George Dassios, Athanassios S. Fokas
Electroencephalography and Magnetoencephalography

De Gruyter Series in Mathematics and Life Sciences

Edited by
Anna Marciniak-Czochra, Heidelberg University, Germany
Benoît Perthame, Sorbonne-Université, France
Jean-Philippe Vert, Mines ParisTech, France

Volume 7

George Dassios, Athanassios S. Fokas

Electroencephalography and Magnetoencephalography

An Analytical–Numerical Approach

DE GRUYTER

Mathematics Subject Classification 2010
Primary: 35R30, 35Q60, 65M32, 78A46; Secondary: 45Q05, 65N21, 78A70, 78M25, 92C55

Authors
Prof. Dr. George Dassios
University of Patras
Dept. of Chemical Engineering
Caratheodory St. 1
265 04 Patras
Greece
gdassios@chemeng.upatras.gr

Prof. Dr. Athanassios S. Fokas
University of Cambridge
Department of Applied Mathematics
and Theoretical Physics
Cambridge CB3 0WA
UK
T.Fokas@damtp.cam.ac.uk

ISBN 978-3-11-054583-8
e-ISBN (PDF) 978-3-11-054753-5
e-ISBN (EPUB) 978-3-11-054578-4
ISSN 2195-5530

Library of Congress Control Number: 2020934398

Bibliographic information published by the Deutsche Nationalbibliothek
The Deutsche Nationalbibliothek lists this publication in the Deutsche Nationalbibliografie; detailed bibliographic data are available on the Internet at http://dnb.dnb.de.

Typesetting: Integra Software Services Pvt. Ltd.
Printing and binding: CPI books GmbH, Leck

www.degruyter.com

Contents

Introduction

The fundamental cells of the nervous tissue are called neurons.[1] A *specific* mental process is associated with brain activation of a *unique* form, which in turn expresses itself via the generation of a *specific* neuronal electric current. Michael Faraday was the first to establish a relation between electricity and magnetism. The precise mathematical form of this relation is expressed by Maxwell's equations of electromagnetism. According to these equations, an electric current gives rise to both a magnetic field and an electric potential. Once a cortical neuronal current is activated, a magnetic field and an electric potential are generated. Magnetoencephalography (MEG) and Electroencephalography (EEG) are based on measurements of the magnetic field outside the head and of the electric potential on the scalp, respectively. The neuronal current is mainly due to the summation of tiny currents (called ionic currents) flowing across certain pores (called sodium and potassium ionic gates), which are located at the membrane of parts of neurons. The particular type of neurons that are mostly responsible for the current affecting EEG and MEG measurements is called pyramidal cells.

In most clinical applications, EEG uses 19 electrodes to measure the electric potential on the scalp (for research purposes there exist high-density, 256-channel EEG, sensory arrays). The measured potential is in the microvolt range. Scalp voltage recordings were first measured in humans in 1924 by Hans Berger (1873–1941), who invented, what he called, the electroencephalogram; Berger extended earlier work by Richard Caton (1842–1926). EEG is an inexpensive, minimally invasive technique, which can provide real time information about the brain's function. This important modality is widely used in both normal and clinical conditions. For example, it is an indispensable tool in sleep studies as well as for the diagnosis of epilepsy. The latter neurological disorder includes *focal seizures* and *generalized seizures* which are further subdivided into *tonic-clonic seizures* and *absence seizures*. Tonic-clonic seizures (previously called grand mal) are the seizures most people associate with epilepsy; absence seizures (previously called petit mal) are more common in children than in adults.[2] The electroencephalograms obtained during different types of seizure exhibit characteristic patterns, which can aid the diagnosis of epilepsy. For example, in generalized seizures there is a pattern of unison

1 The nervous tissue also contains the so-called glia cells. These cells are also crucial for the function of the central nervous system; however, they do not propagate electric impulses, thus are not directly relevant to EEG and MEG.

2 In absence seizures a child may suddenly stop talking or cease walking; the child stares emptily and does not respond to stimuli. After one to several seconds following these short "absences", the child regains consciousness and has no memory of the episode.

https://doi.org/10.1515/9783110547535-001

hyperactivity, which can be contrasted with the highly complex pattern of a normal electroencephalogram.[3]

In general, EEG recordings depend on the location of the electrodes and on the specific mental function performed by the subject during the recordings. In normal subjects there exist several characteristic rhythms that are distinguished by their frequencies: *delta* rhythm involves high amplitude, slow waves, namely waves of frequency less that 4 Hz (i.e., four cycles per second); this is the characteristic rhythm occurring during deep sleep (these waves are best obtained from recordings on the frontal part of the head). Waves of the *theta* rhythm have frequency between 4 and 7 Hz; they occur when subjects try to repress a specific response or action (these waves are found in regions not related to the task at hand). The frequency of *alpha* waves is between 8 and 15 Hz; such waves occur during relaxation. The *beta* waves have frequency in the range of 16 to 32 Hz; they are associated with active thinking. Finally, the *gamma* rhythm involves waves of high frequency, namely waves of frequency higher than 32 Hz, typically occur during the processing of at least two different types of sensory information.

The magnetic field generated by neuronal currents is extremely weak; it is of the order of femtotesla, that is, approximately 10^9 times smaller than the magnetic field of the earth field at its surface. Thus, MEG requires the use of an extremely sensitive magnetometer. In the early 1970s, David Cohen built a magnetically shielded room at MIT and measured MEG signals at a single point outside the head using the exquisitely sensitive superconducting quantum interference device (SQUID). This device had just been developed at Ford Motor Company by James Zimmerman [88]. Remarkably, MEG signals are almost as clear as those obtained via EEG. Present-day technology allows the measurement of the magnetic flux at about 300 points in a helmet-shaped apparatus. In a very recent technological breakthrough, a new type of magnetometer has been developed that takes the form of a portable helmet.[4] In this new device, the sensors are on the scalp instead of being a few millimeters away from the head as they were in the SQUID, and more importantly, the subject can move.

As noted earlier, recordings obtained from EEG can be used to aid the diagnosis of several neurological conditions. Similarly, MEG also has several clinical applications. Later, it was realized that MEG could have an additional, more significant use: it could be employed in order to *determine the electric current* that gave rise to

3 This suggests that the loss of consciousness in a generalized seizure is the result of a dramatic reduction of complexity of the brain's neural state; this indicates the inability of the brain to integrate the diverse repertoire of neural substates that are normally available.

4 This device is called *spin exchange relaxation-free magnetometer* and does not require a bulky cooling system for its operation, thus it is small and portable. Also, the proximity of the sensors to the brain improves the so-called signal to noise ratio.

the MEG recordings. Assuming that the brain can be modelled as an appropriate conductor, the above involves the solution of the following *mathematical inverse problem*: given the magnetic field outside a conductor, determine the electric current generating this field. Similarly, the analogous problem for EEG involves the determination of the electric current from the knowledge of the electric potential measured on the outer surface of a conductor.

The main advantage of both MEG and EEG in comparison to other functional imaging techniques is their very high temporal resolution, which is better than 1 millisecond; that is, EEG and MEG essentially provide *real-time* information about mental processes. This is to be contrasted with the time resolution of other functional imaging techniques, like fMRI, PET and SPECT, which due to slow hemodynamic response, is of the order of 2 s. However, the Achilles heel of MEG and EEG is their spatial resolution. Actually, as it will be discussed in detail in this volume, it is *impossible to obtain the continuous distribution of the cortical current from MEG or EEG data without additional physiological or mathematical assumptions*.

However, there exist particular physiological situations where the current *can* be computed uniquely from either EEG or MEG data. Such a situation, for example, arises when the current is so well localized that it can be adequately approximated by the assumption that the current is supported only at a single point, or a sector of a curve, or a patch of a surface. In this idealized case, the current can be represented in terms of Dirac's delta functions supported on zero, one or two-dimensional subsets [1, 29, 30–34, 37, 38]. The associated idealized point-current is called a *dipole*. Given a set of MEG data, there exists a commercial software that can determine a suitable dipole from the data [73]. The dipole approximation is very useful for the study of focal epilepsy using MEG: in patients with drug-resistant epilepsy,[5] a viable treatment option is brain surgery, namely, the removal of the part of the brain that generates seizures. In order to identify this epileptogenic focus, the usual preoperative evaluation is the combination of long-term video EEG and MRI. For patients with no concordance of the EEG results and a lesion on MRI, MEG (under the assumption of a single dipole) is used to identify the focus.[6] Due to a recent technological development, it is possible to combine MEG with a newer MRI (called ultra-low field MRI), and this improves significantly the localization capability of MEG [85].

In addition to its important role in the investigation and treatment of epilepsy, MEG also has numerous applications in the study of physiology, psychiatry and cognition: from the investigation of the somatosensory system and the relation of brain-rhythms with cognitive processes, to the study of Alzheimer's disease and

---

5 Drug-resistant epilepsy is defined as the failure of adequate control of epilepsy after trials of two appropriately chosen drug schedules with acceptable side effects [70].

6 In a recent study, 77% of 57 patients who underwent surgery (and continued to receive pharmacological treatment) became seizure-free after one year. In comparison, 93% of those receiving only pharmacological treatment continued to have seizures [44].

the identification of neural abnormalities in various psychiatric and neurological diseases [51].

Taking into consideration that the conductivities of the cortex, cerebrospinal fluid and scalp are similar, it turns out that the formulation of MEG gives rise to an *approximate formalism*, which is much simpler than that of EEG. Namely, as it will be explained in this volume, this formalism does not involve the determination of certain auxiliary complicated functions needed for the formulation of EEG; these functions depend on the geometry of the four compartments modelling the brain-head system (cerebrum, cerebrospinal fluid, skull and scalp) as well as the electrical conductivities of these compartments. However, if the current is approximated by a single dipole, the EEG inverse problem can be solved uniquely [29, 30]. Thus, in addition to its well-established use for the diagnosis of epilepsy, EEG can also be used (via the one dipole approximation) for the determination of the precise location of an epileptic focus.[7]

The main purpose of this volume is to elucidate mathematically the nature of the inverse problems arising in EEG and MEG. In addition, efficient numerical algorithms are presented for computing the cortical neuronal current in terms of EEG or MEG data. The emphasis is on the realistic situation of a *continuous distribution of current*; only one chapter of the volume is dedicated to the analysis of dipoles (Chapter 6).

This volume could be of interest to both practitioners of EEG-MEG and *affectionatos* of exact results. With respect to the former, the most important Chapters are 2, 7, 8 and 9. Chapters 2, 7, and 9, which are based mainly on the work of the second author, present: (i) The definitive answer to the non uniqueness question; namely, in chapter 2, the part of the neuronal current that affects the EEG and MEG measurements is identified. (ii) The solution of the minimisation problem for EEG for the particular cases of the spherical and ellipsoidal geometries; namely, it is shown in chapter 7 that in the case of EEG the assumption of minimising the L_2 norm of the current, yields a unique solution. (iii) The definitive solution of the inverse problems for both MEG and EEG in the case that the current is confined to the cerebral surface and is normal to it. Chapter 8, which are based on the joint work of the second author and Parham Hashemzadeh, presents: (i) Numerical algorithms for constructing the components of the current affecting the EEG and MEG measurements in the case of ellipsoidal geometry. (ii) For EEG, an effective numerical algorithm for computing the current in arbitrary geometry by imposing the L_2 minimisation. This result is the most complete result presented in this book; the analogous result for MEG will be presented in the near future. Regarding exact results, based mainly on works of the first author and collaborators, the relevant chapters are 3–6; in these chapters' exact results are presented for the particular cases of spherical and ellipsoidal geometries. More details about the content of the book are presented below.

7 Regarding the determination of epileptic foci via the dipole approximation of the inverse problems of MEG and EEG, comparisons are presented in [75].

Results for an arbitrary geometry

This book is based on the standard four compartment brain-head system model: the cerebrum is modelled by a domain denoted by Ω_c which is bounded by the cortical surface denoted by S_c; it is surrounded by cerebrospinal fluid modelled by the domain Ω_f which is bounded by the external surface S_f; the skull (bone) is modeled by the domain Ω_b which is bounded by the external surface S_b; finally, all the above domains are enclosed by the scalp, modeled by the domain Ω_s, which is bounded by the external surface S_s. The above domains are distinguished by different electrical conductivities denoted, respectively, by σ_c, σ_f, σ_b, σ_s. The constant μ_0 denotes the permeability of the aforementioned four compartments, as well as of the domain exterior to the head, which is denoted by Ω_e.

In Chapter 1, the physical and mathematical foundations of EEG and MEG are discussed. In particular, the validity of the quasistatic approximation for electromagnetoencephalography is demonstrated, and the classical integral formulas of Biot –Savart and Geselowitz are derived.

In Chapter 2, it is shown that the basic equation relating the electric potential measured on S_s with the primary cortical neuronal current denoted by $\mathbf{J}^p$ is the equation

$$u_s(\mathbf{r}) = -\frac{1}{4\pi}\int_{\Omega_c} \upsilon_s(\mathbf{r},\boldsymbol{\tau})(\nabla_{\boldsymbol{\tau}}\cdot\mathbf{J}^p(\boldsymbol{\tau}))\,dV(\boldsymbol{\tau}) + \frac{1}{4\pi}\int_{S_c} \upsilon_s(\mathbf{r},\boldsymbol{\tau})(\hat{\mathbf{n}}(\boldsymbol{\tau})\cdot\mathbf{J}^p(\boldsymbol{\tau}))\,dS(\boldsymbol{\tau}). \quad \text{(i)}$$

In this equation, $\hat{\mathbf{n}}$ denotes the outward unit normal to S_s and υ_s is an auxiliary function (a monopolic potential) that depends on the geometry and conductivities, but it is *independent* of the primary current. This function, as well as the analogous functions $\upsilon_c, \upsilon_f, \upsilon_b$ can be obtained via the solution of a transmission-boundary value problem defined as follows:

for the potential υ_c in Ω_c,

$$\Delta_{\mathbf{r}}\upsilon_c(\mathbf{r},\boldsymbol{\tau}) = 0, \quad \mathbf{r}\in\Omega_c$$

$$\frac{1}{\sigma_c}\left[\frac{1}{|\mathbf{r}-\boldsymbol{\tau}|} + \upsilon_c(\mathbf{r},\boldsymbol{\tau})\right] = \upsilon_f(\mathbf{r},\boldsymbol{\tau}), \quad \mathbf{r}\in S_c$$

$$\left(\frac{\partial}{\partial n(\mathbf{r})}\right)\left[\frac{1}{|\mathbf{r}-\boldsymbol{\tau}|} + \upsilon_c(\mathbf{r},\boldsymbol{\tau})\right] = \sigma_f\left(\frac{\partial}{\partial n(\mathbf{r})}\right)\upsilon_f(\mathbf{r},\boldsymbol{\tau}), \quad \mathbf{r}\in S_c;$$

for the potential υ_f in Ω_f,

$$\Delta_{\mathbf{r}}\upsilon_f(\mathbf{r},\boldsymbol{\tau}) = 0, \quad \mathbf{r}\in\Omega_f$$

$$\upsilon_f(\mathbf{r},\boldsymbol{\tau}) = \upsilon_b(\mathbf{r},\boldsymbol{\tau}), \quad \mathbf{r}\in S_f$$

$$\sigma_f\left(\frac{\partial}{\partial n(\mathbf{r})}\right)\upsilon_f(\mathbf{r},\boldsymbol{\tau})=\sigma_b\left(\frac{\partial}{\partial n(\mathbf{r})}\right)\upsilon_b(\mathbf{r},\boldsymbol{\tau}),\quad \mathbf{r}\in S_f;$$

for the potential υ_b in Ω_b,

$$\Delta_{\mathbf{r}}\upsilon_b(\mathbf{r},\boldsymbol{\tau})=0,\quad \mathbf{r}\in\Omega_b$$
$$\upsilon_b(\mathbf{r},\boldsymbol{\tau})=\upsilon_s(\mathbf{r},\boldsymbol{\tau}),\quad \mathbf{r}\in S_b$$
$$\sigma_b\left(\frac{\partial}{\partial n(\mathbf{r})}\right)\upsilon_b(\mathbf{r},\boldsymbol{\tau})=\sigma_s\left(\frac{\partial}{\partial n(\mathbf{r})}\right)\upsilon_s(\mathbf{r},\boldsymbol{\tau}),\quad \mathbf{r}\in S_b;$$

and finally, for the potential υ_s in Ω_s

$$\Delta_{\mathbf{r}}\upsilon_s(\mathbf{r},\boldsymbol{\tau})=0,\quad \mathbf{r}\in\Omega_s$$
$$\upsilon_s(\mathbf{r},\boldsymbol{\tau})=\upsilon_e(\mathbf{r},\boldsymbol{\tau}),\quad \mathbf{r}\in S_s$$
$$\left(\frac{\partial}{\partial n(\mathbf{r})}\right)\upsilon_s(\mathbf{r},\boldsymbol{\tau})=0,\quad \mathbf{r}\in S_s.$$

Further insight from eq. (i) can be gained by representing the primary current in terms of the Helmholtz decomposition:

$$\mathbf{J}^p(\boldsymbol{\tau})=\nabla_{\boldsymbol{\tau}}\Psi(\boldsymbol{\tau})+\nabla_{\boldsymbol{\tau}}\times\mathbf{A}(\boldsymbol{\tau}),\quad \nabla_{\boldsymbol{\tau}}\cdot\mathbf{A}(\boldsymbol{\tau})=0.$$

Then, eq. (i) becomes

$$u_s(\mathbf{r})=-\frac{1}{4\pi}\int\limits_{\Omega_c}\upsilon_s(\mathbf{r},\boldsymbol{\tau})\Delta_{\boldsymbol{\tau}}\Psi(\boldsymbol{\tau})\,dV(\boldsymbol{\tau})+\frac{1}{4\pi}\int\limits_{S_c}\upsilon_s(\mathbf{r},\boldsymbol{\tau})(\hat{\mathbf{n}}(\boldsymbol{\tau})\cdot\mathbf{J}^p(\boldsymbol{\tau}))\,dS(\boldsymbol{\tau}),\quad \mathbf{r}\in S_s. \tag{ii}$$

In order to decouple the EEG and MEG problems we assume that **A** satisfies the boundary condition:

$$\hat{\mathbf{n}}(\boldsymbol{\tau})\cdot\nabla_{\boldsymbol{\tau}}\times\mathbf{A}(\boldsymbol{\tau})=0,\quad \boldsymbol{\tau}\in S_c. \tag{iii}$$

Then, eq. (ii) becomes

$$u_s(\mathbf{r})=-\frac{1}{4\pi}\int\limits_{\Omega_c}\upsilon_s(\mathbf{r},\boldsymbol{\tau})\Delta_{\boldsymbol{\tau}}\Psi(\boldsymbol{\tau})\,dV(\boldsymbol{\tau})\ \frac{1}{4\pi}\int\limits_{S_c}\upsilon_s(\mathbf{r},\boldsymbol{\tau})(\hat{\mathbf{n}}(\boldsymbol{\tau})\cdot\nabla_{\boldsymbol{\tau}}\Psi(\boldsymbol{\tau}))\,dS(\boldsymbol{\tau}),\quad \mathbf{r}\in S_s. \tag{iv}$$

This equation implies that *only the nonharmonic part of the Ψ component of the current affects the EEG data.* Actually, EEG data provides information only about the value of Ψ and of the divergence of Ψ on the boundary of Ω_c. Indeed, it turns out that the eq. (iv) can be rewritten in the form

$$u_s(\mathbf{r}) = \frac{1}{4\pi} \int_{S_c} \hat{\mathbf{n}}(\boldsymbol{\tau}) \cdot [\Psi(\boldsymbol{\tau}) \nabla_{\boldsymbol{\tau}} \upsilon_s(\mathbf{r}, \boldsymbol{\tau})] dS(\boldsymbol{\tau}), \quad \mathbf{r} \in S_s. \tag{v}$$

This equation captures the definitive nonuniqueness result for EEG.

It is also shown in Chapter 2 that the analogue of eq. (ii) for the radial component of the magnetic flux in Ω_e, is the equation

$$\begin{aligned} \frac{4\pi}{\mu_0} \mathbf{r} \cdot \mathbf{B}(\mathbf{r}) = {} & \frac{1}{4\pi} \int_{\Omega_c} (\Delta_{\boldsymbol{\tau}} \Psi(\boldsymbol{\tau}))(\mathbf{r} \cdot \mathbf{H}(\mathbf{r}, \boldsymbol{\tau})) dV(\boldsymbol{\tau}) - \int_{\Omega_c} \frac{1}{|\mathbf{r} - \boldsymbol{\tau}|} \Delta_{\boldsymbol{\tau}} (\boldsymbol{\tau} \cdot \mathbf{A}(\boldsymbol{\tau})) dV(\boldsymbol{\tau}) \\ & + \int_{S_c} \nabla_{\boldsymbol{\tau}} \cdot \left[|\mathbf{r} - \boldsymbol{\tau}| (\hat{\mathbf{n}}(\boldsymbol{\tau}) \times \mathbf{J}^p(\boldsymbol{\tau})) \right] dS(\boldsymbol{\tau}) - \frac{1}{4\pi} \int_{S_c} (\hat{\mathbf{n}}(\boldsymbol{\tau}) \cdot \mathbf{J}^p(\boldsymbol{\tau}))(\mathbf{r} \cdot \mathbf{H}(\mathbf{r}, \boldsymbol{\tau})) dS(\boldsymbol{\tau}) \\ & - \int_{S_c} \frac{1}{|\mathbf{r} - \boldsymbol{\tau}|} \boldsymbol{\tau} \cdot (\hat{\mathbf{n}}(\boldsymbol{\tau}) \times \mathbf{J}^p(\boldsymbol{\tau})) \, dS(\boldsymbol{\tau}), \quad \mathbf{r} \in \Omega_e, \end{aligned} \tag{vi}$$

where $\mathbf{H}$ is defined in terms of the monopolic potentials υ_l, $l = c, f, b, s$ by the equation

$$\begin{aligned} \mathbf{H}(\mathbf{r}, \boldsymbol{\tau}) = {} & (\sigma_c - \sigma_f) \int_{S_c} \upsilon_f(\boldsymbol{\rho}, \boldsymbol{\tau}) \, \hat{\mathbf{n}}(\boldsymbol{\rho}) \times \left(\nabla_{\rho} \frac{1}{|\mathbf{r} - \boldsymbol{\rho}|} \right) dS(\boldsymbol{\rho}) \\ & + (\sigma_f - \sigma_b) \int_{S_f} \upsilon_b(\boldsymbol{\rho}, \boldsymbol{\tau}) \, \hat{\mathbf{n}}(\boldsymbol{\rho}) \times \left(\nabla_{\boldsymbol{\rho}} \frac{1}{|\mathbf{r} - \boldsymbol{\rho}|} \right) dS(\boldsymbol{\rho}) \\ & + (\sigma_b - \sigma_s) \int_{S_b} \upsilon_s(\boldsymbol{\rho}, \boldsymbol{\tau}) \, \hat{\mathbf{n}}(\boldsymbol{\rho}) \times \left(\nabla_{\boldsymbol{\rho}} \frac{1}{|\mathbf{r} - \boldsymbol{\rho}|} \right) dS(\boldsymbol{\rho}) \\ & + \sigma_s \int_{S_s} \upsilon_s(\boldsymbol{\rho}, \boldsymbol{\tau}) \hat{\mathbf{n}}(\boldsymbol{\rho}) \times \left(\nabla_{\boldsymbol{\rho}} \frac{1}{|\mathbf{r} - \boldsymbol{\rho}|} \right) dS(\boldsymbol{\rho}), \quad \mathbf{r} \in \Omega_s. \end{aligned} \tag{vii}$$

It turns out that for practical purposes the term involving $\mathbf{H}$ can be neglected. Indeed, $\mathbf{H}$ involves explicitly the differences of the various electric conductivities, which are small. Thus, the following equation provides a good approximation of the MEG formulation:

$$\begin{aligned} \frac{4\pi}{\mu_0} \mathbf{r} \cdot \mathbf{B}(\mathbf{r}) \approx {} & - \int_{\Omega_c} \frac{1}{|\mathbf{r} - \boldsymbol{\tau}|} \Delta_{\boldsymbol{\tau}} (\boldsymbol{\tau} \cdot \mathbf{A}(\boldsymbol{\tau})) dV(\boldsymbol{\tau}) + \int_{S_c} \nabla_{\boldsymbol{\tau}} \cdot \left[|\mathbf{r} - \boldsymbol{\tau}| (\hat{\mathbf{n}}(\boldsymbol{\tau}) \times \mathbf{J}^p(\boldsymbol{\tau})) \right] dS(\boldsymbol{\tau}) \\ & - \int_{S_c} \frac{1}{|\mathbf{r} - \boldsymbol{\tau}|} \boldsymbol{\tau} \cdot (\hat{\mathbf{n}}(\boldsymbol{\tau}) \times \mathbf{J}^p(\boldsymbol{\tau})) \, dS(\boldsymbol{\tau}), \quad \mathbf{r} \in \Omega_e. \end{aligned} \tag{viii}$$

Thus, because of (iii) we find

$$\frac{4\pi}{\mu_0}\mathbf{r}\cdot\mathbf{B}(\mathbf{r}) \approx -\int_{S_C}\left[\frac{1}{|\mathbf{r}-\boldsymbol{\tau}|}\nabla_{\boldsymbol{\tau}}(\boldsymbol{\tau}\cdot\mathbf{A}(\boldsymbol{\tau})) - (\boldsymbol{\tau}\cdot\mathbf{A}(\boldsymbol{\tau}))\nabla_{\boldsymbol{\tau}}\frac{1}{|\mathbf{r}-\boldsymbol{\tau}|}\right]\cdot\hat{\mathbf{n}}(\boldsymbol{\tau})\, dS(\boldsymbol{\tau})$$
$$+\int_{S_C}\nabla_{\boldsymbol{\tau}}\cdot\left[|\mathbf{r}-\boldsymbol{\tau}|(\hat{\mathbf{n}}(\boldsymbol{\tau})\times\nabla_{\boldsymbol{\tau}}\Psi(\boldsymbol{\tau}))\right]\, dS(\boldsymbol{\tau}$$
$$-\int_{S_C}\frac{1}{|\mathbf{r}-\boldsymbol{\tau}|}\boldsymbol{\tau}\cdot(\hat{\mathbf{n}}(\boldsymbol{\tau})\times\nabla_{\boldsymbol{\tau}}\Psi(\boldsymbol{\tau}))\, dS(\boldsymbol{\tau}), \quad \mathbf{r}\in\Omega_e. \tag{ix}$$

For simultaneous EEG and MEG measurements Ψ:

$$\frac{4\pi}{\mu_0}\mathbf{r}\cdot\mathbf{B}(\mathbf{r}) \approx -\int_{\Omega_C}\frac{1}{|\mathbf{r}-\boldsymbol{\tau}|}\Delta_{\boldsymbol{\tau}}(\boldsymbol{\tau}\cdot\mathbf{A}(\boldsymbol{\tau}))dV(\boldsymbol{\tau}), \quad \mathbf{r}\in\Omega_e \tag{x}$$

and

$$\frac{4\pi}{\mu_0}\mathbf{r}\cdot\mathbf{B}(\mathbf{r}) \approx -\int_{S_C}\left[\frac{1}{|\mathbf{r}-\boldsymbol{\tau}|}\nabla_{\boldsymbol{\tau}}(\boldsymbol{\tau}\cdot\mathbf{A}(\boldsymbol{\tau})) - (\boldsymbol{\tau}\cdot\mathbf{A}(\boldsymbol{\tau}))\nabla_{\boldsymbol{\tau}}\frac{1}{|\mathbf{r}-\boldsymbol{\tau}|}\right]\cdot\hat{\mathbf{n}}(\boldsymbol{\tau})\, dS(\boldsymbol{\tau}), \quad \mathbf{r}\in\Omega_e. \tag{xi}$$

Thus, the radial component of the magnetic field is affected only by the radial component of **A**. *The aforementioned equations provide the definitive answer to the non-uniqueness question of MEG.* Since the above equations do not involve any of the auxiliary functions, it is clear that the approximate mathematical formulation of MEG is much simpler than that of EEG.

The previously mentioned analysis shows that for a continuously distributed cortical current, EEG and MEG provides some information, respectively, about Ψ and the radial part of **A**.

However, as discussed earlier, there are particular circumstances, where a unique current *can* be determined. In addition to the dipole case discussed before, another physiologically important situation arises as follows: often, EEG and MEG data are dominated by the pyramidal neurons situated perpendicular to the cortical surface. It turns out that under the assumption that the current is *localized on the cortical surface and that it is* also *normal to this surface, this current can be computed uniquely* [51]. In this case the direction of the current vector is known, thus the inverse problem involves determining a single unknown scalar function (the magnitude of the current) supported on a surface.[8] It is shown in Chapter 9 that in this case the following equations for EEG and MEG are valid:

8 It is shown in Chapter 1 that the magnetic field outside the head satisfies Laplace's equation; since the solution of this equation can be obtained from the knowledge of the unknown on the boundary of the domain, it follows that MEG data can always be considered as given on a surface.

$$u_s(\mathbf{r}) = \frac{1}{4\pi}\int_{S_c} J_3(\boldsymbol{\tau})\frac{\partial \upsilon_s(\mathbf{r},\boldsymbol{\tau})}{\partial\lambda}\,dS(\boldsymbol{\tau}), \quad \mathbf{r}\in S_s \tag{xii}$$

and

$$\frac{4\pi}{\mu_0}\mathbf{r}\cdot\mathbf{B}(\mathbf{r}) = \int_{S_c}\boldsymbol{\tau}\cdot\left[\frac{\partial J_3(\boldsymbol{\tau})}{\partial\nu}\boldsymbol{\tau}_1 + \frac{\partial J_3(\boldsymbol{\tau})}{\partial\mu}\boldsymbol{\tau}_2\right]\frac{h_\lambda(\boldsymbol{\tau})}{|\boldsymbol{\tau}_1||\boldsymbol{\tau}_2|}\frac{1}{|\mathbf{r}-\boldsymbol{\tau}|}\,dS(\boldsymbol{\tau})$$

$$-\frac{1}{4\pi}\int_{S_c} J_3(\boldsymbol{\tau})\left(\mathbf{r}\cdot\frac{\partial\mathbf{H}(\mathbf{r},\boldsymbol{\tau})}{\partial\lambda}\right)dS(\boldsymbol{\tau}), \quad \mathbf{r}\in\Omega_e, \tag{xiii}$$

where $\boldsymbol{\tau}_1$, $\boldsymbol{\tau}_2$ and h_λ are defined as follows:

$$\boldsymbol{\tau}_1(\mu,\nu) = \frac{\partial\boldsymbol{\tau}(a,\mu,\nu)}{\partial\mu}, \qquad \boldsymbol{\tau}_2(\mu,\nu) = \frac{\partial\boldsymbol{\tau}(a,\mu,\nu)}{\partial\nu}, \qquad h_\lambda(\mu,\nu) = \left|\frac{\partial\boldsymbol{\tau}(a,\mu,\nu)}{\partial\lambda}\right|.$$

Given EEG data, it is shown in Chapter 9 that the aforementioned equation (xii) gives rise to a simple numerical algorithm for determining the normal component of the current.

Returning to eqs. (iv) and (x) it follows that it is impossible to determine a continuously distributed cortical current. In this connection, it is shown in Chapter 7 that for the particular cases of spherical and ellipsoidal geometries, the additional assumption of minimizing the L_2 norm yields unique answers. Since the brain-system can be approximated by confocal ellipsoidal surfaces, the aforementioned results justify imposing the minimization condition for an arbitrary geometry. For the associated implementation, an effective numerical algorithm is presented in Chapter 8, based on the result of [66] and [67], where: (i) before considering the case of arbitrary geometry (Section 8.2), the case of ellipsoidal geometry is first analyzed (Section 8.1); in this case, it is shown that the direct use of ellipsoidal harmonics is *not* practical. For both the arbitrary and the ellipsoidal geometries the components of current affecting the data is expanded in terms of multiquadric radial basis functions. (ii) Following [65] a novel approach for the fast and accurate numerical determination of the auxiliary function υ_s is employed; actually υ_s can be computed either via the numerical evaluation of a line integral or via the new mathematical technique of *deep learning* (using an appropriate surrogate model).

Hence, both the unknown current and the MEG data are represented as functions supported on a surface, therefore it is not surprising that in this case there exists a unique relation between them.

Results for spherical and ellipsoidal geometries

Spherical geometry corresponds to the case that the surfaces S_c, S_f, S_b, S_s are spheres of radii c, f, b, s. In this case, Ψ and the radial component of A can be expanded in the form

$$\Psi(\mathbf{r}_0) = \sum_{n=1}^{\infty} \sum_{m=-n}^{n} y_n^m(r_0)\, Y_n^m(\hat{\mathbf{r}}_0)$$

and

$$r_0 A^{r_0}(\mathbf{r}_0) = \sum_{n=0}^{\infty} \sum_{m=-n}^{n} a_n^m(r_0)\, Y_n^m(\hat{\mathbf{r}}_0)$$

both for $0 \le r_0 < c$, where Y_n^m denote spherical harmonics. In the aforementioned particular case, it is shown in Chapter 3 that the basic equations of EEG and MEG become:

$$u_s(\mathbf{r}) = -\sum_{n=1}^{\infty} \sum_{m=-n}^{n} \frac{s_n}{2n+1} c^{n+1} \left[c\dot{y}_n^m(c) - n y_n^m(c)\right] Y_n^m(\hat{\mathbf{r}}), \quad r = s \tag{xiv}$$

and

$$\frac{1}{\mu_0} \mathbf{r} \cdot \mathbf{B}(\mathbf{r}) = -\sum_{n=0}^{\infty} \sum_{m=-n}^{n} \frac{c^{n+2}}{2n+1} \frac{1}{r^{n+1}} \left[c\dot{a}_n^m(c) - (n-1) a_n^m(c)\right] Y_n^m(\hat{\mathbf{r}}), \quad r > s. \tag{xv}$$

The determination of the constant s_n requires the explicit computation of υ_s; this is achieved in Chapter 4 where it is shown that:

$$s_n = -\frac{\det \tilde{\mathbf{N}}}{\sigma_s n_{21} (2n+1)^3}, \quad n = 1, 2, \ldots .$$

The matrix $\tilde{\mathbf{N}}$ is a product of specific complicated matrices defined in formulae (4.91)–(4.124) of Chapter 4.

By expanding the left hand side of eqs. (xiv) and (xv) in terms of spherical harmonics, we find:

$$u_s(s\hat{\mathbf{r}}) = \sum_{n=1}^{\infty} \sum_{m=-n}^{n} u_n^m Y_n^m(\hat{\mathbf{r}})$$

and

$$\mathbf{r} \cdot \mathbf{B}(\mathbf{r}) = \mu_0 \sum_{n=0}^{\infty} \sum_{m=-n}^{n} b_n^m \frac{1}{r^{n+1}} Y_n^m(\hat{\mathbf{r}})$$

where u_n^m and b_n^m are known constants.

Using the aforementioned equations in (xiv) and (xv), we find the equations

$$-\frac{s_n}{2n+1}c^{n+1}\left[c\dot{y}_n^m(c)-ny_n^m(c)\right] = u_n^m$$

$$\frac{c^{n+2}}{2n+1}\left[c\dot{a}_n^m(c)-(n-1)a_n^m(c)\right] = b_n^m.$$

It is clear that in order to obtain the unknown functions y_n^m and a_n^m from the aforementioned constants we need some additional information. It is shown in Chapter 7 that in the case of EEG by imposing the minimization of the L_2 norm, we find

$$y_n^m(r_0) = -\frac{(2n+1)}{c_1^{2n+1}}\frac{u_n^m}{k_n^m}r_0^n\left[\ln\frac{r_0}{c_1}-\frac{1}{n}\right]$$

where r_0 denotes the radial component of the location where the current is evaluated.

The numerical implementation of the aforementioned formulae is presented in [7]. Recently, an interesting new treatment of the spherical case is presented in [50], which is not based on Helmholtz decomposition.

In ellipsoidal geometry

$$\Psi(\rho,\mu,\nu) = \sum_{n=1}^{\infty}\sum_{m=1}^{2n+1}\tilde{y}_n^m(\rho)S_n^m(\mu,\nu)$$

where (ρ_0,μ_0,ν_0) are the ellipsoidal coordinates of the point where the current is evaluated and $S_n^m(\mu,\nu)$ are the surface ellipsoidal harmonics.

Following a similar, but more elaborate program than that of the spherical case, we find

$$\tilde{y}_n^m(\rho_o) = \frac{1}{2n+1}\frac{1}{N_n^m}\frac{\tilde{u}_n^m}{C_{sn}^{sm}}\left[E_n^m(\rho_o)\int_{\rho_o}^{c_1}E_n^m(\rho')F_n^m(\rho')d\rho' + F_n^m(\rho_o)\int_{h_2}^{\rho_o}E_n^m(\rho')E_n^m(\rho')d\rho' - E_n^m(\rho_o)\frac{F_n^{m\prime}(c_1)}{E_n^{m\prime}(c_1)}\int_{h_2}^{c_1}E_n^m(\rho')E_n^m(\rho')d\rho'\right].$$

In Chapter 6, it is assumed that the primary cortical current can be approximated by a single dipole; then, the position and the moment of this dipole are determined in terms of either 6 EEG or 6 MEG measured data. This is implemented for both spherical and ellipsoidal geometries.

Chapter 1
The electromagnetics of brain activity

In this chapter we postulate the underlying mathematical model and explain the physical mechanisms that govern the electromagnetic activity of the human brain. We also define the forward and backward problems of electroencephalography (EEG) and magnetoencephalography (MEG). Furthermore, we propose a unified approach for treating these two brain imaging modalities, as well as a generalization of the basic model that includes shells of different conductivities that surround the brain tissue.

1.1 Maxwell's equations

Maxwell's equations provide a system of partial differential equations that describe electromagnetic phenomena. They connect the vector fields of *electric intensity* **E**, *magnetic intensity* **H**, *electric displacement* **D** and *magnetic flux density* **B**, with the *electric charge density* ρ and the *current density* **J**, which constitute the *sources* of the 4 fields **E**, **H**, **D**, **B**. The equations of Maxwell are as follows:

$$\nabla \times \mathbf{E} = -\frac{\partial}{\partial t}\mathbf{E} \tag{1.1}$$

$$\nabla \times \mathbf{H} = \frac{\partial}{\partial t}\mathbf{D} + \mathbf{J} \tag{1.2}$$

$$\nabla \cdot \mathbf{D} = \rho \tag{1.3}$$

$$\nabla \cdot \mathbf{B} = 0. \tag{1.4}$$

Equation (1.1) states *Faraday's law*, (1.2) states *Ampere-Maxwell's law*, (1.3) *Gauss' law* and finally eq. (1.4) ensures the nonexistence of magnetic monopoles. The fields **E** and **H** describe the electric and magnetic fields, respectively, in the absence of any kind of matter, that is, in empty space. On the other hand, the fields **D** and **B** describe the corresponding electric and magnetic fields in the presence of matter and therefore they depend on the particular characteristic of the hosting medium. In a homogeneous and isotropic medium, the parameters that describe its electromagnetic behavior are the *dielectric constant* ε, the *magnetic permeability* μ, and the *conductivity* σ. The parameters ε and μ appear as proportionality constants in the *constitutive equations*,

$$\mathbf{D} = \varepsilon \mathbf{E} \tag{1.5}$$

$$\mathbf{B} = \mu \mathbf{H} \tag{1.6}$$

https://doi.org/10.1515/9783110547535-002

which characterize the medium of propagation. In particular, for empty space we take $\varepsilon_0 = 1$ and $\mu_0 = 1$. Therefore, the parameter ε measures the effect of the medium upon the electric field and μ measures the effect of the medium upon the magnetic field. Regarding conductivity it is noted that within a conductive medium with $\sigma > 0$, an electric field $\mathbf{E}$ generates a secondary current known as *induction current* $\mathbf{J}^i$, which is expressed by the equation

$$\mathbf{J}^i = \sigma \mathbf{E}. \tag{1.7}$$

Thus the higher the conductivity of the medium, the higher the induction current. An electrical *insulator* is a medium for which $\sigma = 0$, while a medium for which $\sigma \to +\infty$ is called a *perfect conductor*. The current $\mathbf{J}$ that appears on the RHS of the Ampere–Maxwell eq. (1.2) can be written in the form

$$\mathbf{J} = \mathbf{J}^p + \mathbf{J}^i = \mathbf{J}^p + \sigma \mathbf{E} \tag{1.8}$$

where $\mathbf{J}^p$ represents the *primary current*, namely the imposed current (as opposed to the current coming from the property of the medium to conduct existing electric charges). Hence, in a homogeneous and isotropic medium characterized by the parameters ε, μ, σ the Maxwell equations assume the form

$$\nabla \times \mathbf{E} = -\mu \frac{\partial}{\partial t} \mathbf{H} \tag{1.9}$$

$$\nabla \times \mathbf{H} = \varepsilon \frac{\partial}{\partial t} \mathbf{E} + \mathbf{J}^p + \sigma \mathbf{E} \tag{1.10}$$

$$\varepsilon \, \nabla \cdot \mathbf{E} = \rho \tag{1.11}$$

$$\nabla \cdot \mathbf{H} = 0. \tag{1.12}$$

Taking the curl of eq. (1.9) and using the vector identify

$$\nabla \times (\nabla \times \mathbf{f}) = \nabla(\nabla \cdot \mathbf{f}) - \Delta \mathbf{f} \tag{1.13}$$

where Δ denotes Laplace's operator, we obtain

$$\begin{aligned} \nabla \times (\nabla \times \mathbf{E}) &= -\mu \frac{\partial}{\partial t} (\nabla \times \mathbf{H}) = -\mu \frac{\partial}{\partial t} \left[\varepsilon \frac{\partial}{\partial t} \mathbf{E} + \mathbf{J}^p + \sigma \mathbf{E} \right] \\ &= -\varepsilon \mu \frac{\partial^2}{\partial t^2} \mathbf{E} - \mu \frac{\partial}{\partial t} \mathbf{J}^p - \mu \sigma \frac{\partial}{\partial t} \mathbf{E} = \nabla(\nabla \cdot \mathbf{E}) - \Delta \mathbf{E} \\ &= \frac{1}{\varepsilon} \nabla \rho - \Delta \mathbf{E}. \end{aligned} \tag{1.14}$$

Thus

$$\Delta\mathbf{E}-\varepsilon\mu\frac{\partial^2}{\partial t^2}\mathbf{E}-\mu\sigma\frac{\partial}{\partial t}\mathbf{E}=\frac{1}{\varepsilon}\nabla\rho+\mu\frac{\partial}{\partial t}\mathbf{J}^p. \tag{1.15}$$

In an analogous way we obtain the corresponding equation for the magnetic field

$$\Delta\mathbf{H}-\varepsilon\mu\frac{\partial^2}{\partial t^2}\mathbf{H}-\mu\sigma\frac{\partial}{\partial t}\mathbf{H}=-\nabla\times\mathbf{J}^p. \tag{1.16}$$

Comparing eqs. (1.15) and (1.16) we see that the medium propagates both the electric and the magnetic field in the same way, but the two fields have different source terms. In fact, the expression

$$\frac{1}{\varepsilon}\nabla\rho+\mu\frac{\partial}{\partial t}\mathbf{J}^p$$

generates the electric field, whereas the magnetic field is generated by the term

$$-\nabla\times\mathbf{J}^p.$$

If the conductivity of the medium vanishes, then eqs. (1.15) and (1.16) reduce to the classical wave equation governing electric and magnetic disturbances, which travel with the *phase velocity*

$$c=(\mu\varepsilon)^{-1/2}. \tag{1.17}$$

The induction current $\mathbf{J}^i$ does not appear as a source for neither field since its appearance in the medium is a consequence and not a prerequisite for the generation of the electric field.

Equations (1.2) and (1.3) imply the *equation of continuity*

$$\nabla\cdot\mathbf{J}+\frac{\partial\rho}{\partial t}=0 \tag{1.18}$$

which provides the mathematical formulation of the conservation of the electric charge. In the absence of sources both fields **E** and **H** solve the homogeneous wave equation

$$\Delta\mathbf{f}-\varepsilon\mu\frac{\partial^2}{\partial t^2}\mathbf{f}-\mu\sigma\frac{\partial}{\partial t}\mathbf{f}=\mathbf{0}. \tag{1.19}$$

This equation implies the *characteristic polynomial* or the *dispersion relation*

$$k^2=\varepsilon\mu\omega^2+i\omega\mu\sigma. \tag{1.20}$$

The vector **f** takes the spectral form

$$\mathbf{f}(\mathbf{r},t)=\alpha e^{i\mathbf{k}\cdot\mathbf{r}-i\omega t} \tag{1.21}$$

where α is the *polarization vector,* $\mathbf{k}$ is the *propagation vector,* $k = |\mathbf{k}|$ is the *wave number,* and ω is the *angular frequency.* In this case, the phase velocity is given by

$$c = \frac{\omega}{\operatorname{Re} k} \tag{1.22}$$

where k is a root of eq. (1.20). We recall the usual relation between the angular frequency ω and the period T

$$\omega = \frac{2\pi}{T} \tag{1.23}$$

as well as the relation between the wave number k and the wavelength λ

$$k = \frac{2\pi}{\lambda}. \tag{1.24}$$

The angular frequency ω defines the temporal density and the wave number k defines the spacial density of the harmonic wave (1.21). Furthermore,

$$c = \frac{\omega}{k} = \frac{2\pi}{T}\frac{\lambda}{2\pi} = \frac{\lambda}{T} \tag{1.25}$$

as it is expected.

1.2 Quasi-static theory

If the propagation space of the wave field is bounded, then we define the *characteristic dimension of* the medium to be the radius a of the smallest sphere that contains the propagation medium. In the special case where the wavelength is much larger than the characteristic dimension of the medium, that is if $a << \lambda$, than the wave does not find enough space to develop and the variation of the corresponding field is relatively small. The spatial dimension is related to the temporal dimension via the constant phase velocity characterizing the medium; thus in the above case the corresponding temporal variation is also relatively small, and hence the time derivatives of the field are also small and can be ignored. This assumption leads to the quasi-static formulation of electromagnetism which is governed by the equations

$$\nabla \times \mathbf{E} = \mathbf{0} \tag{1.26}$$

$$\nabla \times \mathbf{H} = \mathbf{J} \tag{1.27}$$

$$\nabla \cdot \mathbf{D} = \rho \tag{1.28}$$

$$\nabla \cdot \mathbf{B} = 0. \tag{1.29}$$

Note that, the above physical and intuitive arguments can be made rigorous using asymptotical analysis, which implies that for

$$a < \frac{\lambda}{4} \tag{1.30}$$

we can ignore the time-derivative terms in Maxwell's equations. In many practical applications even $a < \frac{\lambda}{2}$ is enough.

Equation (1.26) characterizes the electric field as irrotational; thus there exists an *electric potential* u, such that

$$\mathbf{E} = -\nabla u. \tag{1.31}$$

From (1.27) we obtain

$$\nabla \cdot (\nabla \times \mathbf{H}) = \nabla \cdot \left(\mathbf{J}^p + \mathbf{J}^i\right) = \nabla \cdot \mathbf{J}^p + \sigma\, \nabla \cdot \mathbf{E} = \nabla \cdot \mathbf{J}^p - \sigma\, \Delta u. \tag{1.32}$$

Since the LHS of eq. (1.32) vanishes, we find

$$\sigma\, \Delta u = \nabla \cdot \mathbf{J}^p \tag{1.33}$$

which is the Poisson equation governing the potential u, with the divergence of the primary current defining the source of u. In regions where no primary current exists and $\sigma = 0$, relations (1.27) and (1.29) imply the existence of a *magnetic potential* U which is harmonic, such that

$$\mathbf{B} = \mu\, \nabla U. \tag{1.34}$$

1.3 The brain as a conductor

The human brain is the most complicated material of the known universe. It has a volume of about 1.4 liters shaped in the form of an ellipsoid with average semi axes 6, 6.5, and 9 cm; its weight is close to 2 kg. The human brain continues to evolve. A new generation has about 150,000 neuronal cells more than its parental generation. The complicated structure of the human brain is due to the huge number of neurons it involves. In fact, there are 100 billion $\left(10^{11}\right)$ neuronal cells in the human brain, each one of which is able to exchange information with up to 15,000 neighboring cells. Hence, the human neurons have 10^5 ways of communication among them. It has been estimated that the total length of conducting cable within the brain is approximately 150,000 km. Considering the fact that the length of the Equator is 40,000 km the brain has enough cable to wrap up the Earth almost four times. Each neuron has the *nucleus* of the cell and a *central axon* having a large number of filaments in each one of its ends, knows as dendrites.

Every electrochemical excitation of the brain is represented by an ionic flux across the neuronal membrane which finally generates a very weak *primary* or *neuronal current*. The flux of the current from neuron to neuron is achieved through the release of certain substances known as neurotransmitters, which are released from one cell and are captured by the next one via particular receptors. This relevant structure is called *synapsis*. As mentioned earlier, each neuron has 10^5 possible ways to generate synaptic conducts with neighbor neurons. Via these synaptic conducts and the cellular axons, signals travel through the human body with an average speed of 30 meters per second. The remarkable complexity of this huge electric circuit is not only due to its actual size; in addition, this circuit is not excited by a fixed electric source but via an electrochemically generated source, which is localized at a particular area depending on the corresponding brain activity. Also, the particular flow of the current is not fixed and predetermined but it is decided at each synapsis depending on the particular brain action.

Since the neuronal current runs on individual neurons of the brain tissue, the problem of the electromagnetic brain activity is discrete. On the other hand, the potential generated by the activity of an isolated neuron is so weak that it is impossible to be traced; combining this fact with the high density of neurons, the brain can be modeled as a conductive continuous medium.

Thus, the simplest model is to treat the brain as a conductive homogeneous mass with a density approximately 1.3 times of the water density and a magnetic permeability μ_0 equal to that of the empty space. An improved model is to consider several additional shells. The brain is surrounded by the *cerebrospinal fluid* which has the same permeability μ_0 but a different conductivity of that of the brain tissue; the whole system is contained inside the *skull* which is finally covered with the *scalp* (the exterior skin of the head). Both the skull and the scalp have the same μ_0 but different conductivities. Consequently, an improved model of the head system is a conductive space which is surrounded by 3 shells of different conductivity representing the cerebrospinal fluid, the skull and the scalp.

Inserting the physiological values ε, μ_0 and σ of the brain in the Maxwell equations and calculating the wavelength of the generated electromagnetic wave for frequencies of the order of 100 Hz, (which are typical values for the brain signals), we obtain

$$\lambda \approx 400m. \tag{1.35}$$

Considering the characteristic dimension a of the head to be 20 cm we arrive at

$$\lambda \approx 2.000a. \tag{1.36}$$

Hence, eq. (1.30) implies that the electromagnetic brain activity is indeed governed by the *quasi-static theory of electromagnetism*.

1.4 Electroencephalography (EEG) and Magnetoencephalography (MEG) for a single conductor

In the framework of the quasi-static theory, the electric potential is described by eq. (1.33). Let us assume that $\Omega^- \subset \mathbb{R}^3$ is an open connected subset that is a *domain*; we also assume that Ω^- is bounded having the smooth boundary $\partial\Omega^-$. The level of smoothness of $\partial\Omega^-$ is such that it allows the existence and continuity of the exterior unit normal $\hat{\mathbf{n}}$ over the whole $\partial\Omega^-$ as shown in the following figure.

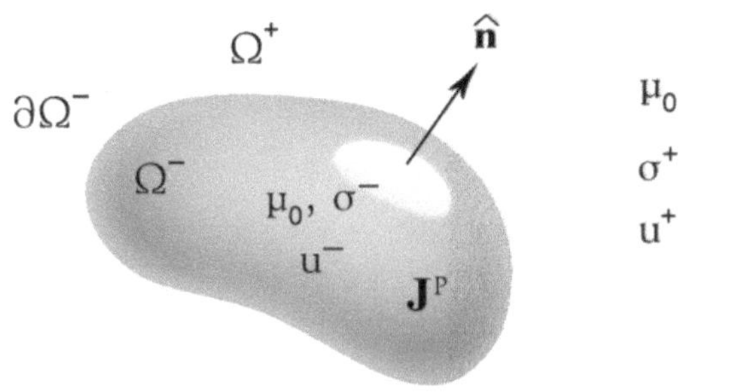

The exterior domain Ω^+ is the complement of the set $\Omega^- \cup \partial\Omega^-$ with respect to $\mathbb{R}^3$. Let σ^- and σ^+ be the constant conductivities of the domains Ω^- and Ω^+ respectively. The permeability μ_0 is constant over the whole space $\mathbb{R}^3$. Finally, we assume that the neuronal current $\mathbf{J}^p$ is supported in Ω^-. Then the electric potential $u^\pm$ satisfies the following transition boundary value problem.

$$\sigma^- \Delta u^-(\mathbf{r}) = \nabla \cdot \mathbf{J}^p(\mathbf{r}), \quad \mathbf{r} \in \Omega^- \tag{1.37}$$

$$\Delta u^+(\mathbf{r}) = 0, \quad \mathbf{r} \in \Omega^+ \tag{1.38}$$

$$u^-(\mathbf{r}) = u^+(\mathbf{r}), \quad \mathbf{r} \in \partial\Omega^- \tag{1.39}$$

$$\sigma^- \frac{\partial}{\partial n} u^-(\mathbf{r}) = \sigma^+ \frac{\partial}{\partial n} u^+(\mathbf{r}), \quad \mathbf{r} \in \partial\Omega^- \tag{1.40}$$

$$u^+(\mathbf{r}) = O\left(\frac{1}{r}\right), \quad r \to \infty \tag{1.41}$$

when $\frac{\partial}{\partial n}$ defines the exterior normal differentiation on $\partial\Omega^-$. The transition conditions (1.39) and (1.40) ensure the continuity as we cross the boundary $\partial\Omega^-$ of the potential and its normal derivative (which is proportional to the normal component of the current). Similarly, the asymptotic condition (1.41) states the behavior of the potential in the neighborhood of infinity. In the case of the head, where we consider the domain Ω^- as the conductive brain and the nonconductive domain Ω^+ as the exterior to the head domain, we have $\sigma^+ = 0$ and $\sigma^- = \sigma \neq 0$. The assumption $\sigma^+ = 0$

splits the transmition problem (1.37)–(1.41) into 2 boundary value problems: the interior Neumann problem

$$\sigma\Delta u^{-}(\mathbf{r}) = \nabla\cdot\mathbf{J}^{p}(\mathbf{r}), \quad \mathbf{r}\in\Omega^{-} \tag{1.42}$$

$$\frac{\partial}{\partial n}u^{-}(\mathbf{r}) = 0, \quad \mathbf{r}\in\partial\Omega^{-} \tag{1.43}$$

and the exterior Dirichlet problem

$$\Delta u^{+}(\mathbf{r}) = 0, \quad \mathbf{r}\in\Omega^{+} \tag{1.44}$$

$$u^{+}(\mathbf{r}) = u^{-}(\mathbf{r}), \quad \mathbf{r}\in\partial\Omega^{-} \tag{1.45}$$

$$u^{+}(\mathbf{r}) = O\left(\frac{1}{r}\right), \quad \mathrm{r}\to\infty. \tag{1.46}$$

Clearly, we need to solve first the Neumann problem (1.42) and (1.43) in order to obtain the Dirichlet data for u^{-} on the boundary, as started in (1.45). The *forward*, or *direct, EEG problem* seeks the calculation of the electric potential u^{-} if the neuronal current $\mathbf{J}^{p}$ is given. The problem of identifying the current $\mathbf{J}^{p}$ from the knowledge of the electric potential u^{-} on the surface $\partial\Omega^{-}$ of the head is known as the *inverse EEG problem.* Once the electric potential is known, the corresponding electric field is obtained via formula (1.31).

The magnetic induction field **B** is calculated as follows: Equation (1.27) is written as

$$\nabla\times\mathbf{B} = \mu_0\mathbf{J} \tag{1.47}$$

and since the field **B** is solenoidal it follows that there exists a vector potential **A** with

$$\nabla\cdot\mathbf{A} = 0 \tag{1.48}$$

such that

$$\mathbf{B} = \nabla\times\mathbf{A}. \tag{1.49}$$

Inserting the representation (1.49) in (1.47) and utilizing condition (1.48) we find the equation

$$\Delta\mathbf{A} = -\mu_0\mathbf{J}. \tag{1.50}$$

Since the fundamental solution F of the Laplacian

$$\Delta_r F(\mathbf{r},\mathbf{r}') = \delta(\mathbf{r}-\mathbf{r}') \tag{1.51}$$

where δ is the Dirac measure, is given by

$$F(\mathbf{r}, \mathbf{r}') = -\frac{1}{4\pi|\mathbf{r}-\mathbf{r}'|} \tag{1.52}$$

with the Euclidean distance

$$|\mathbf{r}| = \sqrt{x_1^2 + x_2^2 + x_3^2} \tag{1.53}$$

it follows that the solution of eq. (1.50) is:

$$\mathbf{A}(\mathbf{r}) = \frac{\mu_0}{4\pi} \int\limits_{\Omega^-} \frac{\mathbf{J}(\mathbf{r}')}{|\mathbf{r}-\mathbf{r}'|} \, dv(\mathbf{r}') \tag{1.54}$$

where the integration is extended over the support of **J**. Note that since $\sigma \neq 0$ in Ω^- the support of **J** is the whole domain Ω^-. Hense,

$$\begin{aligned} \mathbf{B}(\mathbf{r}) = \nabla_{\mathbf{r}} \times \mathbf{A}(\mathbf{r}) &= \frac{\mu_0}{4\pi} \int\limits_{\Omega^-} \left(\nabla_{\mathbf{r}} \frac{1}{|\mathbf{r}-\mathbf{r}'|}\right) \times \mathbf{J}(\mathbf{r}') dv(\mathbf{r}') \\ &= -\frac{\mu_0}{4\pi} \int\limits_{\Omega^-} \frac{\mathbf{r}-\mathbf{r}'}{|\mathbf{r}-\mathbf{r}'|^3} \times \mathbf{J}(\mathbf{r}') dv(\mathbf{r}'). \end{aligned} \tag{1.55}$$

The above equation is the famous Biot–Savart law

$$\mathbf{B}(\mathbf{r}) = \frac{\mu_0}{4\pi} \int\limits_{\Omega^-} \mathbf{J}(\mathbf{r}') \times \frac{\mathbf{r}-\mathbf{r}'}{|\mathbf{r}-\mathbf{r}'|^3} \, dv(\mathbf{r}') \tag{1.56}$$

where the current **J** is the sum of the neuronal plus the induction current.

We know that the electric field generated by a monopole at the point $\mathbf{r}_0$ is proportional to the fundamental solution of the Laplacian given by

$$\Phi_{mon}(\mathbf{r}, \mathbf{r}_0) = \frac{1}{|\mathbf{r}-\mathbf{r}_0|}. \tag{1.57}$$

Similarly, the electric potential due to a dipole at the point $\mathbf{r}_0$, with moment **Q**, is the directional derivate of the monopole field Φ_{mon} in the direction of **Q**,

$$\Phi_{dip}(\mathbf{r}, \mathbf{r}_0) = \mathbf{Q} \cdot \nabla_{\mathbf{r}} \frac{1}{|\mathbf{r}-\mathbf{r}_0|}. \tag{1.58}$$

More generally, a point dipole generates the vector potential

$$\boldsymbol{\Phi}_{dip}(\mathbf{r}, \mathbf{r}_0) = \nabla_{\mathbf{r}} \frac{1}{|\mathbf{r}-\mathbf{r}_0|} = -\frac{\mathbf{r}-\mathbf{r}_0}{|\mathbf{r}-\mathbf{r}_0|^3} \tag{1.59}$$

which gives rise to the magnetic field

$$\mathbf{B}_{dip}(\mathbf{r}, \mathbf{r}_0) = -\mathbf{Q} \times \frac{\mathbf{r} - \mathbf{r}_0}{|\mathbf{r} - \mathbf{r}_0|^3}. \tag{1.60}$$

In view of formula (1.60), which gives the magnetic field of a single dipole, we observe that the Biot–Savart law (1.56) states that the total magnetic field is the sum of the magnetic fields of all dipoles within Ω^-. The calculation of the magnetic field $\mathbf{B}$ in the exterior domain Ω^+ from the knowledge of the total current $\mathbf{J}$ forms the *forward*, or *direct, MEG problem*. The *inverse MEG problem* is the problem of finding the neuronal current $\mathbf{J}^p$ from the knowledge of the magnetic field $\mathbf{B}$ in the exterior domain Ω^+. It is important to note that the goal is to identify the neuronal and not the total current. In other words, during the process of inverting the data of $\mathbf{B}$ we have to exclude the component of the current that is due to the inductive excitation. In fact, this is exactly the main difficulty of the MEG inversion procedure, since the induction current "hides" the primary neuronal current which is the source of the initial excitation.

1.5 The Geselowitz integral representation

Let us start by considering a single dipole at the point $\mathbf{r}_0$ with moment $\mathbf{Q}$, that is

$$\mathbf{J}^P(\mathbf{r}) = \mathbf{Q}\delta(\mathbf{r} - \mathbf{r}_0). \tag{1.61}$$

The electric potential u_d generated by this dipolar current inside a homogeneous conductor with conductivity σ satisfies the Poisson equation

$$\sigma\, \Delta_{\mathbf{r}} u_d(\mathbf{r}; \mathbf{r}_0) = \nabla_{\mathbf{r}} \cdot \mathbf{Q}\delta(\mathbf{r} - \mathbf{r}_0) = \mathbf{Q} \cdot \nabla_{\mathbf{r}}\delta(\mathbf{r} - \mathbf{r}_0). \tag{1.62}$$

Applying the directional derivative operator $\mathbf{Q} \cdot \nabla_{\mathbf{r}}$ on both sides of the equation

$$\Delta_{\mathbf{r}}\left[-\frac{1}{4\pi|\mathbf{r} - \mathbf{r}_0|}\right] = \delta(\mathbf{r} - \mathbf{r}_0) \tag{1.63}$$

we obtain

$$\Delta_{\mathbf{r}}\left[-\frac{1}{4\pi}\mathbf{Q} \cdot \nabla_{\mathbf{r}}\frac{1}{|\mathbf{r} - \mathbf{r}_0|}\right] = \mathbf{Q} \cdot \nabla_{\mathbf{r}}\delta(\mathbf{r} - \mathbf{r}_0). \tag{1.64}$$

Thus the solution of (1.62) is given by

$$u_d(\mathbf{r}; \mathbf{r}_0) = -\frac{1}{4\pi\sigma}\mathbf{Q} \cdot \nabla_{\mathbf{r}}\frac{1}{|\mathbf{r} - \mathbf{r}_0|} = \frac{1}{4\pi\sigma}\mathbf{Q} \cdot \frac{\mathbf{r} - \mathbf{r}_0}{|\mathbf{r} - \mathbf{r}_0|^3}. \tag{1.65}$$

For the magnetic field generated by (1.61) we work as follows. Equation (1.50) gives

$$\Delta_{\mathbf{r}}\mathbf{A}(\mathbf{r};\mathbf{r}_0) = -\mu_0(\mathbf{J}^p + \mathbf{J}^i) = -\mu_0\mathbf{Q}\delta(\mathbf{r}-\mathbf{r}_0) - \mu_0\sigma\mathbf{E}(\mathbf{r};\mathbf{r}_0) \tag{1.66}$$

where $\mathbf{E}(\mathbf{r};\mathbf{r}_0)$ is the electric field associated with the electric potential (1.65). Taking the curl-operator of eq. (1.66) and using the fact that $\mathbf{E}$ is irrotational we find

$$\Delta_{\mathbf{r}}\mathbf{B}_d(\mathbf{r};\mathbf{r}_0) = \mu_0\mathbf{Q}\times\nabla_{\mathbf{r}}\delta(\mathbf{r}-\mathbf{r}_0). \tag{1.67}$$

Following the steps that gave us the solution $u_d(\mathbf{r};\mathbf{r}_0)$ earlier, we finally obtain

$$\mathbf{B}_d(\mathbf{r};\mathbf{r}_0) = \frac{\mu_0}{4\pi}\mathbf{Q}\times\frac{\mathbf{r}-\mathbf{r}_0}{|\mathbf{r}-\mathbf{r}_0|^3}. \tag{1.68}$$

We note that the commutation of the differential operators used in the earlier calculations are justified by the smoothness of the functions and the fact that the linear operators ∇, Δ and $\nabla\times$ in the Cartesian form have constant coefficients.

In the case of a bounded conductor Ω^- with boundary $\partial\Omega^-$, the Biot–Savart law is written as

$$\begin{aligned}\mathbf{B}(\mathbf{r}) &= \frac{\mu_0}{4\pi}\int_{\Omega^-}[\mathbf{Q}\delta(\mathbf{r}-\mathbf{r}_0) - \sigma\,\nabla_{\mathbf{r}'}u^-(\mathbf{r}')]\times\frac{\mathbf{r}-\mathbf{r}'}{|\mathbf{r}-\mathbf{r}'|^3}\,dv(\mathbf{r}')\\ &= \mathbf{B}_{dip}(\mathbf{r};\mathbf{r}_0) - \frac{\mu_0\sigma}{4\pi}\int_{\Omega^-}(\nabla_{\mathbf{r}'}u^-(\mathbf{r}'))\times\frac{\mathbf{r}-\mathbf{r}'}{|\mathbf{r}-\mathbf{r}'|^3}\,dv(\mathbf{r}')\end{aligned} \tag{1.69}$$

where $\mathbf{B}_{dip}(\mathbf{r};\,\mathbf{r}_0)$ is the contribution (1.68) of the localized neuronal current $(\mathbf{r}_0,\mathbf{Q})$. The integral term on the right hand side of (1.69) denotes the contribution of the induction current which is a continuous distribution of dipoles with moments

$$-\sigma\,\nabla_{\mathbf{r}'}u^-(\mathbf{r}')$$

at each point $\mathbf{r}'\in\Omega^-$.

For $\mathbf{r}\neq\mathbf{r}'$ we have

$$\begin{aligned}\nabla_{\mathbf{r}'}\times\left(u^-(\mathbf{r}')\frac{\mathbf{r}-\mathbf{r}'}{|\mathbf{r}-\mathbf{r}'|^3}\right) &= \nabla_{\mathbf{r}'}\times\left(u^-(\mathbf{r}')\nabla_{\mathbf{r}'}\frac{1}{|\mathbf{r}-\mathbf{r}'|}\right)\\ &= (\nabla_{\mathbf{r}'}u^-(\mathbf{r}'))\times\left(\nabla_{\mathbf{r}'}\frac{1}{|\mathbf{r}-\mathbf{r}'|}\right) = (\nabla_{\mathbf{r}'}u^-(\mathbf{r}'))\times\frac{\mathbf{r}-\mathbf{r}'}{|\mathbf{r}-\mathbf{r}'|^3}.\end{aligned} \tag{1.70}$$

The Gauss theorem, for $\mathbf{r}\in\Omega^+$, gives

$$\int_{\Omega^-}(\nabla_{\mathbf{r}'}u^-(\mathbf{r}'))\times\frac{\mathbf{r}-\mathbf{r}'}{|\mathbf{r}-\mathbf{r}'|^3}\,dv(\mathbf{r}') = \int_{\Omega^-}\nabla_{\mathbf{r}'}\times\left(u^-(\mathbf{r}')\frac{\mathbf{r}-\mathbf{r}'}{|\mathbf{r}-\mathbf{r}'|^3}\right)dv(\mathbf{r}')$$

$$= \oint_{\partial\Omega^-} u^-(\mathbf{r}')\hat{\mathbf{n}}(\mathbf{r}') \times \frac{\mathbf{r}-\mathbf{r}'}{|\mathbf{r}-\mathbf{r}'|^3} ds(\mathbf{r}') \tag{1.71}$$

where $\hat{\mathbf{n}}$ is the exterior unit normal on the boundary $\partial\Omega^-$.

In what follows, the symbol "∧" on the top of a vector denotes that the vector has unit length. Substituting relation (1.71) in the formula (1.69) we obtain the *Geselowitz Integral Representation* for the magnetic induction field **B**, namely

$$\mathbf{B}(\mathbf{r};\mathbf{r}_0) = \frac{\mu_0}{4\pi}\mathbf{Q} \times \frac{\mathbf{r}-\mathbf{r}_0}{|\mathbf{r}-\mathbf{r}_0|^3} - \frac{\mu_0 \sigma}{4\pi} \oint_{\partial\Omega^-} u^-(\mathbf{r}')\hat{\mathbf{n}}(\mathbf{r}') \times \frac{\mathbf{r}-\mathbf{r}'}{|\mathbf{r}-\mathbf{r}'|^3} ds(\mathbf{r}'). \tag{1.72}$$

The Geselowitz formula (1.72) replaces the contribution of the volume distribution of dipoles over Ω^- to a surface distribution of normal dipoles over the boundary $\partial\Omega^-$, with moments given by

$$-\sigma u^-(\mathbf{r}')\hat{\mathbf{n}}(\mathbf{r}')$$

for every $\mathbf{r}' \in \partial\Omega^-$. This is depicted in the following figure.

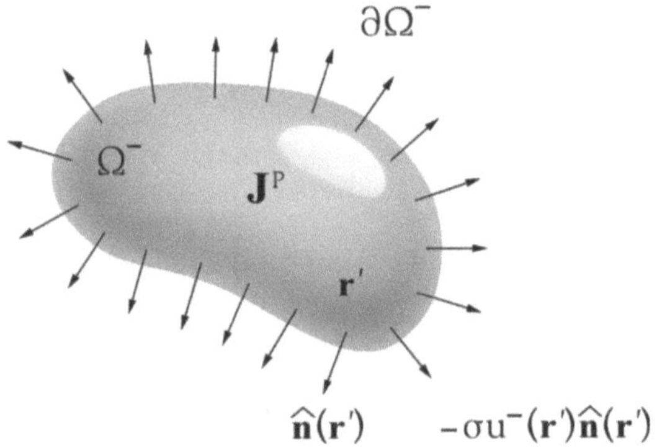

As an equivalent statement we can say that the discontinuity of the conductivity over $\partial\Omega^-$ is expressed via a surface distribution of normal dipolar currents. We finally note that the Geselowitz formula (1.72) holds true for every **r** that does not belong to the boundary $\partial\Omega^-$.

Next we demonstrate the corresponding Geselowitz formula for the electric potential.

We start with Green's second identity

$$\int_{\Omega^-} [f\Delta g - g\Delta f]\, dv = \oint_{\partial\Omega^-} \left[f\frac{\partial}{\partial n}g - g\frac{\partial}{\partial n}f\right] ds. \tag{1.73}$$

Applying this equation for $f = u^-$ and $g = \dfrac{1}{|\mathbf{r}-\mathbf{r}'|}$ we obtain

$$\int_{\Omega^-}\left[u^-(\mathbf{r}')\Delta_{\mathbf{r}'}\frac{1}{|\mathbf{r}-\mathbf{r}'|}-\frac{1}{|\mathbf{r}-\mathbf{r}'|}\Delta_{\mathbf{r}'}u^-(\mathbf{r}')\right]dv(\mathbf{r}')$$
$$=\oint_{\partial\Omega^-}\left[u^-(\mathbf{r}')\frac{\partial}{\partial n'}\frac{1}{|\mathbf{r}-\mathbf{r}'|}-\frac{1}{|\mathbf{r}-\mathbf{r}'|}\frac{\partial}{\partial n'}u^-(\mathbf{r}')\right]ds(\mathbf{r}'). \tag{1.74}$$

Using (1.74) the relations

$$\Delta_{\mathbf{r}'}\frac{1}{|\mathbf{r}-\mathbf{r}'|}=-4\pi\delta(\mathbf{r}-\mathbf{r}') \tag{1.75}$$

$$\Delta_{\mathbf{r}'}u^-(\mathbf{r}')=\frac{1}{\sigma}\nabla\cdot\mathbf{J}^P(\mathbf{r}') \tag{1.76}$$

$$\hat{\mathbf{n}}(\mathbf{r}')\cdot\nabla_{\mathbf{r}'}\frac{1}{|\mathbf{r}-\mathbf{r}'|}=\hat{\mathbf{n}}(\mathbf{r}')\cdot\frac{\mathbf{r}-\mathbf{r}'}{|\mathbf{r}-\mathbf{r}'|^3} \tag{1.77}$$

and

$$\frac{\partial}{\partial n'}u^-(\mathbf{r}')=0,\quad \mathbf{r}'\in\partial\Omega^- \tag{1.78}$$

we obtain

$$u^-(\mathbf{r})=-\frac{1}{4\pi\sigma}\int_{\Omega^-}\frac{\nabla_{\mathbf{r}'}\cdot\mathbf{J}^P(\mathbf{r}')}{|\mathbf{r}-\mathbf{r}'|}dv(\mathbf{r}')$$
$$-\frac{1}{4\pi}\oint_{\partial\Omega^-}u^-(\mathbf{r}')\hat{\mathbf{n}}(\mathbf{r}')\cdot\frac{\mathbf{r}-\mathbf{r}'}{|\mathbf{r}-\mathbf{r}'|^3}ds(\mathbf{r}'). \tag{1.79}$$

The first integral on the right hand side of (1.79) gives

$$\int_{\Omega^-}\frac{\nabla_{\mathbf{r}'}\cdot\mathbf{J}^P(\mathbf{r}')}{|\mathbf{r}-\mathbf{r}'|}dv(\mathbf{r}')=\int_{\Omega^-}\frac{\mathbf{Q}\cdot\nabla_{\mathbf{r}'}\delta(\mathbf{r}'-\mathbf{r}_0)}{|\mathbf{r}-\mathbf{r}'|}dv(\mathbf{r}')$$
$$=-\mathbf{Q}\cdot\nabla_{\mathbf{r}_0}\int_{\Omega^-}\frac{\delta(\mathbf{r}'-\mathbf{r}_0)}{|\mathbf{r}-\mathbf{r}'|}dv(\mathbf{r}')=-\mathbf{Q}\cdot\nabla_{\mathbf{r}_0}\frac{1}{|\mathbf{r}-\mathbf{r}_0|}$$
$$=-\mathbf{Q}\cdot\frac{\mathbf{r}-\mathbf{r}_0}{|\mathbf{r}-\mathbf{r}_0|^3}. \tag{1.80}$$

Therefore, the *Geselowitz formula for the electric potential* is given by

$$u^{-}(\mathbf{r}) = \frac{1}{4\pi\sigma}\mathbf{Q}\cdot\frac{\mathbf{r}-\mathbf{r}_0}{|\mathbf{r}-\mathbf{r}_0|^3} - \frac{1}{4\pi}\oint_{\partial\Omega^-} u^{-}(\mathbf{r}')\,\hat{\mathbf{n}}(\mathbf{r}')\cdot\frac{\mathbf{r}-\mathbf{r}'}{|\mathbf{r}-\mathbf{r}'|^3}ds(\mathbf{r}') \tag{1.81}$$

which holds for every $\mathbf{r} \notin \partial\Omega^-$.

Just as in the case of Magnetoencephalography, the contribution of the induction current to the electric potential is represented by a surface distribution of dipoles, normal to the boundary of the conductor, having a moment which is proportional to the value of the electric potential at the corresponding point.

1.6 Dyadic unification of EEG and MEG

The theory that was presented in this introductory chapter can be unified in a dyadic form in this formulation Electroencephalography and Magnetoencephalography are the associated scalar and vector invariants.

We recall that the governing equation for the electric potential is

$$\sigma\,\Delta u(\mathbf{r}) = \nabla\cdot\mathbf{J}^p(\mathbf{r}) \tag{1.82}$$

whereas the one governing the magnetic field is

$$-\frac{1}{\mu_0}\Delta\mathbf{B}(\mathbf{r}) = \nabla\times\mathbf{J}^p(\mathbf{r}) \tag{1.83}$$

which is obtained by combing eqs. (1.8), (1.26), (1.27) and (1.29). Equations (1.82) and (1.83) indicate that the scalar invariant of the dyadic $\nabla\otimes\mathbf{J}^p$ serves as the source for the electric potential, and the vector invariant of $\nabla\otimes\mathbf{J}^p$serves as the source for the magnetic field.

Equation (1.82) is the scalar invariant of the dyadic equation

$$\nabla\otimes(\nabla\sigma u(\mathbf{r})) = \nabla\otimes\mathbf{J}^p(\mathbf{r}) \tag{1.84}$$

and eq. (1.83) is the vector invariant of

$$\nabla\otimes\left(\nabla\times\frac{1}{\mu_0}\mathbf{B}(\mathbf{r})\right) = \nabla\otimes\mathbf{J}^p(\mathbf{r}) \tag{1.85}$$

where we have utilized the fact that **B** is a solenoidal field. This remark lead us to the following unification of the 2 Geselowitz formulae. We define the neuronal current distribution

$$\mathbf{d}(\mathbf{r}) = -\sigma u^{-}(\mathbf{r})\hat{\mathbf{n}}(\mathbf{r}),\quad \mathbf{r}\in\partial\Omega^- \tag{1.86}$$

and the dyadic field

$$\tilde{\mathbf{D}}(\mathbf{r}) = \mathbf{Q} \otimes \frac{\mathbf{r}-\mathbf{r}_0}{|\mathbf{r}-\mathbf{r}_0|^3} + \oint_{\partial\Omega^-} \mathbf{d}(\mathbf{r}') \otimes \frac{\mathbf{r}-\mathbf{r}'}{|\mathbf{r}-\mathbf{r}'|^3} ds(\mathbf{r}'). \tag{1.87}$$

Then, if we denote by S_D the scalar invariant and by $\mathbf{V}_D$ the vector invariant of the dyadic $\tilde{\mathbf{D}}$, it follows that

$$u^-(\mathbf{r}) = \frac{1}{4\pi\sigma} S_D(\mathbf{r}) \tag{1.88}$$

and

$$\mathbf{B}(\mathbf{r}) = \frac{\mu_0}{4\pi} \mathbf{V}_D(\mathbf{r}). \tag{1.89}$$

Formula (1.88) coincides with the representation (1.81) and formula (1.89) coincides with the representation (1.72).

Hence, EEG and MEG are the 2 fundamental invariants of the same dyadic structure.

1.7 The inhomogeneous shell model

In what follows a more accurate model of the head-brain system is discussed. The geometry of this model is depicted in the following figure.

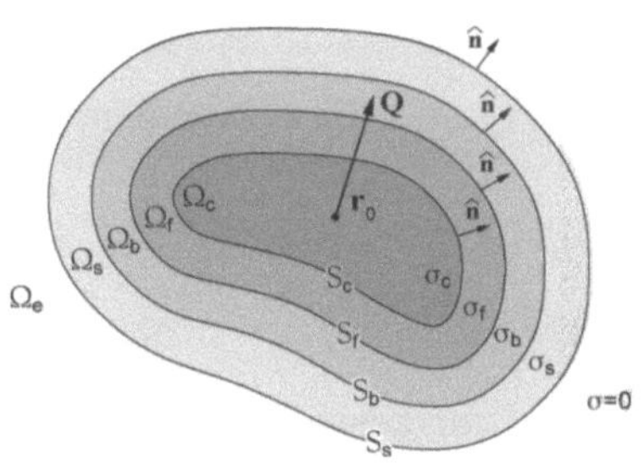

In the above figure, Ω_c represents the actual brain tissue (cerebrum) having the boundary S_c. The region Ω_c is surrounded by the shell Ω_f which contains the cerebrospinal fluid and has the exterior boundary S_f. The next shell Ω_b represents the cranium bones (skull) having the exterior boundary S_b. Finally, the skull is surrounded by the skin (scalp) having the exterior boundary S_s. The exterior to S_s space is denoted by Ω_e.

A dipole $(\mathbf{r}_0, \mathbf{Q})$ is located in the interior of $\Omega_c$giving rise to the neuronal current

$$\mathbf{J}^p(\mathbf{r}) = \mathbf{Q}\delta(\mathbf{r}-\mathbf{r}_0). \tag{1.90}$$

The conductivities of the regions Ω_c, Ω_f, Ω_b, Ω_s are denoted by σ_c, σ_f, σ_b, σ_s respectively, while the conductivity of the exterior space Ω_e is taken to be zero. The magnetic permeability of all regions, including the exterior, is constant and is denoted by μ_0. The electric potential satisfies the following equations:

$$\sigma_c \Delta u_c(\mathbf{r}) = \mathbf{Q} \cdot \nabla \delta(\mathbf{r} - \mathbf{r}_0), \quad \mathbf{r} \in \Omega_c \tag{1.91}$$

$$\Delta u_f(\mathbf{r}) = 0, \quad \mathbf{r} \in \Omega_f \tag{1.92}$$

$$\Delta u_b(\mathbf{r}) = 0, \quad \mathbf{r} \in \Omega_b \tag{1.93}$$

$$\Delta u_s(\mathbf{r}) = 0, \quad \mathbf{r} \in \Omega_s \tag{1.94}$$

$$\Delta u_e(\mathbf{r}) = 0, \quad \mathbf{r} \in \Omega_e \tag{1.95}$$

The electric potential is everywhere harmonic expect in the interior of the brain tissue where the dipolar current is located.

On the interfaces, which support the conductivity jumps, we demand that the electric potential and the normal derivative of the induction current remain continuous.

Since the induction current is proportional to the electric field which is represented by the gradient of the electric potential, the transmition conditions are stated as follows:

$$u_c(\mathbf{r}) = u_f(\mathbf{r}) \tag{1.96}$$

$$\sigma_c \frac{\partial}{\partial n} u_c(\mathbf{r}) = \sigma_f \frac{\partial}{\partial n} u_f(\mathbf{r}), \quad \mathbf{r} \in S_c \tag{1.97}$$

$$u_f(\mathbf{r}) = u_b(\mathbf{r}) \tag{1.98}$$

$$\sigma_f \frac{\partial}{\partial n} u_f(\mathbf{r}) = \sigma_b \frac{\partial}{\partial n} u_b(\mathbf{r}), \quad \mathbf{r} \in S_f \tag{1.99}$$

$$u_b(\mathbf{r}) = u_s(\mathbf{r}) \tag{1.100}$$

$$\sigma_b \frac{\partial}{\partial n} u_b(\mathbf{r}) = \sigma_s \frac{\partial}{\partial n} u_s(\mathbf{r}), \quad \mathbf{r} \in S_b \tag{1.101}$$

and finally on the boundary S_s we have the Neumann condition

$$\frac{\partial}{\partial n} u_s(\mathbf{r}) = 0 \tag{1.102}$$

which ensures that there is no leak of current outside the head because of the nonconductivity of the air and the continuity condition

$$u_s(\mathbf{r}) = u_e(\mathbf{r}). \tag{1.103}$$

Furthermore, the exterior potential has to satisfy the asymptotic condition

$$u_e(\mathbf{r}) = O\left(\frac{1}{r}\right), \quad \mathbf{r} \to \infty. \tag{1.104}$$

Following the steps that led us to the Geselowitz formula (1.72) for the magnetic field **B**, we obtain

$$\begin{aligned}
\mathbf{B}(\mathbf{r}) = {} & \frac{\mu_0}{4\pi}\mathbf{Q} \times \frac{\mathbf{r}-\mathbf{r}_0}{|\mathbf{r}-\mathbf{r}_0|^3} - \frac{\mu_0\sigma_c}{4\pi}\int_{\Omega_c} (\nabla_{\mathbf{r}'} u_c(\mathbf{r}')) \times \frac{\mathbf{r}-\mathbf{r}'}{|\mathbf{r}-\mathbf{r}'|^3}\, dv(\mathbf{r}') \\
& - \frac{\mu_0\sigma_f}{4\pi}\int_{\Omega_f} (\nabla_{\mathbf{r}'} u_f(\mathbf{r}')) \times \frac{\mathbf{r}-\mathbf{r}'}{|\mathbf{r}-\mathbf{r}'|^3}\, dv(\mathbf{r}') - \frac{\mu_0\sigma_b}{4\pi}\int_{\Omega_b} (\nabla_{\mathbf{r}'} u_b(\mathbf{r}')) \times \frac{\mathbf{r}-\mathbf{r}'}{|\mathbf{r}-\mathbf{r}'|^3}\, dv(\mathbf{r}') \\
& - \frac{\mu_0\sigma_s}{4\pi}\int_{\Omega_s} (\nabla_{\mathbf{r}'} u_s(\mathbf{r}')) \times \frac{\mathbf{r}-\mathbf{r}'}{|\mathbf{r}-\mathbf{r}'|^3}\, dv(\mathbf{r}').
\end{aligned} \tag{1.105}$$

In view of Gauss theorem, this equation can be written as

$$\begin{aligned}
\mathbf{B}(\mathbf{r}) = {} & \frac{\mu_0}{4\pi}\mathbf{Q} \times \frac{\mathbf{r}-\mathbf{r}_0}{|\mathbf{r}-\mathbf{r}_0|^3} \\
& - \frac{\mu_0}{4\pi}(\sigma_c - \sigma_f)\oint_{S_c} u_c(\mathbf{r}')\, \hat{\mathbf{n}}(\mathbf{r}') \times \frac{\mathbf{r}-\mathbf{r}'}{|\mathbf{r}-\mathbf{r}'|^3}\, ds(\mathbf{r}') \\
& - \frac{\mu_0}{4\pi}(\sigma_f - \sigma_b)\oint_{S_f} u_f(\mathbf{r}')\, \hat{\mathbf{n}}(\mathbf{r}') \times \frac{\mathbf{r}-\mathbf{r}'}{|\mathbf{r}-\mathbf{r}'|^3}\, ds(\mathbf{r}') \\
& - \frac{\mu_0}{4\pi}(\sigma_b - \sigma_s)\oint_{S_b} u_b(\mathbf{r}')\, \hat{\mathbf{n}}(\mathbf{r}') \times \frac{\mathbf{r}-\mathbf{r}'}{|\mathbf{r}-\mathbf{r}'|^3}\, ds(\mathbf{r}') \\
& - \frac{\mu_0}{4\pi}\sigma_s\oint_{S_s} u_s(\mathbf{r}')\, \hat{\mathbf{n}}(\mathbf{r}') \times \frac{\mathbf{r}-\mathbf{r}'}{|\mathbf{r}-\mathbf{r}'|^3}\, ds(\mathbf{r}'), \quad \mathbf{r} \notin S_c \cup S_f \cup S_b \cup S_s
\end{aligned} \tag{1.106}$$

In an exactly analogous manner we can prove the following generalization of the Geselowitz formula for the electric potential:

$$\begin{aligned}
u(\mathbf{r}) = {} & \frac{1}{4\pi\sigma_c}\mathbf{Q} \cdot \frac{\mathbf{r}-\mathbf{r}_0}{|\mathbf{r}-\mathbf{r}_0|^3} \\
& - \frac{1}{4\pi}\frac{\sigma_c - \sigma_f}{\sigma_c}\oint_{S_c} u_c(\mathbf{r}')\hat{\mathbf{n}}(\mathbf{r}') \cdot \frac{\mathbf{r}-\mathbf{r}'}{|\mathbf{r}-\mathbf{r}'|^3}\, ds(\mathbf{r}')
\end{aligned}$$

$$-\frac{1}{4\pi}\frac{\sigma_f-\sigma_b}{\sigma_c}\oint\limits_{S_f} u_f(\mathbf{r}')\hat{\mathbf{n}}(\mathbf{r}')\cdot\frac{\mathbf{r}-\mathbf{r}'}{|\mathbf{r}-\mathbf{r}'|^3}ds(\mathbf{r}')$$

$$-\frac{1}{4\pi}\frac{\sigma_b-\sigma_s}{\sigma_c}\oint\limits_{S_b} u_b(\mathbf{r}')\hat{\mathbf{n}}(\mathbf{r}')\cdot\frac{\mathbf{r}-\mathbf{r}'}{|\mathbf{r}-\mathbf{r}'|^3}ds(\mathbf{r}'), \quad \mathbf{r}\notin S_c\cup S_f\cup S_b\cup S_s. \qquad (1.107)$$

We observe again that the generalized Geselowitz formulae (1.106) and (1.107) express the conductivity discontinuities with the corresponding surface distributions of dipolar currents, with moments which are normal to the surfaces and have magnitudes proportional to the conductivity jumps across the surfaces. We also mention that the only conductivity which is significantly different than the others is the conductivity of the skull which is the only solid medium. Indicative values of the conductivities are $\sigma_c \approx \sigma_s \approx 0.33(\Omega m)^{-1}$, $\sigma_f \approx 1\ (\Omega m)^{-1}$ and $\sigma_b \approx 4,2\cdot 10^{-3}(\Omega m)^{-1}$. It is evident that the original Geselowitz formulae can be recovered by setting in (1.106) and (1.107) $\sigma_c = \sigma_f = \sigma_b = \sigma_s = \sigma$.

Chapter 2
The definitive nonuniqueness results for deterministic EEG and MEG data

It is well known that most inverse problems suffer from lack of uniqueness and this is indeed the case for the inverse problems of EEG and MEG. Actually, the fact that a current within a homogenous conductor cannot be completely identified from measurements of the magnetic field it generates in the exterior of the conductor was known to Helmholtz since 1853 [21]. However, a first rigorous mathematical proof of this fact appeared only in 1996 [17]. Over the last couple of decades, the authors and their collaborators have investigated extensively the mathematical theory of the direct and inverse problems of EEG and MEG in different geometries [1–4, 5–17, 19, 20, 22]. In these works, a variety of partial results were obtained regarding the dependence of the electric potential and of the magnetic field on parts of the neuronal current, which was assumed to be either a collection of dipoles or a continuous distribution of dipoles. The basic result in this direction was finally obtained by the second author in [16]. It states that in any geometry, the scalar function that defines the irrotational part of the current contributes to both the electric potential and to the magnetic field. Furthermore, it also states that the magnetic field also depends on the radial component of the vector function which defines the solenoidal part of the current. Hence, among the three scalar functions that define the three components of the current vector, one component contributes to both brain imaging modalities of EEG and MEG, and one additional component contributes to MEG. The third component of the current vector is "invisible" by both EEG and MEG.

2.1 The electric potential on the scalp

Using the notation of paragraph 1.7, and assuming that the neuronal currents lives inside the brain tissue Ω_c, it follows that the electric potential $u_c(\mathbf{r})$ in Ω_c satisfies the transmission problem

$$\sigma_c \Delta u_c(\mathbf{r}) = \nabla \cdot \mathbf{J}^p(\mathbf{r}), \quad \mathbf{r} \in \Omega_c \tag{2.1}$$

$$\sigma_c \frac{\partial}{\partial n} u_c(\mathbf{r}) = \sigma_f \frac{\partial}{\partial n} u_f(\mathbf{r}), \quad \mathbf{r} \in S_c. \tag{2.2}$$

The electric potential $u_f(\mathbf{r})$ in the fluid region Ω_f satisfies the transmission problem

$$\Delta u_f(\mathbf{r}) = 0, \quad \mathbf{r} \in \Omega_f \tag{2.3}$$

$$u_f(\mathbf{r}) = u_c(\mathbf{r}), \quad \mathbf{r} \in S_c \tag{2.4}$$

https://doi.org/10.1515/9783110547535-003

$$\sigma_f \frac{\partial}{\partial n} u_f(\mathbf{r}) = \sigma_b \frac{\partial}{\partial n} u_b(\mathbf{r}), \quad \mathbf{r} \in S_f. \tag{2.5}$$

Similarly, $u_b(\mathbf{r})$ in the region Ω_b, occupied by the bone, satisfies

$$\Delta u_b(\mathbf{r}) = 0, \quad \mathbf{r} \in \Omega_b \tag{2.6}$$

$$u_b(\mathbf{r}) = u_f(\mathbf{r}), \quad \mathbf{r} \in S_f \tag{2.7}$$

$$\sigma_b \frac{\partial}{\partial n} u_b(\mathbf{r}) = \sigma_s \frac{\partial}{\partial n} u_s(\mathbf{r}), \quad \mathbf{r} \in S_b. \tag{2.8}$$

Finally, $u_s(\mathbf{r})$ in the skin region Ω_s satisfies

$$\Delta u_s(\mathbf{r}) = 0, \quad \mathbf{r} \in \Omega_s \tag{2.9}$$

$$u_s(\mathbf{r}) = u_b(\mathbf{r}), \quad \mathbf{r} \in S_b \tag{2.10}$$

$$\frac{\partial}{\partial n} u_s(\mathbf{r}) = 0, \quad \mathbf{r} \in S_s. \tag{2.11}$$

The transmission conditions (2.4), (2.7) and (2.10) express the continuity of the electric potential, while the transmission conditions (2.2), (2.5) and (2.8) ensure the continuity of the normal component of the current. Finally, the boundary condition (2.11) states that, since the exterior space Ω_e is not conductive, the current does not leave the region interior to S_s.

Let the primary current $\mathbf{J}^p$ be an isolated dipole with moment $\mathbf{Q}$ which is located at the point τ in Ω_c. If we denote the electric potentials it generates by $u_j(\mathbf{r}, \tau)$, $j = c, f, b, s$, then we obtain the following transmission problems:

For the potential $u_c(\mathbf{r}, \tau)$

$$\sigma_c \Delta_{\mathbf{r}} u_c(\mathbf{r}, \tau) = \nabla_{\mathbf{r}} \cdot (\mathbf{Q}(\tau)\delta(\mathbf{r} - \tau)), \quad \mathbf{r} \in \Omega_c \tag{2.12}$$

$$\sigma_c \left(\frac{\partial}{\partial n(\mathbf{r})}\right) u_c(\mathbf{r}, \tau) = \sigma_f \left(\frac{\partial}{\partial n(\mathbf{r})}\right) u_f(\mathbf{r}, \tau), \quad \mathbf{r} \in S_c. \tag{2.13}$$

For the potential $u_f(\mathbf{r}, \tau)$

$$\Delta_{\mathbf{r}} u_f(\mathbf{r}, \tau) = 0, \quad \mathbf{r} \in \Omega_f \tag{2.14}$$

$$u_f(\mathbf{r}, \tau) = u_c(\mathbf{r}, \tau), \quad \mathbf{r} \in S_c \tag{2.15}$$

$$\sigma_f \left(\frac{\partial}{\partial n(\mathbf{r})}\right) u_f(\mathbf{r}, \tau) = \sigma_b \left(\frac{\partial}{\partial n(\mathbf{r})}\right) u_b(\mathbf{r}, \tau), \quad \mathbf{r} \in S_f. \tag{2.16}$$

For the potential $u_b(\mathbf{r}, \tau)$

$$\Delta_{\mathbf{r}} u_b(\mathbf{r}, \tau) = 0, \quad \mathbf{r} \in \Omega_b \tag{2.17}$$

$$u_b(\mathbf{r},\tau) = u_f(\mathbf{r},\tau), \quad \mathbf{r} \in S_f \tag{2.18}$$

$$\sigma_b\left(\frac{\partial}{\partial n(\mathbf{r})}\right)u_b(\mathbf{r},\tau) = \sigma_s\left(\frac{\partial}{\partial n(\mathbf{r})}\right)u_s(\mathbf{r},\tau), \quad \mathbf{r} \in S_b. \tag{2.19}$$

For the potential $u_s(\mathbf{r},\tau)$

$$\Delta_\mathbf{r} u_s(\mathbf{r},\tau) = 0, \quad \mathbf{r} \in \Omega_s \tag{2.20}$$

$$u_s(\mathbf{r},\tau) = u_b(\mathbf{r},\tau), \quad \mathbf{r} \in S_b \tag{2.21}$$

$$\left(\frac{\partial}{\partial n(\mathbf{r})}\right)u_s(\mathbf{r},\tau) = 0, \quad \mathbf{r} \in S_s. \tag{2.22}$$

In eqs. (2.12)–(2.22) the position of the dipole τ lies in Ω_c.

We know that

$$\Delta_\mathbf{r}\left(-\frac{1}{4\pi|\mathbf{r}-\tau|}\right) = \delta(\mathbf{r}-\tau) \tag{2.23}$$

and that

$$\begin{aligned}\nabla_\mathbf{r}\cdot(\mathbf{Q}(\tau)\delta(\mathbf{r}-\tau)) &= \mathbf{Q}(\tau)\cdot\nabla_\mathbf{r}\delta(\mathbf{r}-\tau)\\ &= -\mathbf{Q}(\tau)\cdot\nabla_\tau\delta(\mathbf{r}-\tau).\end{aligned} \tag{2.24}$$

Utilizing the fact that the operators $(\mathbf{Q}(\tau)\cdot\nabla_\mathbf{r})$ and $\Delta_\mathbf{r}$ in their invariant form commute, and using eqs. (2.23) and (2.24), we can rewrite eq. (2.12) in the form

$$\begin{aligned}\sigma_c\Delta_\mathbf{r} u_c(\mathbf{r},\tau) &= -(\mathbf{Q}(\tau)\cdot\nabla_\tau)\delta(\mathbf{r}-\tau)\\ &= (\mathbf{Q}(\tau)\cdot\nabla_\tau)\Delta_\mathbf{r}\left(\frac{1}{4\pi|\mathbf{r}-\tau|}\right)\\ &= \Delta_\mathbf{r}(\mathbf{Q}(\tau)\cdot\nabla_\tau)\frac{1}{4\pi|\mathbf{r}-\tau|}.\end{aligned} \tag{2.25}$$

Equation (2.25) implies that

$$u_c(\mathbf{r},\tau) = \frac{1}{4\pi\sigma_c}(\mathbf{Q}(\tau)\cdot\nabla_\tau)\left[\frac{1}{|\mathbf{r}-\tau|} + \upsilon_c(\mathbf{r},\tau)\right] \tag{2.26}$$

where, in Ω_c, υ_c is a harmonic function with respect to the variable $\mathbf{r}$.

The above analysis indicates that a dipolar current is represented by the action of the dipole dependent directional derivative $(\mathbf{Q}(\tau)\cdot\nabla_\tau)$ on the monopole source $\delta(\mathbf{r}-\tau)$ which is located at τ. This observation implies that we can reduce all problems (2.12)–(2.22) at the monopole level by introducing the monopolic electric potentials υ_j, $j=c,f,b,s$ which are connected with the corresponding dipolic potentials by eq. (2.26) and by

$$u_j(\mathbf{r},\tau) = \frac{1}{4\pi}\mathbf{Q}(\tau)\cdot\nabla_\tau v_j(\mathbf{r},\tau), \quad j=f,b,s. \tag{2.27}$$

Note that eqs. (2.26) and (2.27) separate the effect of the source $\mathbf{Q}(\tau)\delta(\mathbf{r}-\tau)$ from the effect of the geometry of the problem which is now carried by the potentials v_j, $j=c,f,b,s$. The monopolic potentials satisfy the following problems:

For the potential v_c in Ω_c

$$\Delta_{\mathbf{r}} v_c(\mathbf{r},\tau) = 0, \quad \mathbf{r}\in\Omega_c \tag{2.28}$$

$$\frac{1}{\sigma_c}\left[\frac{1}{|\mathbf{r}-\tau|} + v_c(\mathbf{r},\tau)\right] = v_f(\mathbf{r},\tau), \quad \mathbf{r}\in S_c \tag{2.29}$$

$$\left(\frac{\partial}{\partial n(\mathbf{r})}\right)\left[\frac{1}{|\mathbf{r}-\tau|} + v_c(\mathbf{r},\tau)\right] = \sigma_f\left(\frac{\partial}{\partial n(\mathbf{r})}\right)v_f(\mathbf{r},\tau), \quad \mathbf{r}\in S_c. \tag{2.30}$$

For the potential v_f in Ω_f

$$\Delta_{\mathbf{r}} v_f(\mathbf{r},\tau) = 0, \quad \mathbf{r}\in\Omega_f \tag{2.31}$$

$$v_f(\mathbf{r},\tau) = v_b(\mathbf{r},\tau), \quad \mathbf{r}\in S_f \tag{2.32}$$

$$\sigma_f\left(\frac{\partial}{\partial n(\mathbf{r})}\right)v_f(\mathbf{r},\tau) = \sigma_b\left(\frac{\partial}{\partial n(\mathbf{r})}\right)v_b(\mathbf{r},\tau), \quad \mathbf{r}\in S_f. \tag{2.33}$$

For the potential v_b in Ω_b

$$\Delta_{\mathbf{r}} v_b(\mathbf{r},\tau) = 0, \quad \mathbf{r}\in\Omega_b \tag{2.34}$$

$$v_b(\mathbf{r},\tau) = v_s(\mathbf{r},\tau), \quad \mathbf{r}\in S_b \tag{2.35}$$

$$\sigma_b\left(\frac{\partial}{\partial n(\mathbf{r})}\right)v_b(\mathbf{r},\tau) = \sigma_s\left(\frac{\partial}{\partial n(\mathbf{r})}\right)v_s(\mathbf{r},\tau), \quad \mathbf{r}\in S_b. \tag{2.36}$$

For the potential v_s in Ω_s

$$\Delta_{\mathbf{r}} v_s(\mathbf{r},\tau) = 0, \quad \mathbf{r}\in\Omega_s \tag{2.37}$$

$$v_s(\mathbf{r},\tau) = v_e(\mathbf{r},\tau), \quad \mathbf{r}\in S_s \tag{2.38}$$

$$\left(\frac{\partial}{\partial n(\mathbf{r})}\right)v_s(\mathbf{r},\tau) = 0, \quad \mathbf{r}\in S_s. \tag{2.39}$$

The source location τ belongs again in the interior of Ω_c.

Without loss of generality, a distributed primary current $\mathbf{J}^p$ can be expressed in the form

$$\mathbf{J}^p(\mathbf{r}) = \int_{\Omega_c}\mathbf{Q}(\tau)\delta(\mathbf{r}-\tau)dV(\tau), \quad \mathbf{r}\in\Omega_c. \tag{2.40}$$

In analogy with eq. (2.28) with $j = s$, the corresponding electric potential is given by

$$u_S(\mathbf{r}) = \frac{1}{4\pi} \int\limits_{\Omega_C} \mathbf{J}^p(\tau) \cdot \nabla_\tau \upsilon_S(\mathbf{r}, \tau) dV(\tau) \tag{2.41}$$

for every $\mathbf{r} \in S_S$, assuming that the harmonic field υ_S is uniquely extendable to the boundary S_S of Ω_S.

The divergence theorem implies that

$$\int\limits_{\Omega_C} \nabla_\tau \cdot (\upsilon_S(\mathbf{r}, \tau)\mathbf{J}^p(\tau))\ dV(\tau) = \int\limits_{S_C} \upsilon_S(\mathbf{r}, \tau)\hat{\mathbf{n}}(\tau) \cdot \mathbf{J}^p(\tau)\ dS(\tau) \tag{2.42}$$

and therefore the integral on the right hand side of eq. (2.41) is written as

$$\int\limits_{\Omega_C} \mathbf{J}^p(\tau) \cdot \nabla_\tau \upsilon_S(\mathbf{r}, \tau) dV(\tau) = -\int\limits_{\Omega_C} \upsilon_S(\mathbf{r}, \tau)(\nabla_\tau \cdot \mathbf{J}^p(\tau))\ dV(\tau) + \int\limits_{S_C} \upsilon_S(\mathbf{r}, \tau)(\hat{\mathbf{n}}(\tau) \cdot \mathbf{J}^p(\tau))\ dS(\tau). \tag{2.43}$$

Consequently, (2.40) is written as

$$u_S(\mathbf{r}) = -\frac{1}{4\pi} \int\limits_{\Omega_C} \upsilon_S(\mathbf{r}, \tau)(\nabla_\tau \cdot \mathbf{J}^p(\tau))\ dV(\tau) + \frac{1}{4\pi} \int\limits_{S_C} \upsilon_S(\mathbf{r}, \tau)(\hat{\mathbf{n}}(\tau) \cdot \mathbf{J}^p(\tau))\ dS(\tau), \quad \mathbf{r} \in S_S. \tag{2.44}$$

Next we decompose the current $\mathbf{J}^p$ via the Helmholtz representation

$$\mathbf{J}^p(\tau) = \nabla_\tau \Psi(\tau) + \nabla_\tau \times \mathbf{A}(\tau) \tag{2.45}$$

under the gage condition

$$\nabla_\tau \cdot \mathbf{A}(\tau) = 0. \tag{2.46}$$

Equations (2.45) and (2.46) imply that the current is represented by the scalar potential Ψ and any two components of the vector potential $\mathbf{A}$. The differential representation (2.45) implies that

$$\nabla_\tau \cdot \mathbf{J}^p(\tau) = \Delta_\tau \Psi(\tau) \tag{2.47}$$

thus the source term for the electric potential depends only on the scalar potential Ψ. Then, eq. (2.44) implies

$$\begin{aligned} u_S(\mathbf{r}) = &-\frac{1}{4\pi} \int\limits_{\Omega_C} \upsilon_S(\mathbf{r}, \tau)(\Delta_\tau \Psi(\tau))\ dV(\tau) \\ &+ \frac{1}{4\pi} \int\limits_{S_C} \upsilon_S(\mathbf{r}, \tau)(\hat{\mathbf{n}}(\tau) \cdot \mathbf{J}^p(\tau))\ dS(\tau), \quad \mathbf{r} \in S_S. \end{aligned} \tag{2.48}$$

In summary, the EEG data on the surface of the scalp S_s are affected by the nonharmonic part of the scalar potential Ψ and by the normal component of the current on the boundary S_c of the brain tissue Ω_c. The volume integral on the right hand side of the representation (2.48) can further be expressed as a surface integral S_c. This is achieved by using Green's second identity

$$\int_{\Omega} [u\Delta v - v\Delta u] dV = \int_{\partial\Omega} \hat{\mathbf{n}} \cdot [u\nabla v - v\nabla u] dS \tag{2.49}$$

for $u = v_s$, $v = \Psi$ and $\Omega = \Omega_c$. The equations defining the potentials $v_j(\mathbf{r}, \tau)$, $j = c, f, b, s$ remain invariant under the commutation of $\mathbf{r}$ and τ. Hence, $v_s(\mathbf{r}, \tau)$ is harmonic in τ as well, and (2.49) yields

$$\begin{aligned} &\int_{\Omega_c} v_s(\mathbf{r}, \tau)(\Delta_\tau \Psi(\tau)) \, dV(\tau) \\ &= \int_{S_c} \hat{\mathbf{n}}(\tau) \cdot \left[v_s(\mathbf{r}, \tau)\nabla_\tau \Psi(\tau) - \Psi(\tau)\nabla_\tau v_s(\mathbf{r}, \tau)\right] \, dS(\tau). \end{aligned} \tag{2.50}$$

Consequently, the surface potential $u_s(\mathbf{r})$ has the representation

$$u_s(\mathbf{r}) = \frac{1}{4\pi} \int_{S_c} \hat{\mathbf{n}}(\tau) \cdot \left[\Psi(\tau)\nabla_\tau v_s(\mathbf{r}, \tau) - v_s(\mathbf{r}, \tau)\nabla_\tau \Psi(\tau) + v_s(\mathbf{r}, \tau)\mathbf{J}^p(\tau)\right] \, dS(\tau), \quad \mathbf{r} \in S_s. \tag{2.51}$$

We clarify that the potential $v_s(\mathbf{r}, \tau)$ is defined for $\mathbf{r} \in \Omega_s$ and $\tau \in \Omega_c$.

In view of the decomposition (2.45) formula (2.51) is also written as

$$u_s(\mathbf{r}) = \frac{1}{4\pi} \int_{S_c} \hat{\mathbf{n}}(\tau) \cdot [\Psi(\tau)\nabla_\tau v_s(\mathbf{r}, \tau) + v_s(\mathbf{r}, \tau)\nabla_\tau \times \mathbf{A}(\tau)] \, dS(\tau) \tag{2.52}$$

which exhibits the dependence of the potential u_s on the scalar function Ψ as well as on the tangential components of the vector function. Under the assumption (iii) of the introduction the term involving $\mathbf{A}$ vanishes and we find (v) of the introduction.

2.2 The exterior magnetic field

Equation (1.105), in the first chapter, provides the Geselowitz integral representation for the magnetic field in the exterior of the head, in terms of the electric potentials $u_j(\mathbf{r}, \tau)$, $j = c, f, b, s$ on the corresponding domains Ω_j, $j = c, f, b, s$, when the excitation is due to a dipole $\{\mathbf{r}_0, \mathbf{Q}\}$. If the isolated dipolar source is replaced by the distributed dipolar current $\mathbf{J}^p$ the cerebrum Ω_c the corresponding representation becomes

$$\mathbf{B}(\mathbf{r}) = \frac{\mu_0}{4\pi} \int_{\Omega_c} \mathbf{J}^p(\tau) \times \nabla_\tau \frac{1}{|\mathbf{r}-\tau|} dV(\tau) - \frac{\mu_0}{4\pi} \sum_{j=c,f,b,s} \sigma_j \int_{\Omega_j} (\nabla_\tau u_j(\tau)) \times \left(\nabla_\tau \frac{1}{|\mathbf{r}-\tau|}\right) dV(\tau) \tag{2.53}$$

for every $\mathbf{r}$ in the exterior domain Ω_e. Using the identity

$$(\nabla_\tau u_j(\tau)) \times \left(\nabla_\tau \frac{1}{|\mathbf{r}-\tau|}\right) = \nabla_\tau \times \left(u_j(\tau)\nabla_\tau \frac{1}{|\mathbf{r}-\tau|}\right) \tag{2.54}$$

in the representation (2.53) we obtain

$$\begin{aligned} \mathbf{B}(\mathbf{r}) = {} & \frac{\mu_0}{4\pi} \int_{\Omega_c} \mathbf{J}^p(\rho) \times \nabla_\rho \frac{1}{|\mathbf{r}-\rho|} dV(\rho) \\ & - \frac{\mu_0}{4\pi} \sum_{j=c,f,b,s} \sigma_j \int_{\Omega_j} \nabla_\tau \times \left(u_j(\tau)\nabla_\tau \frac{1}{|\mathbf{r}-\tau|}\right) dV(\tau), \quad \mathbf{r} \in \Omega_e. \end{aligned} \tag{2.55}$$

Utilizing the divergence theorem as well as the transmission and the continuity conditions that ensure the continuity of the electric potential as we cross the interfaces, we rewrite (2.55) as

$$\begin{aligned} \mathbf{B}(\mathbf{r}) = {} & \frac{\mu_0}{4\pi} \int_{\Omega_c} \mathbf{J}^p(\rho) \times \left(\nabla_\rho \frac{1}{|\mathbf{r}-\rho|}\right) dV(\rho) \\ & - \frac{\mu_0}{4\pi}(\sigma_c - \sigma_f) \int_{S_c} u_c(\rho)\ \hat{\mathbf{n}}(\rho) \times \left(\nabla_\rho \frac{1}{|\mathbf{r}-\rho|}\right) dS(\rho) \\ & - \frac{\mu_0}{4\pi}(\sigma_f - \sigma_b) \int_{S_c} u_f(\rho)\ \hat{\mathbf{n}}(\rho) \times \left(\nabla_\rho \frac{1}{|\mathbf{r}-\rho|}\right) dS(\rho) \\ & - \frac{\mu_0}{4\pi}(\sigma_b - \sigma_s) \int_{S_c} u_b(\rho)\ \hat{\mathbf{n}}(\rho) \times \left(\nabla_\rho \frac{1}{|\mathbf{r}-\rho|}\right) dS(\rho) \\ & - \frac{\mu_0 \sigma_s}{4\pi} \int_{S_s} u_s(\rho)\hat{\mathbf{n}}(\rho) \times \left(\nabla_\rho \frac{1}{|\mathbf{r}-\rho|}\right) dS(\rho), \quad \mathbf{r} \in \Omega_e. \end{aligned} \tag{2.56}$$

If we insert, in (2.56), the point source

$$\mathbf{J}^p(\rho) = \mathbf{Q}(\rho)\delta(\rho - \tau) \tag{2.57}$$

and denote the relative potentials by

$$u_j(\rho) = u_j(\rho, \tau), \quad j = c, f, b, s \tag{2.58}$$

where

$$u_c(\rho,\tau) = \frac{1}{4\pi\sigma_c}(\mathbf{Q}(\tau)\cdot\nabla_\tau)\left[\frac{1}{|\rho-\tau|} + \upsilon_c(\rho,\tau)\right] \tag{2.59}$$

and

$$u_j(\rho,\tau) = \frac{1}{4\pi}(\mathbf{Q}(\tau)\cdot\nabla_\tau)\upsilon_j(\rho,\tau), \quad j=f,b,s \tag{2.60}$$

we arrive at the following representation of the exterior magnetic field:

$$\begin{aligned}\mathbf{B}(\mathbf{r},\tau) = {} & \frac{\mu_0}{4\pi}\mathbf{Q}(\tau)\times\left(\nabla_\tau\frac{1}{|\mathbf{r}-\tau|}\right)\\ & - \frac{\mu_0}{(4\pi)^2}(\mathbf{Q}(\tau)\cdot\nabla_\tau)\mathbf{H}(\mathbf{r},\tau), \quad \mathbf{r}\in\Omega_e\end{aligned} \tag{2.61}$$

where $\mathbf{H}$ is defined by

$$\begin{aligned}\mathbf{H}(\mathbf{r},\tau) = {} & \int_{S_c}\left[\frac{1}{|\rho-\tau|} + \upsilon_c(\rho,\tau) - \sigma_f\upsilon_f(\rho,\tau)\right]\hat{\mathbf{n}}(\rho)\times\left(\nabla_\rho\frac{1}{|\mathbf{r}-\rho|}\right)dS(\rho)\\ & + \int_{S_f}(\sigma_f\upsilon_f(\rho,\tau) - \sigma_b\upsilon_b(\rho,\tau))\,\hat{\mathbf{n}}(\rho)\times\left(\nabla_\rho\frac{1}{|\mathbf{r}-\rho|}\right)dS(\rho)\\ & + \int_{S_b}(\sigma_b\upsilon_b(\rho,\tau) - \sigma_s\upsilon_s(\rho,\tau))\,\hat{\mathbf{n}}(\rho)\times\left(\nabla_\rho\frac{1}{|\mathbf{r}-\rho|}\right)dS(\rho)\\ & + \sigma_s\int_{S_s}\upsilon_s(\rho,\tau)\hat{\mathbf{n}}(\rho)\times\left(\nabla_\rho\frac{1}{|\mathbf{r}-\rho|}\right)dS(\rho), \quad \mathbf{r}\in\Omega_s.\end{aligned} \tag{2.62}$$

This equation can be written in the form

$$\begin{aligned}\mathbf{H}(\mathbf{r},\tau) = {} & (\sigma_c-\sigma_f)\int_{S_c}\upsilon_f(\rho,\tau)\,\hat{\mathbf{n}}(\rho)\times\left(\nabla_\rho\frac{1}{|\mathbf{r}-\rho|}\right)dS(\rho)\\ & + (\sigma_f-\sigma_b)\int_{S_f}\upsilon_b(\rho,\tau)\,\hat{\mathbf{n}}(\rho)\times\left(\nabla_\rho\frac{1}{|\mathbf{r}-\rho|}\right)dS(\rho)\\ & + (\sigma_b-\sigma_s)\int_{S_b}\upsilon_s(\rho,\tau)\,\hat{\mathbf{n}}(\rho)\times\left(\nabla_\rho\frac{1}{|\mathbf{r}-\rho|}\right)dS(\rho)\\ & + \sigma_s\int_{S_s}\upsilon_s(\rho,\tau)\hat{\mathbf{n}}(\rho)\times\left(\nabla_\rho\frac{1}{|\mathbf{r}-\rho|}\right)dS(\rho), \quad \mathbf{r}\in\Omega_s.\end{aligned} \tag{2.62a}$$

Going back to the distributed current $\mathbf{J}^p$, eq. (2.61) is now written as

$$\mathbf{B}(\mathbf{r}) = \frac{\mu_0}{4\pi}\int_{\Omega_C} \mathbf{J}^p(\tau)\times\left(\nabla_\tau \frac{1}{|\mathbf{r}-\tau|}\right) dV(\tau) - \frac{\mu_0}{(4\pi)^2}\int_{\Omega_C} \left(\mathbf{J}^p(\tau)\cdot\nabla_\tau\right)\,\mathbf{H}(\mathbf{r},\tau) dV(\tau) \tag{2.63}$$

where $\mathbf{H}(\mathbf{r},\tau)$ is given by (2.62). Using the divergence theorem once more we obtain

$$\begin{aligned}\int_{\Omega_C}\left(\nabla_\tau\times\frac{\mathbf{J}^p(\tau)}{|\mathbf{r}-\tau|}\right) dV(\tau) &= \int_{\Omega_C}\left(\nabla_\tau\frac{1}{|\mathbf{r}-\tau|}\right)\times\mathbf{J}^p(\tau) dV(\tau) + \int_{\Omega_C}\frac{1}{|\mathbf{r}-\tau|}\left(\nabla_\tau\times\mathbf{J}^p(\tau)\right) dV(\tau)\\ &= \int_{S_C}\frac{1}{|\mathbf{r}-\tau|}\left(\hat{\mathbf{n}}(\tau)\times\mathbf{J}^p(\tau)\right) dS(\tau), \quad \mathbf{r}\in\Omega_e.\end{aligned} \tag{2.64}$$

Hence, (2.63) can be written as

$$\begin{aligned}\mathbf{B}(\mathbf{r}) = \frac{\mu_0}{4\pi}\int_{\Omega_C}\frac{1}{|\mathbf{r}-\tau|}\left(\nabla_\tau\times\mathbf{J}^p(\tau)\right) dV(\tau) - \frac{\mu_0}{4\pi}\int_{\Omega_C}\left(\nabla_\tau\times\frac{\mathbf{J}^p(\tau)}{|\mathbf{r}-\tau|}\right) dV(\tau)\\ - \frac{\mu_0}{(4\pi)^2}\int_{\Omega_C}\left(\mathbf{J}^p(\tau)\cdot\nabla_\tau\right)\mathbf{H}(\mathbf{r},\tau) dV(\tau), \quad \mathbf{r}\in\Omega_e.\end{aligned} \tag{2.65}$$

Next we prove the dyadic identity

$$\nabla\cdot(\mathbf{J}\otimes\mathbf{H}) = (\nabla\cdot\mathbf{J})\mathbf{H} + (\mathbf{J}\cdot\nabla)\mathbf{H}. \tag{2.66}$$

Indeed, if $\mathbf{J} = (J_1, J_2, J_3)$ and $\mathbf{H} = (H_1, H_2, H_3)$, then for $i = 1, 2, 3$

$$\nabla\cdot(\mathbf{J}H_i) = (\nabla\cdot\mathbf{J})H_i + (\mathbf{J}\cdot\nabla H_i). \tag{2.67}$$

Multiplying (2.67) by the Cartesian unit vector $\hat{\mathbf{x}}_i$, $i = 1, 2, 3$ and adding up the three resulting equations we arrive at (2.66). Furthermore, integrating the identity (2.66) over a domain Ω and applying the divergence theorem we obtain the integral formula

$$\int_\Omega (\nabla\cdot\mathbf{J})\,\mathbf{H}\,dV + \int_\Omega (\mathbf{J}\cdot\nabla)\,\mathbf{H}\,dV = \int_\Omega \nabla\cdot(\mathbf{J}\mathbf{H})\,dV = \int_{\partial\Omega} (\hat{\mathbf{n}}\cdot\mathbf{J})\,\mathbf{H}\,dS. \tag{2.68}$$

In particular, we can apply (2.68) to Ω_c, use it to replace the third integral on the right hand side of (2.65), and apply the divergence theorem to the second integral on the right hand side of (2.65) to arrive finally at the equation

$$\begin{aligned}\mathbf{B}(\mathbf{r}) = &\frac{\mu_0}{4\pi}\int_{\Omega_C}\frac{1}{|\mathbf{r}-\boldsymbol{\tau}|}(\nabla_\tau \times \mathbf{J}^p(\boldsymbol{\tau}))\, dV(\boldsymbol{\tau})\\ &+\frac{\mu_0}{(4\pi)^2}\int_{\Omega_C}(\nabla_\tau \cdot \mathbf{J}^p(\boldsymbol{\tau}))\,\mathbf{H}(\mathbf{r},\boldsymbol{\tau})\, dV(\boldsymbol{\tau})\\ &-\frac{\mu_0}{4\pi}\int_{S_C}\frac{1}{|\mathbf{r}-\boldsymbol{\tau}|}(\hat{\mathbf{n}}(\boldsymbol{\tau}) \times \mathbf{J}^p(\boldsymbol{\tau}))\, dS(\boldsymbol{\tau})\\ &-\frac{\mu_0}{(4\pi)^2}\int_{S_C}(\hat{\mathbf{n}}(\boldsymbol{\tau}) \cdot \mathbf{J}^p(\boldsymbol{\tau}))\,\mathbf{H}(\mathbf{r},\boldsymbol{\tau})\, dS(\boldsymbol{\tau}), \quad \mathbf{r}\in\Omega_e.\end{aligned} \tag{2.69}$$

The actual purpose for transforming the initial integral representation for **B** to the form (2.69) is to allow the operators of divergence and rotation to act on the primary current $\mathbf{J}^p$. The reason for this will be evident in the sequel. From the representations (2.45) and (2.46) we obtain

$$\nabla_\tau \cdot \mathbf{J}^p(\boldsymbol{\tau}) = \Delta_\tau \Psi(\boldsymbol{\tau}) \tag{2.70}$$

and

$$\nabla_\tau \times \mathbf{J}^p(\boldsymbol{\tau}) = -\Delta_\tau \mathbf{A}(\boldsymbol{\tau}) + \nabla_\tau(\nabla_\tau \cdot \mathbf{A}(\boldsymbol{\tau})) = -\Delta_\tau \mathbf{A}(\boldsymbol{\tau}) \tag{2.71}$$

which, if inserted in (2.69), yield the formula

$$\begin{aligned}\mathbf{B}(\mathbf{r}) = &-\frac{\mu_0}{4\pi}\int_{\Omega_C}\frac{1}{|\mathbf{r}-\boldsymbol{\tau}|}(\Delta_\tau \mathbf{A}(\boldsymbol{\tau}))\, dV(\boldsymbol{\tau})\\ &+\frac{\mu_0}{(4\pi)^2}\int_{\Omega_C}(\Delta_\tau \Psi(\boldsymbol{\tau}))\,\mathbf{H}(\mathbf{r},\boldsymbol{\tau}) dV(\boldsymbol{\tau})\\ &-\frac{\mu_0}{4\pi}\int_{S_C}\frac{1}{|\mathbf{r}-\boldsymbol{\tau}|}(\hat{\mathbf{n}}(\boldsymbol{\tau}) \times \mathbf{J}^p(\boldsymbol{\tau}))\, dS(\boldsymbol{\tau})\\ &-\frac{\mu_0}{(4\pi)^2}\int_{S_C}(\hat{\mathbf{n}}(\boldsymbol{\tau}) \cdot \mathbf{J}^p(\boldsymbol{\tau}))\,\mathbf{H}(\mathbf{r},\boldsymbol{\tau}) dS(\boldsymbol{\tau}), \quad \mathbf{r}\in\Omega_e\end{aligned} \tag{2.72}$$

2.3 The radial component of the exterior magnetic field

It turns out that using eq. (2.72), it is possible to obtain the following convenient equation for the radial component of $\mathbf{B}(\mathbf{r})$

$$\mathbf{r}\cdot\mathbf{B}(\mathbf{r}) = -\frac{\mu_0}{4\pi}\int\limits_{\Omega_C}\frac{1}{|\mathbf{r}-\tau|}\Delta_\tau(\tau\cdot\mathbf{A}(\tau))\;dV(\tau)$$
$$+\frac{\mu_0}{(4\pi)^2}\int\limits_{\Omega_C}(\Delta_\tau\Psi(\tau))(\mathbf{r}\cdot\mathbf{H}(\mathbf{r},\tau))dV(\tau)$$
$$-\frac{\mu_0}{4\pi}\mathbf{r}\cdot\int\limits_{S_C}\frac{1}{|\mathbf{r}-\tau|}(\hat{\mathbf{n}}(\tau)\times\mathbf{J}^p(\tau))\;dS(\tau)$$
$$+\frac{\mu_0}{4\pi}\int\limits_{S_C}(\hat{\mathbf{n}}(\tau)\cdot\Delta_\tau\mathbf{A}(\tau))|\mathbf{r}-\tau|\;dS(\tau)$$
$$-\frac{\mu_0}{(4\pi)^2}\int\limits_{S_C}(\hat{\mathbf{n}}(\tau)\cdot\mathbf{J}^p(\tau))(\mathbf{r}\cdot\mathbf{H}(\mathbf{r},\tau))dS(\tau),\quad \mathbf{r}\in\Omega_e. \quad (2.73)$$

In order to prove the representation (2.73) we utilize the following steps:

Step 1. Using the identity

$$\frac{\mathbf{r}-\tau}{|\mathbf{r}-\tau|} = |\mathbf{r}-\tau|^2\left(\nabla_\tau\frac{1}{|\mathbf{r}-\tau|}\right) \quad (2.74)$$

we find that

$$\begin{aligned}(\Delta_\tau\mathbf{A}(\tau))\cdot\frac{\mathbf{r}-\tau}{|\mathbf{r}-\tau|} &= (\Delta_\tau\mathbf{A}(\tau))|\mathbf{r}-\tau|^2\cdot\left(\nabla_\tau\frac{1}{|\mathbf{r}-\tau|}\right)\\ &= \nabla_\tau\cdot\left[(\Delta_\tau\mathbf{A}(\tau))|\mathbf{r}-\tau|^2\frac{1}{|\mathbf{r}-\tau|}\right] - \nabla_\tau\cdot\left[(\Delta_\tau\mathbf{A}(\tau))|\mathbf{r}-\tau|^2\right]\frac{1}{|\mathbf{r}-\tau|}\\ &= \nabla_\tau\cdot[(\Delta_\tau\mathbf{A}(\tau))|\mathbf{r}-\tau|] - [\nabla_\tau\cdot(\Delta_\tau\mathbf{A}(\tau))]|\mathbf{r}-\tau|\\ &\quad + 2(\Delta_\tau\mathbf{A}(\tau))\cdot\frac{\mathbf{r}-\tau}{|\mathbf{r}-\tau|}\end{aligned} \quad (2.75)$$

where in the last term we have used the identity

$$\begin{aligned}\nabla_\tau|\mathbf{r}-\tau|^2 &= -\nabla_{(\mathbf{r}-\tau)}|\mathbf{r}-\tau|^2 = -2|\mathbf{r}-\tau|\nabla_{(\mathbf{r}-\tau)}|\mathbf{r}-\tau|\\ &= -2|\mathbf{r}-\tau|\frac{\mathbf{r}-\tau}{|\mathbf{r}-\tau|} = -2(\mathbf{r}-\tau).\end{aligned} \quad (2.76)$$

Step 2. Consider the integral

$$I(\mathbf{r}) = \mathbf{r}\cdot\int\limits_{\Omega_C}\frac{1}{|\mathbf{r}-\tau|}(\Delta_\tau\mathbf{A}(\tau))dV(\tau). \quad (2.77)$$

then

$$I(\mathbf{r}) = \int_{\Omega_C} \frac{\mathbf{r}-\tau}{|\mathbf{r}-\tau|} \cdot (\Delta_\tau \mathbf{A}(\tau)) dV(\tau) + \int_{\Omega_C} \frac{1}{|\mathbf{r}-\tau|} \tau \cdot (\Delta_\tau \mathbf{A}(\tau)) dV(\tau). \tag{2.78}$$

Inserting (2.75) in (2.78) and using the divergence theorem, as well as the property

$$\nabla_\tau \cdot (\Delta_\tau \mathbf{A}(\tau)) = \Delta_\tau (\nabla_\tau \cdot \mathbf{A}(\tau)) = 0 \tag{2.79}$$

we obtain

$$\begin{aligned} I(\mathbf{r}) = & \int_{\Omega_C} \nabla_\tau \cdot \left[(\Delta_\tau \mathbf{A}(\tau)) |\mathbf{r}-\tau| + 2(\Delta_\tau \mathbf{A}(\tau)) \cdot \frac{\mathbf{r}-\tau}{|\mathbf{r}-\tau|} \right. \\ & \left. + \frac{1}{|\mathbf{r}-\tau|} \tau \cdot (\Delta_\tau \mathbf{A}(\tau)) \right] dV(\tau) \\ = & \int_{\Omega_C} \frac{1}{|\mathbf{r}-\tau|} \tau \cdot (\Delta_\tau \mathbf{A}(\tau)) dV(\tau) + 2 \int_{\Omega_C} \frac{\mathbf{r}-\tau}{|\mathbf{r}-\tau|} \cdot (\Delta_\tau \mathbf{A}(\tau)) dV(\tau) \\ & + \int_{S_C} (\hat{\mathbf{n}}(\tau) \cdot \Delta_\tau \mathbf{A}(\tau)) |\mathbf{r}-\tau| dS(\tau), \quad \mathbf{r} \in \Omega_e \end{aligned} \tag{2.80}$$

Step 3. Equation (2.80) can be written as

$$\begin{aligned} I(\mathbf{r}) = & \int_{\Omega_C} \frac{1}{|\mathbf{r}-\tau|} \tau \cdot (\Delta_\tau \mathbf{A}(\tau)) dV(\tau) + 2I(\mathbf{r}) \\ & - 2 \int_{\Omega_C} \frac{1}{|\mathbf{r}-\tau|} \tau \cdot (\Delta_\tau \mathbf{A}(\tau)) dV(\tau) \\ & + \int_{S_C} (\hat{\mathbf{n}}(\tau) \cdot \Delta_\tau \mathbf{A}(\tau)) |\mathbf{r}-\tau| dS(\tau) \end{aligned} \tag{2.81}$$

or

$$\begin{aligned} \mathbf{r} \cdot \int_{\Omega_C} \frac{1}{|\mathbf{r}-\tau|} (\Delta_\tau \mathbf{A}(\tau)) dV(\tau) = & \int_{\Omega_C} \frac{1}{|\mathbf{r}-\tau|} \tau \cdot (\Delta_\tau \mathbf{A}(\tau)) dV(\tau) \\ & - \int_{S_C} (\hat{\mathbf{n}}(\tau) \cdot \Delta_\tau \mathbf{A}(\tau)) |\mathbf{r}-\tau| dS(\tau). \end{aligned} \tag{2.82}$$

Finally, we use the identity

$$\Delta_\tau (\tau \cdot \mathbf{A}(\tau)) = \tau \cdot (\Delta_\tau \mathbf{A}(\tau)) + 2(\nabla_\tau \cdot \mathbf{A}(\tau)) = \tau \cdot (\Delta_\tau \mathbf{A}(\tau)) \tag{2.83}$$

in the expression (2.82) to obtain

$$\mathbf{r}\cdot\int_{\Omega_C}\frac{1}{|\mathbf{r}-\tau|}(\Delta_\tau\mathbf{A}(\tau))dV(\tau)=\int_{\Omega_C}\frac{1}{|\mathbf{r}-\tau|}\Delta_\tau(\tau\cdot\mathbf{A}(\tau))dV(\tau)$$
$$-\int_{S_C}(\hat{\mathbf{n}}(\tau)\cdot\Delta_\tau\mathbf{A}(\tau))|\mathbf{r}-\tau|dS(\tau). \qquad (2.84)$$

Step 4. Replacing the first integral on the right hand side of eq. (2.72) by (2.84) we immediately obtain the representation (2.73), holding for all exterior points $\mathbf{r}\in\Omega_e$.

In what follows we will express the representation (2.73) in terms of surface integrals alone.

Indeed, applying Green's second identity (2.49) to the functions $(\tau\cdot\mathbf{A}(\tau))$ and $|\mathbf{r}-\tau|^{-1}$ and utilizing the fact that, since $\mathbf{r}\in\Omega_e$ and $\tau\in\Omega_C$, the function $|\mathbf{r}-\tau|^{-1}$ is harmonic, we obtain

$$I_{\mathbf{A}}(\mathbf{r})=\int_{S_C}\left[\frac{1}{|\mathbf{r}-\tau|}\nabla_\tau(\tau\cdot\mathbf{A}(\tau))-(\tau\cdot\mathbf{A}(\tau))\nabla_\tau\frac{1}{|\mathbf{r}-\tau|}\right]\cdot\hat{\mathbf{n}}(\tau)dS(\tau) \qquad (2.85)$$

where

$$I_{\mathbf{A}}(\mathbf{r})=\int_{\Omega_C}\frac{1}{|\mathbf{r}-\tau|}\Delta_\tau(\tau\cdot\mathbf{A}(\tau))dV(\tau),\quad \mathbf{r}\in\Omega_e \qquad (2.86)$$

Similarly, the function $\mathbf{H}(\mathbf{r},\tau)$ is harmonic with respect to the variable τ, because it involves only the function $|\rho-\tau|^{-1}$ and the potentials υ_j, $j=c,f,b,s$, and therefore if we apply (2.49) to $\mathbf{r}\cdot\mathbf{H}(\mathbf{r},\tau)$ and $(\Delta_\tau\Psi(\tau))$ we obtain the formula

$$I_{\mathbf{H}}(\mathbf{r})=\int_{S_C}[(\mathbf{r}\cdot\mathbf{H}(\mathbf{r},\tau))(\nabla_\tau\Psi(\tau))-\Psi(\tau)\nabla_\tau(\mathbf{r}\cdot\mathbf{H}(\mathbf{r},\tau))]\cdot\hat{\mathbf{n}}(\tau)dS(\tau) \qquad (2.87)$$

where

$$I_{\mathbf{H}}(\mathbf{r})=\int_{\Omega_C}(\Delta_\tau\Psi(\tau))(\mathbf{r}\cdot\mathbf{H}(\mathbf{r},\tau))dV(\tau),\quad \mathbf{r}\in\Omega_e. \qquad (2.88)$$

Inserting expressions (2.85) and (2.87) in (2.73) we obtain the surface representation

$$\mathbf{r}\cdot\mathbf{B}(\mathbf{r})=-\frac{\mu_0}{4\pi}I_{\mathbf{A}}(\mathbf{r})+\frac{\mu_0}{(4\pi)^2}I_{\mathbf{H}}(\mathbf{r})$$
$$-\frac{\mu_0}{(4\pi)^2}\mathbf{r}\cdot\int_{S_C}(\hat{\mathbf{n}}(\tau)\cdot\mathbf{J}^p(\tau))\mathbf{H}(\mathbf{r},\tau)dS(\tau)$$

$$
\begin{aligned}
&-\frac{\mu_0}{4\pi}\mathbf{r}\cdot\int_{S_C}\frac{1}{|\mathbf{r}-\tau|}(\hat{\mathbf{n}}(\tau)\times\mathbf{J}^p(\tau))\,dS(\tau)\\
&+\frac{\mu_0}{4\pi}\int_{S_C}(\hat{\mathbf{n}}(\tau)\cdot\Delta_\tau\mathbf{A}(\tau))|\mathbf{r}-\tau|\,dS(\tau),\quad \mathbf{r}\in\Omega_e
\end{aligned}
\tag{2.89}
$$

Elaborating further on the above representation we observe that

$$
\begin{aligned}
&\nabla_\tau\cdot[|\mathbf{r}-\tau|(\hat{\mathbf{n}}(\tau)\times\mathbf{J}^p(\tau))]\\
&\quad=|\mathbf{r}-\tau|[\nabla_\tau\cdot(\hat{\mathbf{n}}(\tau)\times\mathbf{J}^p(\tau))]+(\nabla_\tau|\mathbf{r}-\tau|)\cdot(\hat{\mathbf{n}}(\tau)\times\mathbf{J}^p(\tau))\\
&\quad=|\mathbf{r}-\tau|[\nabla_\tau\cdot(\hat{\mathbf{n}}(\tau)\times\mathbf{J}^p(\tau))]-\frac{\mathbf{r}-\tau}{|\mathbf{r}-\tau|}\cdot(\hat{\mathbf{n}}(\tau)\times\mathbf{J}^p(\tau)).
\end{aligned}
\tag{2.90}
$$

This immediately implies

$$
\begin{aligned}
&\mathbf{r}\cdot\int_{S_C}\frac{1}{|\mathbf{r}-\tau|}(\hat{\mathbf{n}}(\tau)\times\mathbf{J}^p(\tau))\,dS(\tau)\\
&\quad=\int_{S_C}\frac{1}{|\mathbf{r}-\tau|}\tau\cdot(\hat{\mathbf{n}}(\tau)\times\mathbf{J}^p(\tau))\,dS(\tau)\\
&\qquad+\int_{S_C}|\mathbf{r}-\tau|[\nabla_\tau\cdot(\hat{\mathbf{n}}(\tau)\times\mathbf{J}^p(\tau))]\,dS(\tau)\\
&\qquad-\int_{S_C}\nabla_\tau\cdot[|\mathbf{r}-\tau|(\hat{\mathbf{n}}(\tau)\times\mathbf{J}^p(\tau))]\,dS(\tau).
\end{aligned}
\tag{2.91}
$$

If we insert (2.91) in (2.89) we arrive at

$$
\begin{aligned}
\frac{4\pi}{\mu_0}\mathbf{r}\cdot\mathbf{B}(\mathbf{r})=&-I_{\mathbf{A}}(\mathbf{r})+\frac{1}{4\pi}I_{\mathbf{H}}(\mathbf{r})\\
&+\int_{S_C}\nabla_\tau\cdot[|\mathbf{r}-\tau|(\hat{\mathbf{n}}(\tau)\times\mathbf{J}^p(\tau))]\,dS(\tau)\\
&-\frac{1}{4\pi}\int_{S_C}(\hat{\mathbf{n}}(\tau)\cdot\mathbf{J}^p(\tau))(\mathbf{r}\cdot\mathbf{H}(\mathbf{r},\tau))dS(\tau)\\
&-\int_{S_C}\frac{1}{|\mathbf{r}-\tau|}\tau\cdot(\hat{\mathbf{n}}(\tau)\times\mathbf{J}^p(\tau))\,dS(\tau)\\
&-\int_{S_C}|\mathbf{r}-\tau|[\nabla_\tau\cdot(\hat{\mathbf{n}}(\tau)\times\mathbf{J}^p(\tau))-\hat{\mathbf{n}}(\tau)\cdot(\Delta_\tau\mathbf{A}(\tau))]\,dS(\tau).
\end{aligned}
\tag{2.92}
$$

From the Helmholtz decomposition with a divergence free vector potential $\mathbf{A}$ we obtain the identity

$$\nabla \times \mathbf{J}^p = \nabla \times (\nabla \Psi + \nabla \times \mathbf{A}) = -\Delta \mathbf{A} + \nabla(\nabla \cdot \mathbf{A}) = -\Delta \mathbf{A}. \tag{2.93}$$

This implies

$$\begin{aligned}
&\nabla_\tau \cdot (\hat{\mathbf{n}}(\tau) \times \mathbf{J}^p(\tau)) - \hat{\mathbf{n}}(\tau) \cdot (\Delta_\tau \mathbf{A}(\tau)) \\
&\quad = \nabla_\tau \cdot (\hat{\mathbf{n}}(\tau) \times \mathbf{J}^p(\tau)) + \hat{\mathbf{n}}(\tau) \cdot (\nabla_\tau \times \mathbf{J}^p(\tau)) \\
&\qquad = \mathbf{J}^p(\tau) \cdot (\nabla_\tau \times \hat{\mathbf{n}}(\tau)) - \hat{\mathbf{n}}(\tau) \cdot (\nabla_\tau \times \mathbf{J}^p(\tau)) + \hat{\mathbf{n}}(\tau) \cdot (\nabla_\tau \times \mathbf{J}^p(\tau)) \\
&\qquad = \mathbf{J}^p(\tau) \cdot (\nabla_\tau \times \hat{\mathbf{n}}(\tau)).
\end{aligned} \tag{2.94}$$

However, the rotation of any unit vector vanishes

$$\nabla_\tau \times \hat{\mathbf{n}}(\tau) = \mathbf{0} \tag{2.95}$$

and therefore the last integral on the right hand side of (2.92) disappears. The resulting expression for the radial component of the exterior magnetic field is given in terms of surface integrals:

$$\begin{aligned}
&\frac{4\pi}{\mu_0} \mathbf{r} \cdot \mathbf{B}(\mathbf{r}) \\
&= \frac{1}{4\pi} \int_{S_C} [(\mathbf{r} \cdot \mathbf{H}(\mathbf{r}, \tau))(\nabla_\tau \Psi(\tau)) - \Psi(\tau) \nabla_\tau (\mathbf{r} \cdot \mathbf{H}(\mathbf{r}, \tau))] \cdot \hat{\mathbf{n}}(\tau)\, dS(\tau) \\
&\quad - \int_{S_C} \left[\frac{1}{|\mathbf{r} - \tau|} \nabla_\tau (\tau \cdot \mathbf{A}(\tau)) - (\tau \cdot \mathbf{A}(\tau)) \nabla_\tau \frac{1}{|\mathbf{r} - \tau|} \right] \cdot \hat{\mathbf{n}}(\tau)\, dS(\tau) \\
&\quad + \int_{S_C} \nabla_\tau \cdot [|\mathbf{r} - \tau| (\hat{\mathbf{n}}(\tau) \times \mathbf{J}^p(\tau))]\, dS(\tau) \\
&\quad - \frac{1}{4\pi} \int_{S_C} (\hat{\mathbf{n}}(\tau) \cdot \mathbf{J}^p(\tau))(\mathbf{r} \cdot \mathbf{H}(\mathbf{r}, \tau)) dS(\tau) \\
&\quad - \int_{S_C} \frac{1}{|\mathbf{r} - \tau|} \tau \cdot (\hat{\mathbf{n}}(\tau) \times \mathbf{J}^p(\tau))\, dS(\tau).
\end{aligned} \tag{2.96}$$

An alternative expression involving both surface and volume integrals is

$$\begin{aligned}\frac{4\pi}{\mu_0}\mathbf{r}\cdot\mathbf{B}(\mathbf{r}) = \frac{1}{4\pi}\int_{\Omega_C}(\Delta_\tau\Psi(\tau))(\mathbf{r}\cdot\mathbf{H}(\mathbf{r},\tau))dV(\tau)\\ - \int_{\Omega_C}\frac{1}{|\mathbf{r}-\tau|}\Delta_\tau(\tau\cdot\mathbf{A}(\tau))dV(\tau)\\ + \int_{S_C}\nabla_\tau\cdot[|\mathbf{r}-\tau|(\hat{\mathbf{n}}(\tau)\times\mathbf{J}^p(\tau))]\,dS(\tau)\\ - \frac{1}{4\pi}\int_{S_C}(\hat{\mathbf{n}}(\tau)\cdot\mathbf{J}^p(\tau))(\mathbf{r}\cdot\mathbf{H}(\mathbf{r},\tau))dS(\tau)\\ - \int_{S_C}\frac{1}{|\mathbf{r}-\tau|}\tau\cdot(\hat{\mathbf{n}}(\tau)\times\mathbf{J}^p(\tau))\,dS(\tau), \quad \mathbf{r}\in\Omega_e.\end{aligned} \tag{2.97}$$

2.4 Calculation of the exterior magnetic field

It was shown in Chapter 1 that outside the head the magnetic flux can be expressed in the form

$$\mathbf{B}(\mathbf{r}) = \mu_0\nabla U(\mathbf{r}), \quad \mathbf{r}\in\Omega_e \tag{2.98}$$

where the scalar function U is called the *magnetic potential*. Hence

$$\mathbf{r}\cdot\mathbf{B}(\mathbf{r}) = \mu_0\ \mathbf{r}\cdot\nabla U(\mathbf{r}), \quad \mathbf{r}\in\Omega_e. \tag{2.99}$$

Let ∇_t denote the tangential part of the gradient operator, that is,

$$\nabla = \hat{\mathbf{r}}\frac{\partial}{\partial r} + \nabla_t, \hat{\mathbf{r}}\cdot\nabla_t = 0 \tag{2.100}$$

where $\hat{\mathbf{r}}$ is the unit vector in the direction of $\mathbf{r}$. Employing eq. (2.100) in (2.99) we obtain

$$\mathbf{r}\cdot\mathbf{B}(\mathbf{r}) = \mu_0 r\frac{\partial}{\partial r}U(\mathbf{r}) \tag{2.101}$$

or

$$\frac{\partial}{\partial r}U(\mathbf{r}) = \frac{1}{\mu_0 r}\mathbf{r}\cdot\mathbf{B}(\mathbf{r}), \quad \mathbf{r}\in\Omega_e. \tag{2.102}$$

Letting in this equation $\mathbf{r}=\rho\ \hat{\mathbf{r}}$, integrating the resulting equation with respect to ρ and using the condition that the potential $U(\mathbf{r})$ vanishes as $r\to\infty$, we obtain

$$U(\mathbf{r}) = -\frac{1}{\mu_0}\int_r^{+\infty} \hat{\mathbf{r}}\cdot\mathbf{B}(t\hat{\mathbf{r}})\,dt. \tag{2.103}$$

In order to calculate the integral in (2.103) we utilize the representation (2.62), which gives the following representation of the radial component of the field $\mathbf{B}$, generated by the distributed current $\mathbf{J}^p$ over the cerebrum Ω_c:

$$\begin{aligned}\hat{\mathbf{r}}\cdot\mathbf{B}(\mathbf{r}) &= \frac{\mu_0}{4\pi}\int_{\Omega_c} \hat{\mathbf{r}}\cdot\mathbf{J}^p(\tau)\times\left(\nabla_\tau\frac{1}{|\mathbf{r}-\tau|}\right)dV(\tau)\\ &\quad - \frac{\mu_0}{(4\pi)^2}\int_{\Omega_c} (\mathbf{J}^p(\tau)\cdot\nabla_\tau)(\hat{\mathbf{r}}\cdot\mathbf{H}(\mathbf{r},\tau))dV(\tau).\end{aligned} \tag{2.104}$$

Observe that the $t=|\mathbf{r}|$ dependence of $\mathbf{B}$ comes from the factor

$$\nabla_\tau\frac{1}{|t\hat{\mathbf{r}}-\tau|}$$

in the first integral and from $\mathbf{H}(t\hat{\mathbf{r}},\tau)$ in the second integral on the right hand side of (2.104).

However, in view of the definition (2.62), the t dependence of $\mathbf{H}(t\hat{\mathbf{r}},\tau)$ comes from the factor

$$\nabla_\rho\frac{1}{|t\hat{\mathbf{r}}-\rho|}$$

which appears in all four surface integrals.

Consequently, all we need to calculate is the following line integral:

$$\begin{aligned}\int_r^{+\infty}\nabla_\mathbf{y}\frac{1}{|t\hat{\mathbf{r}}-\mathbf{y}|}dt &= \int_r^{+\infty}\frac{t\hat{\mathbf{r}}-\mathbf{y}}{|t\hat{\mathbf{r}}-\mathbf{y}|^3}dt\\ &= \hat{\mathbf{r}}\int_r^{+\infty}\frac{t}{(t^2-2ty\cos\theta+y^2)^{3/2}}dt - \mathbf{y}\int_r^{+\infty}\frac{1}{(t^2-2ty\cos\theta+y^2)^{3/2}}dt\end{aligned} \tag{2.105}$$

with

$$\cos\theta = \hat{\mathbf{r}}\cdot\hat{\mathbf{y}} \tag{2.106}$$

when $\mathbf{y}=\tau$ and $\mathbf{y}=\rho$.

If we denote by $R(t)$ the expression

$$R(t) = |t\hat{\mathbf{r}}-\mathbf{y}|^2 = t^2 - 2ty\cos\theta + y^2 \tag{2.107}$$

then we need to calculate the indefinite integrals of the functions $R(t)^{-3/2}$ and $tR(t)^{-3/2}$. Calculations by standard transformations lead to the formulae

$$\int \frac{dt}{\sqrt{R(t)}^3} = \frac{t - y\cos\theta}{y^2\sin^2\theta\sqrt{R(t)}} \tag{2.108}$$

$$\int \frac{tdt}{\sqrt{R(t)}^3} = \frac{t\cos\theta - y}{y\sin^2\theta\sqrt{R(t)}}. \tag{2.109}$$

It is also easy to evaluate the following limits:

$$\lim_{t\to+\infty} \frac{t - y\cos\theta}{\sqrt{R(t)}} = \lim_{t\to+\infty} \frac{1 - \frac{y}{t}\cos\theta}{\sqrt{1 - 2\frac{y}{t}\cos\theta + \frac{y^2}{t^2}}} = 1 \tag{2.110}$$

and

$$\lim_{t\to+\infty} \frac{t\cos\theta - y}{\sqrt{R(t)}} = \lim_{t\to+\infty} \frac{\cos\theta - \frac{y}{t}}{\sqrt{1 - 2\frac{y}{t}\cos\theta + \frac{y^2}{t^2}}} = \cos\theta. \tag{2.111}$$

Hence, we obtain

$$\int_r^{+\infty} \frac{dt}{\sqrt{R(t)}^3} = \frac{1}{y^2\sin^2\theta}\left[1 - \frac{r - y\cos\theta}{\sqrt{R(r)}}\right] \tag{2.112}$$

and

$$\int_r^{+\infty} \frac{tdt}{\sqrt{R(t)}^3} = \frac{1}{y\sin^2\theta}\left[\cos\theta - \frac{r\cos\theta - y}{\sqrt{R(r)}}\right]. \tag{2.113}$$

Thus, the integral (2.105) has the closed form

$$\begin{aligned} &\int_r^{+\infty} \nabla_{\mathbf{y}} \frac{1}{|t\hat{\mathbf{r}} - \mathbf{y}|dt} \\ &= \frac{1}{y\sin^2\theta}\left[\cos\theta - \frac{r\cos\theta - y}{|\mathbf{r} - \mathbf{y}|}\right]\hat{\mathbf{r}} - \frac{1}{y\sin^2\theta}\left[1 - \frac{r - y\cos\theta}{|\mathbf{r} - \mathbf{y}|}\right]\hat{\mathbf{y}} \end{aligned} \tag{2.114}$$

where $\mathbf{y} = \tau$ or $\mathbf{y} = \rho$.

In particular,

$$\begin{aligned} &\int_r^{+\infty} \hat{\mathbf{r}} \cdot \mathbf{J}^p(\tau) \times \left(\nabla_\tau \frac{1}{|t\hat{\mathbf{r}} - \tau|}\right) dt = -\int_r^{+\infty} \mathbf{J}^p(\tau) \cdot \hat{\mathbf{r}} \times \left(\nabla_\tau \frac{1}{|t\mathbf{r} - \tau|}\right) dt \\ &= -\frac{1}{\tau\sin^2\theta} \mathbf{J}^p(\tau) \cdot \hat{\mathbf{r}} \times \left[\left(\cos\theta - \frac{r\cos\theta - \tau}{|\mathbf{r} - \tau|}\right)\hat{\mathbf{r}} - \left(1 - \frac{r - \tau\cos\theta}{|\mathbf{r} - \tau|}\right)\hat{\tau}\right] \end{aligned}$$

$$= \frac{1}{\tau \sin^2\theta}\left(1 - \frac{r - \tau\cos\theta}{|\mathbf{r}-\tau|}\right)\mathbf{J}^p(\tau)\cdot\hat{\mathbf{r}}\times\hat{\tau} = \left(1 - \hat{\mathbf{r}}\cdot\frac{\mathbf{r}-\tau}{|\mathbf{r}-\tau|}\right)\mathbf{J}^p(\tau)\cdot\frac{\hat{\mathbf{r}}\times\tau}{|\hat{\mathbf{r}}\times\tau|^2}. \quad (2.115)$$

Similarly,

$$\int_r^{+\infty} \hat{\mathbf{r}}\cdot\hat{\mathbf{n}}(\rho)\times\left(\nabla_\rho \frac{1}{|t\hat{\mathbf{r}}-\rho|}\right)dt = \left(1-\hat{\mathbf{r}}\cdot\frac{\mathbf{r}-\rho}{|\mathbf{r}-\rho|}\right)\hat{\mathbf{n}}(\rho)\cdot\frac{\hat{\mathbf{r}}\times\rho}{|\hat{\mathbf{r}}\times\rho|^2}. \quad (2.116)$$

Define

$$\begin{aligned}
h(\mathbf{r},\tau) &= \int_r^{+\infty} \hat{\mathbf{r}}\cdot\mathbf{H}(t\hat{\mathbf{r}},\rho)\, dt \\
&= \int_{S_c}\left[\frac{1}{|\rho-\tau|} + \upsilon_c(\rho,\tau) - \sigma_f\upsilon_f(\rho,\tau)\right]\left(1-\hat{\mathbf{r}}\cdot\frac{\mathbf{r}-\rho}{|\mathbf{r}-\rho|}\right)\hat{\mathbf{n}}(\rho)\cdot\frac{\hat{\mathbf{r}}\times\rho}{|\hat{\mathbf{r}}\times\rho|^2}dS(\rho) \\
&+ \int_{S_f}\left[\sigma_f\upsilon_f(\rho,\tau) - \sigma_b\upsilon_b(\rho,\tau)\right]\left(1-\hat{\mathbf{r}}\cdot\frac{\mathbf{r}-\rho}{|\mathbf{r}-\rho|}\right)\hat{\mathbf{n}}(\rho)\cdot\frac{\hat{\mathbf{r}}\times\rho}{|\hat{\mathbf{r}}\times\rho|^2}dS(\rho) \\
&+ \int_{S_b}\left[\sigma_b\upsilon_b(\rho,\tau) - \sigma_s\upsilon_s(\rho,\tau)\right]\left(1-\hat{\mathbf{r}}\cdot\frac{\mathbf{r}-\rho}{|\mathbf{r}-\rho|}\right)\hat{\mathbf{n}}(\rho)\cdot\frac{\hat{\mathbf{r}}\times\rho}{|\hat{\mathbf{r}}\times\rho|^2}dS(\rho) \\
&+ \sigma_s\int_{S_s}\left[\upsilon_s(\rho,\tau)\right]\left(1-\hat{\mathbf{r}}\cdot\frac{\mathbf{r}-\rho}{|\mathbf{r}-\rho|}\right)\hat{\mathbf{n}}(\rho)\cdot\frac{\hat{\mathbf{r}}\times\rho}{|\hat{\mathbf{r}}\times\rho|^2}dS(\rho), \quad \mathbf{r}\in\Omega_e. \qquad (2.117)
\end{aligned}$$

Then, in view of (2.103) and (2.104) we obtain

$$\begin{aligned}
U(\mathbf{r}) &= -\frac{1}{4\pi}\int_{\Omega_c}\left(1-\hat{\mathbf{r}}\cdot\frac{\mathbf{r}-\tau}{|\mathbf{r}-\tau|}\right)\mathbf{J}^p(\tau)\cdot\frac{\hat{\mathbf{r}}\times\tau}{|\hat{\mathbf{r}}\times\tau|^2}dV(\tau) \\
&+ \frac{1}{(4\pi)^2}\int_{\Omega_c}(\mathbf{J}^p(\tau)\cdot\nabla_\tau)h(\mathbf{r},\tau)dV(\tau)\,\mathbf{r}\in\Omega_e. \qquad (2.118)
\end{aligned}$$

Finally, the exterior magnetic flux is given by (2.98) with U as in (2.118) and h as in (2.117).

It turns out that the second integral in the right hand side of (2.118) is much smaller than the first integral in this side. This is due to the fact that the conductivities of the different shells do not differ much and therefore the same is true for the corresponding electric potentials. Consequently, the value of functions $\mathbf{H}$ and h are small, as it can be seen in (1.107) and (1.108). The above facts can also be verified numerically.

2.5 Convenient Boundary Conditions for A and Ψ

By employing the boundary conditions (iii) for **A**, eq (2.48) becomes eq (iv) of the introduction. Furthermore, if we impose on Ψ the boundary condition

$$\hat{\mathbf{n}}(\tau)\cdot\nabla_\tau\Psi(\tau)=0,\quad \tau\in S_c,$$

we find

$$u_s^o(\mathbf{r})=-\frac{1}{4\pi}\int_{\Omega_c} \upsilon_s(\mathbf{r},\tau)(\Delta_\tau\Psi(\tau))dV(\tau). \tag{2.119}$$

Furthermore if the cross product of $\hat{\mathbf{n}}(\tau)$ with the $\nabla\times\mathbf{A}$ also vanishes, formulae (2.86), (2.88) and (2.89) yield

$$\begin{aligned}\mathbf{r}\cdot\mathbf{B}(\mathbf{r})=&-\frac{\mu_0}{4\pi}\int_{\Omega_c}\frac{1}{|\mathbf{r}-\tau|}\Delta_\tau(\tau\cdot\mathbf{A}(\tau))dV(\tau)\\ &+\frac{\mu_0}{(4\pi)^2}\int_{\Omega_c}(\mathbf{r}\cdot\mathbf{H}(\mathbf{r},\tau))(\Delta_\tau\Psi(\tau))dV(\tau).\end{aligned} \tag{2.120}$$

This way we recover the results obtained in [47]. In fact, since the radial component of **B** provides the magnetic potential U (as shown in (2.103)), it follows that the electric potential depends on Ψ alone, while the magnetic flux density **B** depends both on Ψ and on the radial component of **A**.

We note that even if $\mathbf{J}^p$ is localized, it does follow that the scalar potential Ψ and the vector potential **A** also vanish on S_c. In this connection we note that eqs. (2.45) and (2.46) imply

$$\Delta\Psi=\nabla\cdot\mathbf{J}^p,\Delta\mathbf{A}=-\nabla\times\mathbf{J}^p. \tag{2.121}$$

Hence, particular solutions of the above equations are given by the following expressions:

$$\Psi(\mathbf{r})=-\frac{1}{4\pi}\int_{\Omega_c}\frac{\nabla_\tau\cdot\mathbf{J}^p(\tau)}{|\tau-\mathbf{r}|}dV(\tau) \tag{2.122}$$

and

$$\mathbf{A}(\mathbf{r})=\frac{1}{4\pi}\int_{\Omega_c}\frac{\nabla_\tau\times\mathbf{J}^p(\tau)}{|\tau-\mathbf{r}|}dV(\tau). \tag{2.123}$$

As an example, we compute Ψ and **A** in the particular case that $\mathbf{J}^p$ consists of the isolated dipole

$$\mathbf{J}^p(\tau)=\mathbf{Q}\delta(\tau-\mathbf{r}_0). \tag{2.124}$$

Then, using the identities

$$\int_{\Omega_C} \frac{\nabla_\tau \cdot (\mathbf{Q}\delta(\tau - \mathbf{r}_0))}{|\tau - \mathbf{r}|} dV(\tau) = \mathbf{Q} \cdot \int_{\Omega_C} \frac{\nabla_\tau \delta(\tau - \mathbf{r}_0)}{|\tau - \mathbf{r}|} dV(\tau)$$

$$= -\mathbf{Q} \cdot \int_{\Omega_C} \delta(\tau - \mathbf{r}_0) \left(\nabla_\tau \frac{1}{|\mathbf{r} - \tau|} \right) dV(\tau) \tag{2.125}$$

we obtain

$$\Psi(\mathbf{r}) = \frac{1}{4\pi} \mathbf{Q} \cdot \frac{\mathbf{r} - \mathbf{r}_0}{|\mathbf{r} - \mathbf{r}_0|^3}. \tag{2.126}$$

Similarly, the identity

$$\int_{\Omega_C} \frac{\nabla_\tau \times (\mathbf{Q}\delta(\tau - \mathbf{r}_0))}{|\tau - \mathbf{r}|} dV(\tau) = -\mathbf{Q} \times \int_{\Omega_C} \frac{\nabla_\tau \delta(\tau - \mathbf{r}_0)}{|\tau - \mathbf{r}|} dV(\tau) \tag{2.127}$$

implies

$$\mathbf{A}(\mathbf{r}) = \frac{1}{4\pi} \mathbf{Q} \times \frac{\mathbf{r} - \mathbf{r}_0}{|\mathbf{r} - \mathbf{r}_0|^3}. \tag{2.128}$$

Note that the right hand sides of eqs. (2.126) and (2.128) are proportional to the principal contribution of the dipolic source (2.124) to the electric potential u^p and to the magnetic flux density $\mathbf{B}^p$, respectively. Indeed,

$$\sigma \Delta u^p = \nabla \cdot \mathbf{J}^p = \Delta \Psi \tag{2.129}$$

and

$$\nabla \times (\nabla \times \mathbf{B}^p) = \mu_0 \nabla \times \mathbf{J}^p = -\mu_0 \Delta \mathbf{A}. \tag{2.130}$$

Thus,

$$\Delta \mathbf{B}^p = \mu_0 \Delta \mathbf{A}. \tag{2.131}$$

The above imply

$$u^p(\mathbf{r}) = \frac{1}{4\pi\sigma} \mathbf{Q} \cdot \frac{\mathbf{r} - \mathbf{r}_0}{|\mathbf{r} - \mathbf{r}_0|^3} \tag{2.132}$$

and

$$\mathbf{B}^p(\mathbf{r}) = \frac{\mu_0}{4\pi}\mathbf{Q} \times \frac{\mathbf{r}-\mathbf{r}_0}{|\mathbf{r}-\mathbf{r}_0|^3}. \tag{2.133}$$

2.6 Surface geometry

In what follows we will demonstrate that all previous results, that give the EEG and MEG data in terms of surface integrals, can also be expressed in intrinsic surface geometry.

In this connection, we define the boundary S_c in terms of its parametric representation

$$\tau = \tau(u, v). \tag{2.134}$$

We assume that the corresponding local system$\hat{\tau}_1$, $\hat{\tau}_2$, $\hat{\mathbf{n}}(\tau)$, is orthogonal, where

$$\tau_1 = \tau_1\hat{\tau}_1 = \frac{\partial}{\partial u}\tau(u,v) \tag{2.135}$$

$$\tau_2 = \tau_2\hat{\tau}_2 = \frac{\partial}{\partial v}\tau(u,v) \tag{2.136}$$

$$\hat{\mathbf{n}}(\tau) = \frac{\tau_1 \times \tau_2}{|\tau_1 \times \tau_2|}. \tag{2.137}$$

The fundamental quantities E, G, and H are defined by

$$E = \tau_1^2, \quad G = \tau_2^2, \quad H = \tau_1\tau_2. \tag{2.138}$$

In terms of this local system, we have the following formulae: If $\mathbf{F}$ is a vector field defined on the surface S_c, then it can be represented as

$$\mathbf{F}(\tau) = P\tau_1 + Q\tau_2 + R\hat{\mathbf{n}}(\tau) \tag{2.139}$$

The divergence of $\mathbf{F}$ multiplied by the scalar field f gives

$$\nabla \cdot (f\ \mathbf{F}) = \frac{1}{H}\left[\frac{\partial}{\partial u}(HfP) + \frac{\partial}{\partial v}(HfQ)\right] - JfR \tag{2.140}$$

where J is the mean curvature at the point τ defined as the sum of the two principal curvatures of the surface at the point τ. If we rewrite (2.126) as

$$\mathbf{F}(\tau) = \mathbf{F}_T + R\hat{\mathbf{n}} \tag{2.141}$$

where $\mathbf{F}_T$ is the projection of $\mathbf{F}$ on the tangent plane, then since the surface gradient is written as

$$\nabla_s = \frac{\tau_1}{E}\frac{\partial}{\partial u} + \frac{\tau_2}{G}\frac{\partial}{\partial v} \tag{2.142}$$

we obtain

$$\begin{aligned}\mathbf{F}_T \cdot \nabla_s f &= (P\tau_1 + Q\tau_2) \cdot \left(\frac{\tau_1}{E}\frac{\partial}{\partial u}f + \frac{\tau_2}{G}\frac{\partial}{\partial v}f\right) \\ &= P\frac{\tau_1 \cdot \tau_1}{E}\frac{\partial}{\partial u}f + Q\frac{\tau_2 \cdot \tau_2}{G}\frac{\partial}{\partial v}f = P\frac{\partial}{\partial u}f + Q\frac{\partial}{\partial v}f.\end{aligned} \tag{2.143}$$

Therefore

$$\nabla \cdot (f\ \mathbf{F}) = f(\nabla \cdot \mathbf{F}) + \mathbf{F}_T \cdot \nabla_s f. \tag{2.144}$$

Using formula (2.143) with f replaced by $|\mathbf{r} - \tau|^{-1}$ and with $\mathbf{F}$ replaced by $|\mathbf{r} - \tau|^2\mathbf{F}_T$, we obtain

$$\begin{aligned}&|\mathbf{r} - \tau|^2\mathbf{F}_T \cdot \nabla_\tau \frac{1}{|\mathbf{r} - \tau|} \\ &\quad = \nabla_\tau \cdot (\mathbf{F}_T|\mathbf{r} - \tau|) - \frac{1}{|\mathbf{r} - \tau|}\left[|\mathbf{r} - \tau|^2(\nabla_\tau \cdot \mathbf{F}_T) + \mathbf{F}_T \cdot \left(\nabla_\tau|\mathbf{r} - \tau|^2\right)\right] \\ &\quad = \nabla_\tau \cdot (\mathbf{F}_T|\mathbf{r} - \tau|) - (\nabla_\tau \cdot \mathbf{F}_T)|\mathbf{r} - \tau| - \frac{1}{|\mathbf{r} - \tau|}\mathbf{F}_T \cdot \nabla_\tau|\mathbf{r} - \tau|^2\end{aligned} \tag{2.145}$$

where

$$\mathbf{F}_T \cdot \nabla_\tau|\mathbf{r} - \tau|^2 = -2\mathbf{F}_T \cdot (\mathbf{r} - \tau). \tag{2.146}$$

We can now use the earlier calculations to verify the equality below, and at the same time to rewrite this equality in an intrinsic surface form:

$$\int_{S_C} (\hat{\mathbf{n}}(\tau) \times \mathbf{J}^p(\tau)) \cdot \frac{\mathbf{r} - \tau}{|\mathbf{r} - \tau|} dS(\tau) = \int_{S_C} (\hat{\mathbf{n}}(\tau) \times \mathbf{J}^p(\tau))|\mathbf{r} - \tau|^2 \cdot \left[\nabla_\tau \frac{1}{|\mathbf{r} - \tau|}\right] dS(\tau). \tag{2.147}$$

Using the identity

$$\tilde{I} = \frac{\tau_1 \otimes \tau_1}{E} + \frac{\tau_2 \otimes \tau_2}{G} \tag{2.148}$$

where $\tilde{\mathbf{I}}$ is the identity dyadic, we obtain

$$\nabla_\tau \frac{1}{|\mathbf{r} - \tau|} = \frac{1}{|\mathbf{r} - \tau|^3}\left[\frac{\tau_1}{E}((\mathbf{r} - \tau) \cdot \tau_1) + \frac{\tau_2}{G}((\mathbf{r} - \tau) \cdot \tau_2)\right]$$

$$
\begin{aligned}
&= \frac{1}{|\mathbf{r}-\tau|^3}\left[\frac{\tau_1}{E}(\mathbf{r}\cdot\tau_1) + \frac{\tau_2}{G}(\mathbf{r}\cdot\tau_2)\right] \\
&\quad - \frac{1}{|\mathbf{r}-\tau|^3}\left[\frac{\tau_1}{E}(\tau\cdot\tau_1) + \frac{\tau_2}{G}(\tau\cdot\tau_2)\right].
\end{aligned}
\tag{2.149}
$$

Expanding the current in the form

$$
\mathbf{J}^p(\tau) = P\tau_1 + Q\tau_2 + R\hat{\mathbf{n}}(\tau) \tag{2.150}
$$

we obtain

$$
\hat{\mathbf{n}}(\tau)\times\mathbf{J}^p(\tau) = P\hat{\mathbf{n}}(\tau)\times\tau_1 + Q\hat{\mathbf{n}}(\tau)\times\tau_2. \tag{2.151}
$$

Therefore the right hand side of (2.134) gives the following:

$$
\begin{aligned}
RHS &= \int_{S_C} (P\hat{\mathbf{n}}(\tau)\times\tau_1 + Q\hat{\mathbf{n}}(\tau)\times\tau_2)\frac{1}{|\mathbf{r}-\tau|} \\
&\quad \cdot\left[\frac{\tau_1}{E}(\mathbf{r}\cdot\tau_1) + \frac{\tau_2}{G}(\mathbf{r}\cdot\tau_2) - \frac{\tau_1}{E}(\tau\cdot\tau_1) - \frac{\tau_2}{G}(\tau\cdot\tau_2)\right] dS(\tau) \\
&= \int_{S_C} \left[\frac{P}{G}(\hat{\mathbf{n}}(\tau)\times\tau_1\cdot\tau_2)(\mathbf{r}-\tau)\cdot\tau_2 + \frac{Q}{E}(\hat{\mathbf{n}}(\tau)\times\tau_2\cdot\tau_1)(\mathbf{r}-\tau)\cdot\tau_1\right]\frac{1}{|\mathbf{r}-\tau|} dS(\tau) \\
&= \int_{S_C} \left[P\frac{\tau_1}{\tau_2}(\mathbf{r}-\tau)\cdot\tau_2 + Q\frac{\tau_2}{\tau_1}(\mathbf{r}-\tau)\cdot\tau_1\right]\frac{1}{|\mathbf{r}-\tau|} dS(\tau) \\
&= \int_{S_C} \left[P\frac{\tau_1}{\tau_2}\tau_2 + Q\frac{\tau_2}{\tau_1}\tau_1\right]\cdot\frac{\mathbf{r}-\tau}{|\mathbf{r}-\tau|} dS(\tau) \\
&= \int_{S_C} [P\tau_1(\hat{\mathbf{n}}(\tau)\times\hat{\tau}_1) + Q\tau_2(\hat{\mathbf{n}}(\tau)\times\hat{\tau}_2)]\cdot\frac{\mathbf{r}-\tau}{|\mathbf{r}-\tau|} dS(\tau) \\
&= \int_{S_C} [\hat{\mathbf{n}}(\tau)\times(P\tau_1 + Q\tau_2)]\cdot\frac{\mathbf{r}-\tau}{|\mathbf{r}-\tau|} dS(\tau) \\
&= \int_{S_C} (\hat{\mathbf{n}}(\tau)\times\mathbf{J}^p(\tau))\cdot\frac{\mathbf{r}-\tau}{|\mathbf{r}-\tau|} dS(\tau)
\end{aligned}
\tag{2.152}
$$

which is equal to the left hand side of (2.134).

As a second example of rewriting the surface integrals in an intrinsic surface form we demonstrate the equality

$$
(\hat{\mathbf{n}}(\tau)\times\mathbf{J}^p(\tau))\cdot\nabla_\tau|\mathbf{r}-\tau|^2 = -2(\hat{\mathbf{n}}(\tau)\times\mathbf{J}^p(\tau))\cdot(\mathbf{r}-\tau). \tag{2.153}
$$

Since

$$\hat{\mathbf{n}}(\tau) \times \mathbf{J}^p(\tau) = P\hat{\mathbf{n}}(\tau) \times \tau_1 + Q\hat{\mathbf{n}}(\tau) \times \tau_2 = P\tau_1\hat{\tau}_2 - Q\tau_2\hat{\tau}_1, \tag{2.154}$$

and since ∇_s is given by (2.141), we obtain the following:

$$\begin{aligned}
&(\hat{\mathbf{n}}(\tau) \times \mathbf{J}^p(\tau)) \cdot \nabla_\tau |\mathbf{r} - \tau|^2 \\
&= (P\tau_1\hat{\tau}_2 - Q\tau_2\hat{\tau}_1) \cdot \left(\frac{\tau_1}{E}\frac{\partial}{\partial u} + \frac{\tau_2}{G}\frac{\partial}{\partial v}\right)(r^2 - 2\mathbf{r} \cdot \tau + \tau^2) \\
&= (P\tau_1\hat{\tau}_2 - Q\tau_2\hat{\tau}_1) \cdot \left[\frac{\tau_1}{E}(-2\mathbf{r} \cdot \tau_1 + 2\tau \cdot \tau_1) + \frac{\tau_2}{G}(-2\mathbf{r} \cdot \tau_2 + 2\tau \cdot \tau_2)\right] \\
&= -2(P\tau_1\hat{\tau}_2 - Q\tau_2\hat{\tau}_1) \cdot [(\mathbf{r} \cdot \hat{\tau}_1)\hat{\tau}_1 - (\tau \cdot \hat{\tau}_1)\hat{\tau}_1 + (\mathbf{r} \cdot \hat{\tau}_2)\hat{\tau}_2 - (\tau \cdot \hat{\tau}_2)\hat{\tau}_2] \\
&= -2P\tau_1[(\mathbf{r} - \tau) \cdot \hat{\tau}_2] + 2Q\tau_2[(\mathbf{r} - \tau) \cdot \hat{\tau}_1] \\
&= -2(P\tau_1\hat{\tau}_2 - Q\tau_2\hat{\tau}_1) \cdot (\mathbf{r} - \tau). \\
&= -2(\hat{\mathbf{n}}(\tau) \times \mathbf{J}^p(\tau)) \cdot (\mathbf{r} - \tau).
\end{aligned} \tag{2.155}$$

In a similar way, we can realize all surface integrals in an intrinsic surface form.

Chapter 3
Distributed current in spherical and ellipsoidal geometry

In this chapter, we consider the 3-shell model of the inhomogeneous head-brain system and assume that the neuronal current vanishes on S_c. For this case, we present details for the spherical and ellipsoidal geometries.

3.1 Statement of the problem

Some of the results obtained in Chapter 2 in the case that $\mathbf{J}^p$ vanishes on S_c are summarized in Proposition 3.1.

Proposition 3.1 Assume the 3-shell model of the head-brain system and the relative notation $\Omega_i, S_i, \sigma_i, u_i, v_\iota$ with $i = c, f, b, s, e$. Assume that the current which is supported in Ω_c vanishes on S_c. Let the current be represented in the form

$$\mathbf{J}(\tau) = \nabla\Psi(\tau) + \nabla\times\mathbf{A}(\tau) \tag{3.1}$$

where

$$\nabla\cdot\mathbf{A}(\tau) = 0, \quad \tau\in\Omega_c \tag{3.2}$$

Then the following integral representations hold

$$u_s(\mathbf{r}) = -\frac{1}{4\pi}\int\limits_{\Omega_c}(\Delta\Psi(\tau))\ v_s(\mathbf{r},\tau)dv(\tau), \quad \mathbf{r}\in S_s \tag{3.3}$$

and

$$\frac{4\pi}{\mu_0}\mathbf{B}(\mathbf{r}) = -\int\limits_{\Omega_c}(\Delta\mathbf{A}(\tau))\ \frac{1}{|\mathbf{r}-\tau|}dv(\tau) + \frac{1}{4\pi}\int\limits_{\Omega_c}(\Delta\Psi(\tau))\ \mathbf{H}(\mathbf{r},\tau)dv(\tau), \quad \mathbf{r}\in\Omega_c. \tag{3.4}$$

Also,

$$\frac{4\pi}{\mu_0}\mathbf{r}\cdot\mathbf{B}(\mathbf{r}) = -\int\limits_{\Omega_c}\Delta(\tau\cdot\mathbf{A}(\tau))\ \frac{1}{|\mathbf{r}-\tau|}dv(\tau) + \frac{1}{4\pi}\int\limits_{\Omega_c}(\Delta\Psi(\tau))\mathbf{r}\cdot\ \mathbf{H}(\mathbf{r},\tau)dv(\tau), \ \mathbf{r}\in\Omega_c \tag{3.5}$$

The constant μ_0 denotes the magnetic permeability which is the same in every region and the function $\mathbf{H}(\mathbf{r},\tau)$ is given in (2.62). As it can be seen in (3.3) and (3.5) the electric potential depends only on the nonharmonic part of Ψ and the magnetic field depends only on the nonharmonic parts of Ψ and of the radial part of $\mathbf{A}$.

https://doi.org/10.1515/9783110547535-004

3.2 Spherical geometry

In the case of spherical geometry the relative regions are specified as follows

$$\left.\begin{aligned} &\Omega_c : 0 \le r < c \\ &\Omega_f : c < r < f \\ &\Omega_b : f < r < b \\ &\Omega_s : b < r < s \\ &\Omega_e : s < r < +\infty \end{aligned}\right\} \tag{3.6}$$

We recall the Laplace expansion

$$\frac{1}{|\mathbf{r}-\mathbf{r}_0|} = \sum_{n=0}^{\infty} \frac{r^n}{r^{n+1}} P_n(\hat{\mathbf{r}} \cdot \hat{\mathbf{r}}_0) = 4\pi \sum_{n=0}^{\infty} \sum_{m=-n}^{n} \frac{1}{2n+1} \frac{r_0{}^n}{r^{n+1}} Y_n^{m*}(\hat{\mathbf{r}}_0) Y_n^m(\hat{\mathbf{r}}) \tag{3.7}$$

which holds for $r > r_0$,P_n are the Legendre polynomials and Y_n^m are the complex form of the normalized surface spherical harmonics.

For an isolated dipolar current at $\mathbf{r}_0$ with moment, $\mathbf{J}(\mathbf{r}_0)$ we can solve the transmission problem for the harmonic functions v_i, $i=c,f,b,s,e$ and obtain the corresponding electric potentials u_i, $i=c,f,b,s,e$ via the expressions (2.26) and (2.27). This program, which will be applied in detail in the next chapter, provides the values of v_s on the boundary $r=s$ in the form

$$v_s(\mathbf{r}, \mathbf{r}_0) = \sum_{n=1}^{\infty} s_n r_0{}^n P_n(\hat{\mathbf{r}} \cdot \hat{\mathbf{r}}_0) \tag{3.8}$$

with $r_0 < c$, where as we have explained earlier, the $n=0$ term in (3.8) will be annihilated by the action of the operator $\frac{1}{4\pi}\mathbf{J}(\mathbf{r}_0) \cdot \nabla_{\mathbf{r}_0}$. The coefficients s_n, which depend on the radii c, f, b, s and the conductivities $\sigma_c, \sigma_f, \sigma_b, \sigma_s$, will be calculated explicitly in the next chapter.

For the spherical model, the dot product $\mathbf{r} \cdot \mathbf{H}(\mathbf{r}, \mathbf{r}_0)$ appearing in eq. (3.5) vanishes, since on the surfaces S_i, $i=c,f,b,s$, we have

$$\mathbf{r} \cdot \hat{\mathbf{n}}(\mathbf{r}') \times \nabla_{\mathbf{r}'} \frac{1}{|\mathbf{r}-\mathbf{r}'|} = \mathbf{r} \cdot \frac{\mathbf{r}'}{r_i} \times \frac{\mathbf{r}-\mathbf{r}'}{|\mathbf{r}-\mathbf{r}'|^3} = 0 \tag{3.9}$$

for every $r_i = c, f, b, s$.

The main results of this section are stated and proved in the next proposition.

Proposition 3.2 Consider the 3-shell spherical model described in (3.6) and the distributed current $\mathbf{J}(\mathbf{r}_0)$, $0 \le r_0 < c$. Then, for $r=s$, we have

$$u_s(\mathbf{r}) = \frac{1}{4\pi} \int_{\Omega_c} \mathbf{J}(\mathbf{r}_0) \cdot \nabla_{\mathbf{r}_0} v_s(\mathbf{r}, \mathbf{r}_0) dv(\mathbf{r}_0) \tag{3.10}$$

where $v_s(\mathbf{r},\mathbf{r}_0)$ is given by (3.8).
Hence

$$u_s(\mathbf{r}) = -\frac{1}{4\pi}\int_{\Omega_c}\sum_{n=1}^{\infty} s_n(\Delta\Psi(\mathbf{r}_0)){r_0}^n P_n(\hat{\mathbf{r}},\hat{\mathbf{r}}_0)dv(\mathbf{r}_0). \tag{3.11}$$

Also, we have

$$\frac{4\pi}{\mu_0}\mathbf{r}\cdot\mathbf{B}(\mathbf{r}) = -\int_{\Omega_c}\sum_{n=0}^{\infty}\Delta(r_0A^{r_0}(\mathbf{r}_0))\frac{{r_0}^n}{r^{n+1}}P_n(\hat{\mathbf{r}},\hat{\mathbf{r}}_0)dv(\mathbf{r}_0), \quad r>s \tag{3.12}$$

where

$$\mathbf{J}(\mathbf{r}_0) = \nabla\Psi(\mathbf{r}_0) + \nabla\times\mathbf{A}(\mathbf{r}_0) \tag{3.13}$$

with $\nabla\cdot\mathbf{A}(\mathbf{r}_0) = 0$, A^{r_0} denotes the radial part of the vector potential $\mathbf{A}$ and s_n are known constants.

Furthermore, the expansions (3.11) and (3.12) can be simplified into the following expressions

$$u_s(\mathbf{r}) = -\sum_{n=1}^{\infty}\sum_{m=-n}^{n}\frac{s_n}{2n+1}c^{n+1}\left[c\dot{y}_n^m(c) - ny_n^m(c)\right]Y_n^m(\hat{\mathbf{r}}), \quad r=s \tag{3.14}$$

and

$$\frac{1}{\mu_0}\mathbf{r}\cdot\mathbf{B}(\mathbf{r}) = -\sum_{n=0}^{\infty}\sum_{m=-n}^{n}\frac{c^{n+2}}{2n+1}\frac{1}{r^{n+1}}\left[c\dot{a}_n^m(c) - (n-1)a_n^m(c)\right]Y_n^m(\hat{\mathbf{r}}), \quad r>s, \tag{3.15}$$

where $\dot{y}_n^m$, $\dot{a}_n^m$ denote derivatives with respect to r_0 and $y_n^m(r_0)$, $a_n^m(r_0)$ are the r_0 – dependent parts of the expansions of $\Psi(\mathbf{r}_0)$ and $A^{r_0}(\mathbf{r}_0)$ in spherical harmonics, namely, $0\le r<c$

$$\Psi(\mathbf{r}_0) = \sum_{n=1}^{\infty}\sum_{m=-n}^{n} y_n^m(r_0)Y_n^m(\hat{\mathbf{r}}_0) \tag{3.16}$$

and

$$A^{r_0}(\mathbf{r}_0) = \sum_{n=0}^{\infty}\sum_{m=-n}^{n} a_n^m(r_0)Y_n^m(\hat{\mathbf{r}}_0). \tag{3.17}$$

Proof. First we observed that the expansion for Ψ starts with $n=1$ since the expansion of v_s starts from $n=1$, while the expansion for A^{r_0} starts from $n=0$. Utilizing (3.8), we obtain expression (3.14). Then using (3.7) and (3.9), we obtain (3.12). Inserting (3.16) in (3.11), we find

$$u_s(\mathbf{r}) = -\frac{1}{4\pi}\sum_{n'=1}^{\infty}\sum_{n=1}^{\infty}\sum_{m=-n}^{n} s_{n'} \int_{\Omega_c} \left(\Delta y_n^m(r_0)Y_n^m(\hat{\mathbf{r}}_0)\right) r_0^{n'} P_{n'}(\hat{\mathbf{r}}, \hat{\mathbf{r}}_0) dv(\mathbf{r}_0). \tag{3.18}$$

If $\mathbb{B}$ denotes the Betrami operator (the angular part of the Laplacian), then

$$\begin{aligned}\Delta y_n^m(r_0)Y_n^m(\hat{\mathbf{r}}_0) &= \frac{1}{r_0^2}\frac{d}{dr_0}\left(r_0^2\frac{d}{dr_0}y_n^m(r_0)\right)Y_n^m(\hat{\mathbf{r}}_0) + y_n^m(r_0)\frac{1}{r_0^2}\mathbb{B}Y_n^m(\hat{\mathbf{r}}_0)\\ &= \frac{1}{r_0^2}\left[\frac{d}{dr_0}\left(r_0^2\frac{d}{dr_0}y_n^m(r_0)\right) - n(n+1)y_n^m(r_0)\right]Y_n^m(\hat{\mathbf{r}}_0).\end{aligned} \tag{3.19}$$

Furthermore, we have the integral mapping

$$\begin{aligned}&\oint_{|\mathbf{r}_0|=1} Y_n^m(\hat{\mathbf{r}}_0)P_{n'}(\hat{\mathbf{r}}, \hat{\mathbf{r}}_0)d\omega(\hat{\mathbf{r}}_0)\\ &= \sum_{m'=-n'}^{n'} \frac{4\pi}{2n'+1}Y_{n'}^{m'}(\hat{\mathbf{r}}) \oint_{|\hat{\mathbf{r}}_0|=1} Y_n^m(\hat{\mathbf{r}}_0)Y_{n'}^{m'}(\hat{\mathbf{r}}_0)d\omega(\hat{\mathbf{r}}_0)\\ &= \sum_{m'=-n'}^{n'} \frac{4\pi}{2n'+1}Y_{n'}^{m'}(\hat{\mathbf{r}})\delta_{nn'}\delta_{mm'} = \frac{4\pi}{2n+1}Y_{n'}^m(\hat{\mathbf{r}})\delta_{nn'}.\end{aligned} \tag{3.20}$$

Substituting (3.19) and (3.20) in (3.18), we obtain

$$u_s(\mathbf{r}) = -\sum_{n=1}^{\infty}\sum_{m=-n}^{n} \frac{s_n}{2n+1}Y_n^m(\hat{\mathbf{r}}) \int_0^c \left[\frac{d}{dr_0}\left(r_0^2\frac{d}{dr_0}y_n^m(r_0)\right) - n(n+1)y_n^m(r_0)\right] r_0^n dr_0. \tag{3.21}$$

Next, we integrate by parts twice to obtain

$$\int_0^c \left[\frac{d}{dr_0}\left(r_0^2\frac{d}{dr_0}y_n^m(r_0)\right)\right] r_0^n dr_0 = c^{n+1}\left[c\dot{y}_n^m(c) - ny_n^m(c)\right] + n(n+1)\int_0^c y_n^m(r_0)\, r_0^n dr_0 \tag{3.22}$$

which implies that the integral in (3.21) is equal to

$$c^{n+1}\left[c\dot{y}_n^m(c) - ny_n^m(c)\right].$$

Finally, substituting this value of the integral in (3.21), we obtain (3.14).

The proof of expansion (3.15) follows exactly the same steps used above where y_n^m is replaced by $r_0 a_n^m$. Hence, the proof of Proposition 3.2 is completed.

The aforementioned proposition shows that the electric potential, as given by (3.14), depends on the boundary expressions

$$c\dot{y}_n^m(c) - ny_n^m(c) \tag{3.23}$$

whereas the radial component of the magnetic field $\mathbf{r} \cdot \mathbf{B}$, as given by (3.15), depends on the boundary expressions

$$c\dot{a}_n^m(c) - (n-1)\, a_n^m(c) \tag{3.24}$$

where the functions $y_n^m(r_0)$ provide the $r_0 -$ dependence of the expansions for Ψ and A^{r_0}, respectively, in spherical harmonics. Since the vector potential is considered to be divergence free it only has two independent functional components. One of these is the radial component of $\mathbf{A}$ and the other is a tangential component of $\mathbf{A}$ that does not affect the MEG recordings.

Corollary 3.1 Given that

$$\frac{1}{\mu_0}\mathbf{r} \cdot \mathbf{B}(\mathbf{r}) = \mathbf{r} \cdot \nabla U(\mathbf{r}) = r\frac{\partial}{\partial r} U(\mathbf{r}) \tag{3.25}$$

where U is the magnetic potential, which vanishes at infinity, we find the following expansions of U in the spherical geometry of Proposition 3.2:

$$U(\mathbf{r}) = -\frac{1}{4\pi}\sum_{n=0}^{\infty}\frac{1}{(n+1)r^{n+1}}\int_{\Omega_c} \Delta(r_0 A^{r_0}(\mathbf{r}_0)) {r_0}^n P_n(\hat{\mathbf{r}} \cdot \hat{\mathbf{r}}_0) dv(\mathbf{r}_0), \quad r > s \tag{3.26}$$

and

$$U(\mathbf{r}) = -\sum_{n=0}^{\infty}\sum_{m=-n}^{n}\frac{c^{n+2}}{(n+1)(2n+1)}\left[c\dot{a}_n^m(c) - (n-1)a_n^m(c)\right]\frac{1}{r^{n+1}} Y_n^m(\hat{\mathbf{r}}), \quad r > s. \tag{3.27}$$

Proof. Integrate expressions (3.12) and (3.15) with respect to r along the ray that goes from r to ∞.

3.3 Ellipsoidal geometry

The ellipsoidal model provides a much better approximation to the realistic brain geometry. However, the mathematical analysis is much harder than the corresponding analysis of the spherical model.

In the present chapter, we present details of the mathematical analysis of the ellipsoidal model using a confocal ellipsoidal system. This system is the only ellipsoidal system that admits spectral analysis for the Laplace's operator.

The results obtained for the general triaxial ellipsoid can be reduced to the corresponding results for the prorate spheroid, the oblate spheroid, their asymptotic forms of needle and disc, as well as for the sphere. However, the implementation of this reduction is quite complicated. For the geometry as well as for the analysis of ellipsoidal harmonic functions we refer to [26].

We start with a short description of the ellipsoidal system and the related ellipsoidal harmonics.

The spherical coordinate system represents quantities exhibiting isotropic structure, whereas the ellipsoidal system can be adjusted to represent quantities with anisotropic structure. In this sense, the ellipsoid can be considered as the sphere of the anisotropic space. In an isotropic space, distances from a center are measured by the radius of a sphere. On the other hand, in an anisotropic space, distances from a center are measured in terms of the three different semiaxes of the ellipsoid that incorporates the given special anisotropy.

The coordinate system that allows spectral decomposition of harmonic functions is the one that preserves the position of the six foci of the system. Such a system is termed *confocal ellipsoidal system.* In analogy with the spherical system, where we start with a *reference sphere* involving its center and its unit radius in the ellipsoidal system, we start with the *reference ellipsoid* specified by its center, three semiaxes and three Euler angles, defining the ellipsoidal analogue of the unit sphere.

Given the reference ellipsoid

$$\frac{x_1^2}{a_1^2}+\frac{x_2^2}{a_2^2}+\frac{x_3^2}{a_3^2}=1, \tag{3.28}$$

in its canonical form, with $0<a_3<a_2<a_1<+\infty$, three semifocal distances $h_1,\ h_2,\ h_3$ are defined by

$$h_1^2=a_2^2-a_3^2,\ h_2^2=a_1^2-a_3^2,\ h_3^2=a_1^2-a_2^2. \tag{3.29}$$

The semifocal distances satisfy the relation

$$h_1^2-h_2^2+h_3^2=0. \tag{3.30}$$

The six foci are located at the pvoints

$$(\pm h_2,0,0),\ (\pm h_3,0,0),\ (0,\pm h_1,0). \tag{3.31}$$

By extending eq. (3.28) to one parameter family of equations

$$\frac{x_1^2}{a_1^2-\lambda}+\frac{x_2^2}{a_2^2-\lambda}+\frac{x_3^2}{a_3^2-\lambda}=1 \tag{3.32}$$

we obtain a family of *confocal ellipsoids* when $\lambda\in(-\infty,a_3^2)$. The focal ellipse

$$\frac{x_1^2}{h_2^2}+\frac{x_2^2}{h_1^2}=1,\quad x_3=0 \tag{3.33}$$

is the particular surface where $\lambda = a_3^2$. If $a_3^2 < \lambda < a_2^2$ we obtain a family of *confocal hyperboloids of one sheet*. The focal hyperbola

$$\frac{x_1^2}{h_3^2} - \frac{x_3^2}{h_1^2} = 1, \quad x_2 = 0 \tag{3.34}$$

is the particular surface where $\lambda = a_2^2$. If $a_2^2 < \lambda < a_1^2$, we obtain a family of *confocal hyperboloids of two sheets*.

Finally, when $\lambda \geq a_1^2$ eq. (3.32) does not represents a real surface.

We define the ellipsoidal coordinates by (ρ, μ, ν) where $\rho \in (h_2, +\infty)$ represents the family of ellipsoids. Similarly, $\mu \in (h_3, h_2)$ represents the family of hyperboloids of one sheet and $\nu \in (-h_3, h_3)$ represents the family of hyperboloids of two sheets.

The ellipsoidal system (3.32) is written in the axes system defined by its principal directions. Therefore, the eight octants of the Cartesian system (x_1, x_2, x_3) represent similar ellipsoidal characteristics. In particular, in the first octant, where $x_1 > 0$, $x_2 > 0$, $x_3 > 0$, the ellipsoidal coordinates are related with the Cartesian ones the following equations:

$$x_1 = \frac{\rho\mu\nu}{h_2 h_3} \tag{3.35}$$

$$x_2 = \frac{\sqrt{\rho^2 - h_3^2}\sqrt{\mu^2 - h_3^2}\sqrt{h_3^2 - \nu^2}}{h_1 h_3} \tag{3.36}$$

$$x_3 = \frac{\sqrt{\rho^2 - h_2^2}\sqrt{h_2^2 - \mu^2}\sqrt{h_2^2 - \nu^2}}{h_1 h_2} \tag{3.37}$$

where $h_2 < \rho < +\infty$, $h_3 < \mu < h_2$, $0 < \nu < h_3$.

The confocal ellipsoidal system allows separation of variables for the Laplace equation. All three separated functions satisfy the same Lamé equation

$$\begin{aligned} &(x^2 - h_3^2)(x^2 - h_2^2)E''(x) + x(2x^2 - h_3^2 - h_2^2)E'(x) \\ &\left[(h_2^2 + h_3^2)p - n(n+1)x^2\right]E(x) = 0 \end{aligned} \tag{3.38}$$

where $n = 0, 1, 2, \ldots$ is the *first separation constant* and p is the *second separation constant* which is specified through the solvability of a linear algebraic system, different for each n. In particular, there are $2n+1$ different values of p for each n denoted by p_n^m, $m = 1, 2, 3, \ldots, 2n+1$. For any pair (n, m), where n defines the *degree* and m the *order* of the eq. (3.38), we obtain two linearly independent solutions $E_n^m(x)$ and $F_n^m(x)$ of the *first (interior) kind* and the *second (exterior) kind*. When $x = \rho \in (h_2, +\infty)$, we obtain the Lamé functions $E_n^m(\rho)$ and $F_n^m(\rho)$ which define the $\rho-$ dependence of the generated harmonic functions. Similarly, when $x = \mu \in (h_3, h_2)$, $E_n^m(\mu)$ and $F_n^m(\mu)$ define the $\mu-$ dependence, and when $x = \nu \in (-h_3, h_3)$, $E_n^m(\nu)$ and $F_n^m(\nu)$ define the $\nu-$ dependence of the corresponding harmonic function. The function $F(\rho)$ is defined by

$$F_n^m(\rho) = (2n+1)E_n^m(\rho)I_n^m(\rho) \tag{3.39}$$

where $I_n^m(x)$ is the elliptic integral

$$I_n^m(\rho) = \int_\rho^{+\infty} \frac{dx}{\left(E_n^m(x)\right)^2 \sqrt{\left(x^2-h_3^2\right)\left(x^2-h_2^2\right)}}. \tag{3.40}$$

In terms of these Lamé functions, we define the *interior ellipsoidal harmonics*

$$\mathbb{E}_n^m(\rho,\mu,\nu) = E_n^m(\rho)E_n^m(\mu)E_n^m(\nu), \tag{3.41}$$

and the *exterior ellipsoidal harmonics*

$$\mathbb{F}_n^m(\rho,\mu,\nu) = F_n^m(\rho)E_n^m(\mu)E_n^m(\nu). \tag{3.42}$$

The reason why we consider only the $F-$ functions of the variable ρ is due to the fact that interior and exterior harmonics are defined with respect to domains inside and outside the ellipsoids. On the ellipsoid $\rho =$ constant, the angular part of the harmonic functions E and F is given by the *surface ellipsoidal harmonics*

$$S_n^m(\mu,\nu) = E_n^m(\mu)E_n^m(\nu). \tag{3.43}$$

The surface harmonics S_n^m are orthogonal on the surface of any ellipsoid with respect to the weighting function

$$l_\rho(\mu,\nu) = \frac{1}{\sqrt{\rho^2-\mu^2}\sqrt{\rho^2-\nu^2}} \tag{3.44}$$

where ρ denotes the ellipsoidal surface. In fact, the solid angle element in the ellipsoidal system is given by

$$d\Omega(\mu,\nu) = l_\rho(\mu,\nu)dS_\rho(\mu,\nu) \tag{3.45}$$

where dS_ρ is the surface element on the ellipsoid ρ. The solid angle element satisfies the identity

$$\oint_{S_\rho} d\Omega(\mu,\nu) = 4\pi \tag{3.46}$$

for any $\rho \in (h_2, +\infty)$. The orthgonality relations satisfied by $S_n^m(\mu,\nu)$ are given by

$$\oint_{S_0} S_n^m(\mu,\nu)S_{n'}^{m'}(\mu,\nu)d\Omega(\mu,\nu) = \gamma_n^m \delta_{nn'}\delta_{mm'} \tag{3.47}$$

where S_0 represents the full solid angle and γ_n^m are the normalization constants. It is impossible to obtain all of the ellipsoidal harmonics in closed form. Analytic solutions can be obtained only for degree $n \leq 7$. This is due to the lack of recurrence formulae for the Lamé functions, which in turn is due to the fact that as we move from one n to the next, all values of the p_n^m $m = 1, 2, \ldots, 2n+1$, change in an irregular way.

The metric coefficients of the ellipsoidal system are given as follows:

$$h_\rho = \left|\frac{\partial \mathbf{r}}{\partial \rho}\right| = \frac{\sqrt{\rho^2-\mu^2}\sqrt{\rho^2-\nu^2}}{\sqrt{\rho^2-h_3^2}\sqrt{\rho^2-h_2^2}} \tag{3.48}$$

$$h_\mu = \left|\frac{\partial \mathbf{r}}{\partial \mu}\right| = \frac{\sqrt{\rho^2-\mu^2}\sqrt{\mu^2-\nu^2}}{\sqrt{\mu^2-h_3^2}\sqrt{h_2^2-\mu^2}} \tag{3.49}$$

$$h_\nu = \left|\frac{\partial \mathbf{r}}{\partial \nu}\right| = \frac{\sqrt{\rho^2-\nu^2}\sqrt{\mu^2-\nu^2}}{\sqrt{h_3^2-\nu^2}\sqrt{h_2^2-\nu^2}}. \tag{3.50}$$

The corresponding unit vectors of the local system are given by

$$\hat{\boldsymbol{\rho}} = \frac{\rho}{h_\rho}\sum_{i=1}^{3}\frac{x_i}{\rho^2-a_1^2+a_i^2}\hat{\mathbf{x}}_i \tag{3.51}$$

$$\hat{\boldsymbol{\mu}} = \frac{\mu}{h_\mu}\sum_{i=1}^{3}\frac{x_i}{\mu^2-a_1^2+a_i^2}\hat{\mathbf{x}}_i \tag{3.52}$$

$$\hat{\boldsymbol{\nu}} = \frac{\nu}{h_\nu}\sum_{i=1}^{3}\frac{x_i}{\nu^2-a_1^2+a_i^2}\hat{\mathbf{x}}_i. \tag{3.53}$$

Also, the position vector is represented as

$$\mathbf{r} = \frac{\rho}{h_\rho}\hat{\boldsymbol{\rho}} + \frac{\mu}{h_\mu}\hat{\boldsymbol{\mu}} + \frac{\nu}{h_\nu}\hat{\boldsymbol{\nu}} \tag{3.54}$$

and the gradient operator as

$$\nabla = \frac{\hat{\boldsymbol{\rho}}}{h_\rho}\frac{\partial}{\partial \rho} + \frac{\hat{\boldsymbol{\mu}}}{h_\mu}\frac{\partial}{\partial \mu} + \frac{\hat{\boldsymbol{\nu}}}{h_\nu}\frac{\partial}{\partial \nu}. \tag{3.55}$$

Consequently, the outward normal derivative on the ellipsoid ρ is given by

$$\frac{\partial}{\partial n} = \hat{\boldsymbol{\rho}}\cdot\nabla = \frac{1}{h_\rho}\frac{\partial}{\partial \rho} \tag{3.56}$$

which, on the surface of the reference ellipsoid becomes

$$\left.\frac{\partial}{\partial n}\right|_{\rho=a_1} = \frac{a_2 a_3}{\sqrt{a_1^2-\mu^2}\sqrt{a_1^2-\nu^2}}\frac{\partial}{\partial \rho} = a_2 a_3 l_{a_1}(\mu,\nu)\frac{\partial}{\partial \rho}. \tag{3.57}$$

We now turn to the 3-shell ellipsoidal model of the head-brain system.

The surfaces S_j, $j=c,f,b,s$ are confocal ellipsoidal surfaces with the following characteristics:

$$\begin{aligned}
&S_c\colon\ 0<c_3<c_2<c_1<+\infty\\
&S_f\colon\ 0<f_3<f_2<f_1<+\infty\\
&S_b\colon\ 0<b_3<b_2<b_1<+\infty\\
&S_s\colon\ 0<s_3<s_2<s_1<+\infty
\end{aligned} \tag{3.58}$$

These surfaces define the regions

$$\begin{aligned}
\Omega_c &= \{(\rho,\mu,\nu)|h_2\le\rho<c_1\}\\
\Omega_f &= \{(\rho,\mu,\nu)|c_1<\rho<f_1\}\\
\Omega_b &= \{(\rho,\mu,\nu)|f_1<\rho<b_1\}\\
\Omega_s &= \{(\rho,\mu,\nu)|b_1<\rho<s_1\}\\
\Omega_e &= \{(\rho,\mu,\nu)|s_1<\rho<+\infty\}
\end{aligned} \tag{3.59}$$

This means that the surface S_c is defined by the equation

$$\frac{x_1^2}{c_1^2}+\frac{x_2^2}{c_2^2}+\frac{x_3^2}{c_3^2}=1 \tag{3.60}$$

and similarly for the other surfaces.

The aforementioned surfaces have the same semifocal distances h_1, h_2, h_3, where

$$\begin{aligned}
h_1^2 &= c_2^2-c_3^2 = f_2^2-f_3^2 = b_2^2-b_3^2 = s_2^2-s_3^2\\
h_2^2 &= c_1^2-c_3^2 = f_1^2-f_3^2 = b_1^2-b_3^2 = s_1^2-s_3^2.\\
h_3^2 &= c_1^2-c_2^2 = f_1^2-f_2^2 = b_1^2-b_2^2 = s_1^2-s_2^2
\end{aligned} \tag{3.61}$$

The Cartesian coordinates are related to their ellipsoidal coordinates by eqs. (3.35)–(3.37).

If we set $\mathbf{r}_0=(\rho_0,\mu_0,\nu_0)$ and $\mathbf{r}=(\rho,\mu,\nu)$, then the fundamental solution has the expansion

$$\frac{1}{|\mathbf{r}-\mathbf{r}_0|} = \sum_{n=0}^{\infty}\sum_{m=1}^{2n+1}\frac{4\pi}{2n+1}\frac{1}{\gamma_n^m}\mathbb{E}_n^m(\mathbf{r}_0)\mathbb{F}_n^m(\mathbf{r}), \quad \rho_0<c_1. \tag{3.62}$$

$I\!E_n^m$ and $I\!F_n^m$ are the interior and exterior ellipsoidal harmonics as given in (3.41) and (3.42), respectively. The numbers γ_n^m are the normalization constants defined in (3.47).

By solving the transmission problem, where v_i, $i=c,f,b,s,e$ are harmonic functions, we obtain five expansions in ellipsoidal harmonics for the aforementioned five monopolic potential fields. The exact form of these fields are derived in Chapter 5.

In contrast to the case of spherical geometry, the dot product of $\mathbf{r}$ and $\mathbf{H}$ does not vanish in the ellipsoidal case. Thus, it is now necessary to evaluate the surface integrals appearing in the definition of $\mathbf{H}$, given in (3.62).

In this connection, we need the form of the monopolic potentials v_i, $i=c,f,b,s$ on the surfaces S_k which are given by

$$v_i(\mathbf{r},\mathbf{r}_0)=\sum_{n=1}^{\infty}\sum_{m=1}^{2n+1}\frac{4\pi}{\gamma_n^m}C_{in}^{km}I\!E_n^m(\mathbf{r}_0)S_n^m(\mu,\nu) \tag{3.63}$$

where, i and k take the values $c,f,b,s,\ \tau\in\Omega_c$. The constants C_{in}^{km} depend only on the geometry, via c_1,f_1,b_1,s_1, and the physics, via $\sigma_c,\sigma_f,\sigma_b,\sigma_s$. The explicit values of C_{in}^{km} will be calculated in Chapter 5.

Lemma 3.1 The function $\mathbf{H}(\mathbf{r},\mathbf{r}_0)$, defined in (2.62) is written as

$$\mathbf{H}(r,r_0)=(4\pi)^2\sum_{n=1}^{\infty}\sum_{m=1}^{2n+1}\frac{I\!F_n^m(r)}{(2n+1)\gamma_n^m}\sum_{\substack{l=1\\ l\neq m}}^{2n+1}\mathbf{H}_n^{ml}I\!E_n^l(r_0) \tag{3.64}$$

where $\mathbf{r}\in\Omega_e$, $\mathbf{r}_0\in\Omega_c$ and

$$\begin{aligned}\hat{\mathbf{x}}_i\cdot\mathbf{H}_n^{ml}=H_{ni}^{ml}&=c_1c_2c_3(\sigma_c-\sigma_f)C_{cn}^{cl}c_{ni}^{ml}\\&+f_1f_2f_3(\sigma_f-\sigma_b)C_{fn}^{fl}f_{ni}^{ml}\\&+b_1b_2b_3(\sigma_b-\sigma_s)C_{bn}^{bl}b_{ni}^{ml}\\&+s_1s_2s_3\sigma_sC_{sn}^{sl}s_{ni}^{ml}\end{aligned} \tag{3.65}$$

with $i=1,2,3$. The constants c_{ni}^{ml}, f_{ni}^{ml}, b_{ni}^{ml} and s_{ni}^{ml} are specified by the following formulae:

$$\begin{aligned}&\hat{\mathbf{n}}(\mathbf{r})\times\nabla I\!E_n^m(\mathbf{r})|_{\mathbf{r}\in S_c}\\&=\frac{c_1c_2c_3}{\sqrt{c_1^2-\mu^2}\sqrt{c_1^2-\nu^2}}\sum_{\substack{l=1\\ l\neq m}}^{2n+1}(c_{n1}^{ml}\hat{\mathbf{x}}_1+c_{n2}^{ml}\hat{\mathbf{x}}_2+c_{n3}^{ml}\hat{\mathbf{x}}_3)S_n^l(\mu,\nu)\end{aligned} \tag{3.66}$$

$$\hat{\mathbf{n}}(\mathbf{r}) \times \nabla I\!E_n^m(\mathbf{r})|_{\mathbf{r}\in S_f}$$
$$= \frac{f_1 f_2 f_3}{\sqrt{f_1^2-\mu^2}\sqrt{f_1^2-\nu^2}} \sum_{\substack{l=1\\ l\neq m}}^{2n+1} (f_{n1}^{ml}\hat{\mathbf{x}}_1 + f_{n2}^{ml}\hat{\mathbf{x}}_2 + f_{n3}^{ml}\hat{\mathbf{x}}_3) S_n^l(\mu,\nu) \tag{3.67}$$

$$\hat{\mathbf{n}}(\mathbf{r}) \times \nabla I\!E_n^m(\mathbf{r})|_{\mathbf{r}\in S_b} = \frac{b_1 b_2 b_3}{\sqrt{b_1^2-\mu^2}\sqrt{b_1^2-\nu^2}} \sum_{\substack{l=1\\ l\neq m}}^{2n+1} (b_{n1}^{ml}\hat{\mathbf{x}}_1 + b_{n2}^{ml}\hat{\mathbf{x}}_2 + b_{n3}^{ml}\hat{\mathbf{x}}_3) S_n^l(\mu,\nu) \tag{3.68}$$

$$\hat{\mathbf{n}}(\mathbf{r}) \times \nabla I\!E_n^m(\mathbf{r})|_{\mathbf{r}\in S_s} = \frac{s_1 s_2 s_3}{\sqrt{s_1^2-\mu^2}\sqrt{s_1^2-\nu^2}} \sum_{\substack{l=1\\ l\neq m}}^{2n+1} (s_{n1}^{ml}\hat{\mathbf{x}}_1 + s_{n2}^{ml}\hat{\mathbf{x}}_2 + s_{n3}^{ml}\hat{\mathbf{x}}_3) S_n^l(\mu,\nu) . \tag{3.69}$$

Proof. The proof is based on the identity

$$\hat{\mathbf{n}}(\mathbf{r}) \times \nabla I\!E_n^m(\mathbf{r})|_{\rho=a_1} = \frac{a_1 a_2 a_3}{\sqrt{a_1^2-\mu^2}\sqrt{a_1^2-\nu^2}} \sum_{\substack{l=1\\ l\neq m}}^{2n+1} (a_{n1}^{ml}\hat{\mathbf{x}}_1 + a_{n2}^{ml}\hat{\mathbf{x}}_2 + a_{n3}^{ml}\hat{\mathbf{x}}_3) S_n^l(\mu,\nu) \tag{3.70}$$

where $\rho = a_1$ is the ellipsoidal surface

$$\frac{x_1^2}{a_1^2} + \frac{x_2^2}{a_2^2} + \frac{x_3^2}{a_3^2} = 1 \tag{3.71}$$

(ρ,μ,ν) are the ellipsoidal coordinates of the position pvoint $\mathbf{r}$ and $S_n^m(\mu,\nu)$ are the surface ellipsoidal harmonics. Formulae (3.70) expresses the tangential surface gradient operator on the surface (3.71). The proof of (3.70) is rather technical and can be found in [30].

Using formulae (3.43), (3.44), (3.45), (3.21), (3.62) and (3.63), we can write the first integral on the right hand side of (2.62a) as follows

$$\int_{S_c} v_f(\mathbf{r}',\mathbf{r}_0)\,\hat{\mathbf{n}}(\mathbf{r}') \times \nabla_{\mathbf{r}'} \frac{1}{|\mathbf{r}-\mathbf{r}'|} ds(\mathbf{r}') = \int_{S_c} \left[\sum_{n=1}^{\infty}\sum_{m=1}^{2n+1} \frac{4\pi}{\gamma_n^m} C_{fn}^{cm} I\!E_n^m(\mathbf{r}_0) S_n^m(\mu',\nu')\right]$$
$$\cdot \left[\frac{c_1 c_2 c_3}{\sqrt{c_1^2-\mu'^2}\sqrt{c_1^2-\nu'^2}} \sum_{i=1}^{3} \frac{x'_i}{c_i^2} \hat{\mathbf{x}}_i \times \nabla_{\mathbf{r}'}\right] \left[\sum_{n'=0}^{\infty}\sum_{m'=1}^{2n'+1} \frac{4\pi}{(2n'+1)\gamma_{n'}^{m'}} I\!E_{n'}^{m'}(\mathbf{r}') I\!F_{n'}^{m'}(\mathbf{r})\right]$$
$$\cdot \sqrt{c_1^2-\mu'^2}\sqrt{c_1^2-\nu'^2}\, d\Omega(\mu',\nu')$$
$$= (4\pi)^2 c_1 c_2 c_3 \sum_{n=1}^{\infty}\sum_{m=1}^{2n+1}\sum_{n'=0}^{\infty}\sum_{m'=1}^{2n'+1} \frac{C_{cn}^{cm} I\!E_n^m(\mathbf{r}_0) I\!F_{n'}^{m'}(\mathbf{r})}{(2n'+1)\gamma_n^m \gamma_{n'}^{m'}}$$
$$\cdot \int_{S_c} \left[\sum_{i=1}^{3} \frac{x'_i}{c_i^2} \hat{\mathbf{x}}_i \times \nabla_{\mathbf{r}'} I\!E_{n'}^{m'}(\mathbf{r}')\right] S_n^m(\mu',\nu')\, d\Omega(\mu',\nu') \tag{3.72}$$

where we have used the fact that because of continuity

$$C_{fn}^{cm} = C_{cn}^{cm}. \tag{3.73}$$

Inserting (3.66) in (3.72) and using the orthogonality property (3.47) we write (3.72) as

$$\int_{S_c} \upsilon_f(\mathbf{r}', \mathbf{r}_0)\, \hat{\mathbf{n}}(\mathbf{r}') \times \nabla_{\mathbf{r}'} \frac{1}{|\mathbf{r}-\mathbf{r}'|} ds(\mathbf{r}')$$

$$= (4\pi)^2 c_1 c_2 c_3 \sum_{n=1}^{\infty} \sum_{m=1}^{2n+1} \sum_{n'=0}^{\infty} \sum_{m'=1}^{2n'+1} \sum_{\substack{l=1 \\ l \neq m'}}^{2n'+1} \frac{C_{cn}^{cm} \mathbb{E}_n^m(\mathbf{r}_0) \mathbb{F}_{n'}^{m'}(\mathbf{r})}{(2n'+1)\gamma_n^m \gamma_{n'}^{m'}} (c_{n'1}^{m'l}\, \hat{\mathbf{x}}_1 + c_{n'2}^{m'l}\, \hat{\mathbf{x}}_2 + c_{n'3}^{m'l}\, \hat{\mathbf{x}}_3)$$

$$\cdot \int_{S_c} S_{m'}^{l}(\mu', \nu')\, S_n^m(\mu', \nu') d\Omega(\mu', \nu')$$

$$= (4\pi)^2 c_1 c_2 c_3 \sum_{n=1}^{\infty} \sum_{m=1}^{2n+1} \sum_{n'=0}^{\infty} \sum_{m'=1}^{2n'+1} \sum_{\substack{l=1 \\ l \neq m'}}^{2n'+1} \frac{C_{cn}^{cm} \mathbb{E}_n^m(\mathbf{r}_0) \mathbb{F}_{n'}^{m'}(\mathbf{r})}{(2n'+1)\gamma_n^m \gamma_{n'}^{m'}} \sum_{i=1}^{3} c_{n'i}^{m'l}\, \hat{\mathbf{x}}_i \gamma_n^l \delta_{nn'} \delta_{ml}$$

$$= (4\pi)^2 c_1 c_2 c_3 \sum_{n=1}^{\infty} \sum_{m'=1}^{2n+1} \sum_{\substack{l=1 \\ l \neq m'}}^{2n+1} \frac{C_{cn}^{cl} \mathbb{E}_n^m(\mathbf{r}_0) \mathbb{F}_{n'}^{m'}(\mathbf{r})}{(2n'+1)\gamma_n^{m'}} \sum_{i=1}^{3} c_{n'i}^{m'l}\, \hat{\mathbf{x}}_i. \tag{3.74}$$

Replacing m' by m in (3.74) and analyzing the second, third and fourth integrals on the right hand side of (2.62a) in exactly the same way as the first integral we finally obtain

$$\mathbf{H}(\mathbf{r}, \mathbf{r}_0) = (4\pi)^2 \sum_{n=1}^{\infty} \sum_{m=1}^{2n+1} \sum_{\substack{l=1 \\ l \neq m}}^{2n+1} \frac{\mathbb{E}_n^l(\mathbf{r}_0) \mathbb{F}_n^m(\mathbf{r})}{(2n+1)\gamma_n^m}$$
$$+ \sum_{i=1}^{3} \Big[(\sigma_c - \sigma_f) c_1 c_2 c_3 C_{cn}^{cl} c_{ni}^{ml} + (\sigma_f - \sigma_b) f_1 f_2 f_3 C_{fn}^{fl} f_{ni}^{ml}$$
$$+ (\sigma_b - \sigma_s) b_1 b_2 b_3 C_{bn}^{bl} b_{ni}^{ml} + \sigma_s s_1 s_2 s_3 C_{sn}^{sl} s_{ni}^{ml} \Big]\, \hat{\mathbf{x}}_i \tag{3.75}$$

which coincides with (3.64) and (3.65).

Hence, the proof of Lemma 3.1.

In view of (3.64), we obtain that

$$\frac{1}{\mu_0} \mathbf{B}(\mathbf{r}, \mathbf{r}_0) = \sum_{n=1}^{\infty} \sum_{m=1}^{2n+1} \frac{\mathbb{F}_n^m(\mathbf{r})}{(2n+1)\gamma_n^m} \cdot \left[\mathbf{J}(\mathbf{r}_0) \times \nabla_{\mathbf{r}_0} \mathbb{E}_n^m(\mathbf{r}_0) - \mathbf{J}(\mathbf{r}_0) \cdot \nabla_{\mathbf{r}_0} \sum_{\substack{l=1 \\ l \neq m}}^{2n+1} \mathbf{H}_n^{nl} \mathbb{E}_n^l(\mathbf{r}_0) \right],$$
$$\mathbf{r} \in \Omega_e, \quad \mathbf{r}_0 \in \Omega_c. \tag{3.76}$$

Proposition 3.3 Consider the 3-shell ellipsoidal model specified by the inequalities (3.59) and let the neuronal current $\mathbf{J}(\mathbf{r}_0)$, $\mathbf{r}_0 \in \Omega_c$, be expressed in the form (3.1) and (3.2). Then

$$u_s(\mathbf{r}) = -\frac{1}{4\pi}\sum_{n=1}^{\infty}\sum_{m=1}^{2n+1}\frac{1}{\gamma_n^m}C_{sn}^{sm}S_n^m(\mu,\nu)\int_{\Omega_c}(\Delta\Psi(\mathbf{r}_0))\mathbb{E}_n^m(\mathbf{r}_0)dv(\mathbf{r}_0), \quad \mathbf{r}\in S_s \tag{3.77}$$

and

$$\frac{1}{\mu_0}\mathbf{r}\cdot\mathbf{B}(\mathbf{r}) = -\sum_{n=1}^{\infty}\sum_{m=1}^{2n+1}\frac{1}{(2n+1)\gamma_n^m}\mathbb{F}_n^m(\mathbf{r})$$
$$\cdot\int_{\Omega_c}\left[(\Delta(\mathbf{r}_0\cdot\mathbf{A}(\mathbf{r}_0)))\mathbb{E}_n^m(\mathbf{r}_0) - (\Delta\Psi(\mathbf{r}_0))\sum_{\substack{l=1\\ l\neq m}}^{2n+1}\mathbf{r}\cdot\mathbf{H}_n^{ml}\mathbb{E}_n^l(\mathbf{r}_0)\right]dv(\mathbf{r}_0), \quad \mathbf{r}\in\Omega_e. \tag{3.78}$$

In eq. (3.77), the ellipsoidal coordinates of the pvoint $\mathbf{r}\in S_s$ are (s_1,μ,ν), while the constants C_{sn}^{sm} as well as the constant vectors $\mathbf{H}_n^{ml}$ can be explicitly computed in terms of the conductivities $\sigma_c, \sigma_f, \sigma_b, \sigma_s$ and the ellipsoidal parameters $c_i, f_i, b_i, s_i,\ i=1,2,3$.

The expressions in eqs. (3.77) and (3.78) can be simplified into the following expressions:

$$u_s(\mathbf{r}) = -\frac{c_2c_3}{4\pi}\sum_{n=1}^{\infty}\sum_{m=1}^{2n+1}C_{sn}^{sm}S_n^m(\mu,\nu)\left[\dot{y}_n^m(c_1)E_n^m(c_1) - y_n^m(c_1)\dot{E}_n^m(c_1)\right], \quad \mathbf{r}\in S_s \tag{3.79}$$

and

$$\frac{1}{\mu_0}\mathbf{r}\cdot\mathbf{B}(\mathbf{r}) = -\ c_2c_3\sum_{n=0}^{\infty}\sum_{m=1}^{2n+1}\frac{\mathbb{F}_n^m(\mathbf{r})}{2n+1}\left\{\left[\dot{a}_n^m(c_1)E_n^m(c_1) - a_n^m(c_1)\dot{E}_n^m(c_1)\right]\right.$$
$$\left.-\sum_{\substack{l=1\\ l\neq m}}^{2n+1}\left(\mathbf{r}\cdot\mathbf{H}_n^{ml}\right)\left[\dot{y}_n^l(c_1)E_n^l(c_1) - y_n^l(c_1)\dot{E}_n^l(c_1)\right]\right\}, \quad \mathbf{r}\in\Omega_e \tag{3.80}$$

where $y_n^m(\rho_0)$ and $a_n^m(\rho_0)$ are the $\rho-$ parts of the expansions of $\Psi(\mathbf{r}_0)$ and $\mathbf{r}_0\cdot\mathbf{A}(\mathbf{r}_0)$ in surface ellipsoidal harmonics, namely,

$$\Psi(\mathbf{r}_0) = \sum_{n=0}^{\infty}\sum_{m=1}^{2n+1}y_n^m(\rho_0)S_n^m(\mu_0,\nu_0) \tag{3.81}$$

and

$$\mathbf{r}_0\cdot\mathbf{A}(\mathbf{r}_0) = \sum_{n=0}^{\infty}\sum_{m=1}^{2n+1}a_n^m(\rho_0)S_n^m(\mu_0,\nu_0) \tag{3.82}$$

with $\mathbf{r}_0\in\Omega_c$.

Proof. Replacing in equation (3.3), the function υ_s as it is given by (3.63) with $i=k=s$ we obtain (3.77). Similarly, replacing in eq. (3.5), the function $1/|\mathbf{r}-\mathbf{r}'|$ by the right hand side of (3.62) and the function $\mathbf{H}$ by the right hand side of (3.64), we obtain eq. (3.78).

We recall that

$$d\upsilon = h_\rho h_\mu h_\nu d\rho\, d\mu d\nu = h_\rho d_\rho ds = \frac{(\rho^2-\mu^2)(\rho^2-\nu^2)}{\sqrt{\rho^2-h_3^2}\sqrt{\rho^2-h_2^2}}\, d\rho\, d\Omega(\mu,\nu) \tag{3.83}$$

where $d\Omega$ denotes the elementary solid angle in ellipsoidal coordinates, that is,

$$d\Omega(\mu,\nu) = \frac{\mu^2-\nu^2}{\sqrt{\mu^2-h_3^2}\sqrt{h_2^2-\mu^2}\sqrt{h_3^2-\nu^2}\sqrt{h_2^2-\nu^2}}\, d\mu d\nu. \tag{3.84}$$

The Laplace's operator in ellipsoidal coordinates has the form

$$\begin{aligned}\Delta = {} & \frac{1}{(\rho^2-\mu^2)(\rho^2-\nu^2)}\left(\sqrt{\rho^2-h_3^2}\sqrt{\rho^2-h_2^2}\frac{\partial}{\partial\rho}\right)\left(\sqrt{\rho^2-h_3^2}\sqrt{\rho^2-h_2^2}\frac{\partial}{\partial\rho}\right)\\ & + \frac{1}{(\rho^2-\mu^2)(\rho^2-\nu^2)}\mathbb{B}_e\end{aligned} \tag{3.85}$$

where $\mathbb{B}_e$ is the following *ellipsoidal Beltrami operator* or the ellipsoidal surface Laplacian:

$$\begin{aligned}\mathbb{B}_e = {} & \frac{\rho^2-\nu^2}{\mu^2-\nu^2}\left(\sqrt{\mu^2-h_3^2}\sqrt{h_2^2-\mu^2}\frac{\partial}{\partial\mu}\right)\left(\sqrt{\mu^2-h_3^2}\sqrt{h_2^2-\mu^2}\frac{\partial}{\partial\mu}\right)\\ & + \frac{\rho^2-\mu^2}{\mu^2-\nu^2}\left(\sqrt{h_3^2-\nu^2}\sqrt{h_2^2-\nu^2}\frac{\partial}{\partial\nu}\right)\left(\sqrt{h_3^2-\nu^2}\sqrt{h_2^2-\nu^2}\frac{\partial}{\partial\nu}\right).\end{aligned} \tag{3.86}$$

The Beltrami operator satisfies the spectral equations

$$\mathbb{B}_e S_n^m(\mu,\nu) = \left[(h_3^2+h_2^2)p_n^m - n(n+1)\rho^2\right] S_n^m(\mu,\nu) \tag{3.87}$$

where p_n^m is the second separation constant of the Lamé functions appearing in eq. (3.38). Furthermore, it is straightforward to show that

$$\frac{(\rho^2-\mu^2)(\rho^2-\nu^2)}{\sqrt{\rho^2-h_3^2}\sqrt{\rho^2-h_2^2}}\Delta\left(y_n^m(\rho)S_n^m(\mu,\nu)\right) = \left(L_n^m y_n^m(\rho)\right)S_n^m(\mu,\nu) \tag{3.88}$$

where

$$L_n^m = \frac{d}{d\rho}\sqrt{\rho^2-h_3^2}\sqrt{\rho^2-h_2^2}\frac{d}{d\rho} + \frac{(h_2^2+h_3^2)p_n^m - n(n+1)\rho^2}{\sqrt{\rho^2-h_3^2}\sqrt{\rho^2-h_2^2}} \tag{3.89}$$

is the Lamé operator.

Replacing in eq. (3.77) the expression (3.81) for Ψ, (3.83) for dv, then making use of eq. (3.88) and the orthogonality condition (3.47), we obtain

$$\begin{aligned}
u_s(\mathbf{r}) &= -\frac{1}{4\pi}\sum_{n=1}^{\infty}\sum_{m=1}^{2n+1}\sum_{n'=0}^{\infty}\sum_{m'=1}^{2n'+1}\frac{1}{\gamma_n^m}C_{sn}^{sm}S_n^m(\mu,\nu)\\
&\cdot\int_{h_2}^{c_1}\oint_{S_0}\Delta\left(y_{n'}^{m'}(\rho_0)S_{n'}^{m'}(\mu_0,\nu_0)\right)E_n^m(\rho_0)S_n^m(\mu_0,\nu_0)\frac{(\rho_0{}^2-\mu_0{}^2)(\rho_0{}^2-\nu_0{}^2)}{\sqrt{\rho_0{}^2-h_3^2}\sqrt{\rho_0{}^2-h_2^2}}d\rho_0 d\Omega(\mu_0,\nu_0)\\
&= -\frac{1}{4\pi}\sum_{n=1}^{\infty}\sum_{m=1}^{2n+1}\sum_{n'=0}^{\infty}\sum_{m'=1}^{2n'+1}\frac{1}{\gamma_n^m}C_{sn}^{sm}S_n^m(\mu,\nu)\\
&\cdot\int_{h_2}^{c_1}\oint_{S_0}\left(L_{n'}^{m'}y_{n'}^{m'}(\rho_0)\right)S_{n'}^{m'}(\mu_0,\nu_0)E_{n'}^{m'}(\rho_0)S_n^m(\mu_0,\nu_0)d\rho_0 d\Omega(\mu_0,\nu_0)\\
&= -\frac{1}{4\pi}\sum_{n=1}^{\infty}\sum_{m=1}^{2n+1}\sum_{n'=0}^{\infty}\sum_{m'=1}^{2n'+1}\frac{1}{\gamma_n^m}C_{sn}^{sm}S_n^m(\mu,\nu)\int_{h_2}^{c_1}\left(L_{n'}^{m'}y_{n'}^{m'}(\rho_0)\right)E_{n'}^{m'}(\rho_0)d\rho_0\ \gamma_n^m\delta_{nn'}\delta_{mm'}\\
&= -\frac{1}{4\pi}\sum_{n=1}^{\infty}\sum_{m=1}^{2n+1}C_{sn}^{sm}S_n^m(\mu,\nu)\int_{h_2}^{c_1}(L_n^m y_n^m(\rho_0))E_n^m(\rho_0)d\rho_0.
\end{aligned} \tag{3.90}$$

Two successive integrations by parts give

$$\begin{aligned}
&\int_{h_2}^{c_1}E_n^m(\rho_0)\frac{d}{d\rho_0}\left[\sqrt{\rho_0{}^2-h_3^2}\sqrt{\rho_0{}^2-h_2^2}\frac{d}{d\rho_0}y_n^m(\rho_0)\right]d\rho_0\\
&\quad= c_2c_3\left[E_n^m(c_1)\dot{y}_n^m(c_1)-\dot{E}_n^m(c_1)y_n^m(c_1)\right]\\
&\quad-\int_{h_2}^{c_1}\frac{(h_2^2+h_3^2)p_n^m-n(n+1)\rho_0{}^2}{\sqrt{\rho_0{}^2-h_3^2}\sqrt{\rho_0{}^2-h_2^2}}E_n^m(\rho_0)y_n^m(\rho_0)d\rho_0.
\end{aligned} \tag{3.91}$$

In view of (3.89), we obtain

$$\int_{h_2}^{c_1}(L_n^m y_n^m(\rho_0))E_n^m(\rho_0)d\rho_0 = c_2c_3\left[E_n^m(c_1)\dot{y}_n^m(c_1)-\dot{E}_n^m(c_1)y_n^m(c_1)\right]. \tag{3.92}$$

Inserting eq. (3.92) in (3.90), we obtain (3.79).

Finally, for the proof of formula (3.80), we follow the same steps as with (3.88) and obtain

$$\frac{({\rho_0}^2-{\mu_0}^2)({\rho_0}^2-{\nu_0}^2)}{\sqrt{{\rho_0}^2-h_3^2}\sqrt{{\rho_0}^2-h_2^2}}\Delta(a_n^m(\rho_0)S_n^m(\mu_0,\nu_0))=(L_n^m a_n^m(\rho_0))S_n^m(\mu_0,\nu_0) \tag{3.93}$$

where L_n^m is defined in (3.89) and we have use of the expansion (3.82).

Then,

$$\begin{aligned}\frac{1}{\mu_0}\mathbf{r}\cdot\mathbf{B}(\mathbf{r}) &= -\sum_{n=0}^{\infty}\sum_{m=1}^{2n+1}\frac{I\!F_n^m(\mathbf{r})}{(2n+1)\gamma_n^m}\\
&\quad\cdot\int_{\Omega_c}\Bigg[(\Delta(\mathbf{r}_0\cdot\mathbf{A}(\mathbf{r}_0)))I\!E_n^m(\mathbf{r}_0)dv(\mathbf{r}_0)\\
&\quad-\sum_{\substack{l=1\\l\neq m}}^{2n+1}(\mathbf{r}\cdot\mathbf{H}_n^{ml})\int_{\Omega_c}(\Delta\Psi(\mathbf{r}_0))I\!E_n^l(\mathbf{r}_0)\,dv(\mathbf{r}_0)\Bigg]\\
&= -\sum_{n=0}^{\infty}\sum_{m=1}^{2n+1}\frac{I\!F_n^m(\mathbf{r})}{2n+1}\Bigg[\int_{h_2}^{c_1}(L_n^m a_n^m(\rho_0))E_n^m(\rho_0)d\rho_0\\
&\quad-\sum_{\substack{l=1\\l\neq m}}^{2n+1}(\mathbf{r}\cdot\mathbf{H}_n^{ml})\int_{h_2}^{c_1}(L_n^m y_n^m(\rho_0))E_n^m(\rho_0)d\rho_0\Bigg]\\
&= -c_2c_3\sum_{n=0}^{\infty}\sum_{m=1}^{2n+1}\frac{I\!F_n^m(\mathbf{r})}{2n+1}\Bigg[(E_n^m(c_1)\dot a_n^m(c_1)-\dot E_n^m(c_1)a_n^m(c_1))\\
&\quad-\sum_{\substack{l=1\\l\neq m}}^{2n+1}(\mathbf{r}\cdot\mathbf{H}_n^{ml})(E_n^l(c_1)\dot y_n^l(c_1)-\dot E_n^l(c_1)y_n^l(c_1))\Bigg]\end{aligned} \tag{3.94}$$

which is exactly (3.80).

Hence, the proof of Proposition 3.3 is completed.

Remark 3.1 The general identity (3.70) can also be replaced by its ellipsoidal analogue which is

$$\begin{aligned}\hat{\mathbf{n}}(\mathbf{r})\times\nabla I\!E_n^m(\mathbf{r}) &= \hat\rho\times\left(\frac{\hat\rho}{h_\rho}\frac{\partial}{\partial\rho}+\frac{\hat\mu}{h_\mu}\frac{\partial}{\partial\mu}+\frac{\hat\nu}{h_\nu}\frac{\partial}{\partial\nu}\right)I\!E_n^m(\mathbf{r})\\
&=\left(-\frac{\hat\nu}{h_\mu}\frac{\partial}{\partial\mu}+\frac{\hat\mu}{h_\nu}\frac{\partial}{\partial\nu}\right)I\!E_n^m(\mathbf{r})\\
&=I\!E_n^m(\rho)\left(\frac{\hat\mu}{h_\nu}E_n^m(\mu)\dot E_n^m(\nu)-\frac{\hat\nu}{h_\mu}\dot E_n^m(\mu)E_n^m(\nu)\right).\end{aligned} \tag{3.95}$$

However, this form does not involve the surface functions $S_n^m(\mu, \nu)$ and therefore we can not apply the orthogonality condition as we did in (3.74). We observe here the unusual case where a combination of an ellipsoidal with a Cartesian form is more effective than the case where every expression is written in ellipsoidal form.

Remark 3.2 Using the expansions of the potentials u_s, u_b, u_f and u_c in ellipsoidal harmonics, as they are given analytically in Chapter 5, and the corresponding potentials given in (3.63), after we apply the source dependent operator

$$\frac{1}{4\pi}\mathbf{J}(\mathbf{r}_0) \cdot \nabla_{\mathbf{r}_0}$$

we obtain the relations

$$C_{sn}^{sm} = \frac{1}{s_2 s_3 \dot{E}_n^m(s_1) G_{3,n}^m} \tag{3.96}$$

$$C_{bn}^{bm} = C_{sn}^{bm} = C_{sn}^{sm} + \frac{E_n^m(b_1)}{G_{3,n}^m} I_n^m(b_1, s_1) \tag{3.97}$$

$$C_{fn}^{fm} = C_{bn}^{fm} = C_{bn}^{bm} + \frac{E_n^m(f_1) G_{1,n}^m}{\sigma_b G_{3,n}^m} I_n^m(f_1, b_1) \tag{3.98}$$

$$C_{cn}^{cm} = C_{fn}^{cm} = C_{fn}^{fm} + \frac{E_n^m(c_1) G_{2,n}^m}{\sigma_f G_{3,n}^m} I_n^m(c_1, f_1) \tag{3.99}$$

where

$$I_n^m(x, y) = I_n^m(x) - I_n^m(y) \tag{3.100}$$

and the constants $G_{1,n}^m$, $G_{2,n}^m$, $G_{3,n}^m$ are given in Chapter 5.

Chapter 4
The spherical model

The spherical model of the head–brain system was analyzed in Section 3.2. For the electroencephalogram (EEG) problem, the monopolic potential v_s on the head surface $r=s$ is given by formula (3.8), where the sequence $\{s_n\}_{n=1}^{\infty}$ of coefficients involves the radii c, f, b, s as well as the conductivity constants $\sigma_c, \sigma_f, \sigma_b, \sigma_s$. In this chapter, we calculate analytically the above coefficients. This is achieved by computing all electric potentials in the cerebrum and in all shells that surround the cerebrum. In fact in this chapter we also establish closed-form solutions for the spherical geometry. Related results for magnetoencephalography (MEG) are also included.

4.1 The interior electric potential

In this chapter, we assume that the conductive brain is modeled by the sphere

$$x_1^2 + x_2^2 + x_3^2 = \alpha^2 \tag{4.1}$$

which contain in its interior a dipolar current located at the point $\mathbf{r}_0$, with $|\mathbf{r}_0| < \alpha$ and dipole moment $\mathbf{Q}$. In this case, the interior electric potential u^- solves the Neumann problem

$$\sigma \Delta u^-(\mathbf{r}) = \mathbf{Q} \cdot \nabla_{\mathbf{r}} \delta(\mathbf{r} - \mathbf{r}_0), \quad r < \alpha \tag{4.2}$$

$$\frac{\partial}{\partial n} u^-(\mathbf{r}) = 0, \quad r = \alpha. \tag{4.3}$$

We recall that the solution of the above Neumann problem is unique up to an additive constant. This is a typical nonuniqueness situation for potential problems since the actual physical quantities are defined as potential differences. Consequently, the zero values of the potential can be taken arbitrarily. Here, we choose the center of the sphere as the point of vanishing electric potential, that is,

$$u^-(\mathbf{0}) = 0. \tag{4.4}$$

Since

$$\nabla_{\mathbf{r}} \delta(\mathbf{r} - \mathbf{r}_0) = -\nabla_{\mathbf{r}_0} \delta(\mathbf{r} - \mathbf{r}_0) \tag{4.5}$$

it follows that the solution u^- can be written as

$$u^-(\mathbf{r}) = \mathbf{Q} \cdot \nabla_{\mathbf{r}_0} w^-(\mathbf{r}). \tag{4.6}$$

https://doi.org/10.1515/9783110547535-005

Hence, it can be interpreted as the action of the directional derivative along $\mathbf{Q}$ on the potential w^-, which solves the Poisson equation

$$\sigma\Delta_{\mathbf{r}} w^-(\mathbf{r}) = -\delta(\mathbf{r}-\mathbf{r}_0), \quad r<\alpha. \tag{4.7}$$

A particular solution of eq. (4.7) can be expressed in terms of the fundamental solution of the Laplacian and therefore the general solution of eq. (4.7) is written as

$$w^-(\mathbf{r}) = \frac{1}{4\pi\sigma|\mathbf{r}-\mathbf{r}_0|} + v^-(\mathbf{r}) \tag{4.8}$$

where v^- is a solution of the Laplace equation

$$\Delta_{\mathbf{r}} v^-(\mathbf{r}) = 0, \quad r<\alpha. \tag{4.9}$$

The general solution of eq. (4.9) can be expressed in spherical coordinates in terms of the *interior solid spherical harmonics*

$$r^n Y_n^m(\hat{\mathbf{r}})$$

with $n=0,1,2,\dots$ and $m=-n,-n+1,\dots,n-1,n$, where the *surface spherical harmonics* Y_n^m are given by

$$Y_n^m(\hat{\mathbf{r}}) = \sqrt{\frac{2n+1}{4\pi}\frac{(n-|m|)!}{(n+|m|)!}}\,P_n^{|m|}(\cos\theta)e^{im\varphi} \tag{4.10}$$

with P_n^m being the well-known associated Legendre functions of the first kind. The functions Y_n^m are normalized over the unit sphere. The general solution of eq. (4.9) has the expansion

$$v^-(\mathbf{r}) = \sum_{n=0}^{\infty}\sum_{m=-n}^{n} A_n^m r^n Y_n^m(\hat{\mathbf{r}}) \tag{4.11}$$

where the coefficients A_n^m are chosen by the requirement that the boundary condition (4.3) is also satisfied. In order to utilize the orthogonality of Y_n^m we also need to expand the fundamental solution. In this connection, we start with the Laplace expansion

$$\frac{1}{\sqrt{1-2r\cos\gamma+r^2}} = \sum_{n=0}^{\infty} r^n P_n(\cos\gamma) \tag{4.12}$$

which converges absolutely and almost uniformly for $r<1$. We recall that almost uniform convergence means that the convergence is uniform over any compact subset of the region of convergence. The functions P_n are the *ordinary Legendre polynomials* for every $n=0,1,2,\dots$ which have the function $[1-2r\cos\gamma+r^2]^{-1/2}$ as their *generating function.* From (4.12) we obtain

$$\frac{1}{|\mathbf{r}-\mathbf{r}_0|} = \sum_{n=0}^{\infty}\frac{r_0^n}{r^{n+1}}P_n(\hat{\mathbf{r}}\cdot\hat{\mathbf{r}}_0) \tag{4.13}$$

where $r_0 < r$ and $\hat{\mathbf{r}} \cdot \hat{\mathbf{r}}_0$ defines the cosine of the angle between the direction of observation $\hat{\mathbf{r}}$ and that of the location $\hat{\mathbf{r}}_0$ of the dipole. Next we utilize the *addition theorem* for the Legendre polynomial

$$P_n(\hat{\mathbf{r}} \cdot \hat{\mathbf{r}}_0) = \sum_{m=-n}^{n} \frac{4\pi}{2n+1} Y_n^m(\hat{\mathbf{r}}) Y_n^{m*}(\hat{\mathbf{r}}_0) \tag{4.14}$$

which expresses every Legendre polynomial, involving the cosine $\hat{\mathbf{r}} \cdot \hat{\mathbf{r}}_0$, in terms of the surface spherical harmonics of the independent directions $\hat{\mathbf{r}}$ and $\hat{\mathbf{r}}_0$. Inserting the representation (4.14) in expansion (4.13), we obtain

$$\frac{1}{|\mathbf{r} - \mathbf{r}_0|} = \sum_{n=0}^{\infty} \sum_{m=-n}^{n} \frac{4\pi}{2n+1} \frac{r_0^n}{r^{n+1}} Y_n^m(\hat{\mathbf{r}}) Y_n^{m*}(\hat{\mathbf{r}}_0) \tag{4.15}$$

which holds for $r > r_0$, and hence in a neighborhood of the boundary $r = \alpha$.

Therefore, the boundary condition (4.3) is written as

$$\begin{aligned} &\mathbf{Q} \cdot \nabla_{\mathbf{r}_0} \sum_{n=0}^{\infty} \sum_{m=-n}^{n} A_n^m \left(\frac{d}{dr} r^n \right) Y_n^m(\hat{\mathbf{r}}) \\ &= -\frac{1}{\sigma} \mathbf{Q} \cdot \nabla_{\mathbf{r}_0} \sum_{n=0}^{\infty} \sum_{m=-n}^{n} \frac{r_0^n}{2n+1} \left(\frac{d}{dr} \frac{1}{r^{n+1}} \right) Y_n^m(\hat{\mathbf{r}}) Y_n^{m*}(\hat{\mathbf{r}}_0) \end{aligned} \tag{4.16}$$

which has to be evaluated at $r = \alpha$. Utilizing the orthogonality of the surface spherical harmonics, we obtain

$$A_0^0 = \text{arbitrary} \tag{4.17}$$

$$A_n^m = \frac{1}{\sigma} \frac{n+1}{n(2n+1)} \mathbf{Q} \cdot \nabla_{\mathbf{r}_0} \frac{r_0^n}{\alpha^{2n+1}} Y_n^{m*}(\hat{\mathbf{r}}_0) \tag{4.18}$$

for every $n \geq 1$ and $|m| \leq n$. The above arguments are valid because expansion (4.15) converges in the spherical shell $r_0 < r < \alpha$. Inserting expressions (4.17) and (4.18) into expansion (4.11) and substituting the result, as well as eq. (4.15), in representations (4.8) and (4.6) we obtain

$$u^-(\mathbf{r}) = c + \frac{1}{\sigma} \mathbf{Q} \cdot \nabla_{\mathbf{r}_0} \sum_{n=1}^{\infty} \sum_{m=-n}^{n} \frac{r_0^n}{2n+1} \left(\frac{n+1}{n} \frac{r^n}{\alpha^{2n+1}} + \frac{1}{r^{n+1}} \right) Y_n^{m*}(\hat{\mathbf{r}}_0) Y_n^m(\hat{\mathbf{r}}), \quad r \in (r_0, \alpha) \tag{4.19}$$

where c is the arbitrary constant of the Neumann problem. An analogous expression that holds for every $r < \alpha$ is given by

$$\begin{aligned} u^-(\mathbf{r}) &= c + \frac{1}{4\pi\sigma} \mathbf{Q} \cdot \nabla_{\mathbf{r}_0} \frac{1}{|\mathbf{r} - \mathbf{r}_0|} \\ &+ \frac{1}{\sigma} \mathbf{Q} \cdot \nabla_{\mathbf{r}_0} \sum_{n=1}^{\infty} \sum_{m=-n}^{n} \frac{n+1}{n(2n+1)} \frac{r_0^n r^n}{\alpha^{2n+1}} Y_n^{m*}(\hat{\mathbf{r}}_0) Y_n^m(\hat{\mathbf{r}}), \quad r < \alpha \end{aligned} \tag{4.20}$$

and the choice (4.4) implies that

$$c = \frac{\mathbf{Q}\cdot\mathbf{r}_0}{4\pi\sigma r_0^3}. \tag{4.21}$$

Utilizing the addition theorem (4.14) in (4.20) we arrive at

$$\begin{aligned} u^-(\mathbf{r}) &= \frac{1}{4\pi\sigma}\mathbf{Q}\cdot\nabla_{\mathbf{r}_0}\left[\frac{1}{|\mathbf{r}-\mathbf{r}_0|} - \frac{1}{r_0}\right] \\ &+ \frac{1}{4\pi\sigma}\mathbf{Q}\cdot\nabla_{\mathbf{r}_0}\sum_{n=1}^{\infty}\frac{n+1}{n}\frac{r_0^n r^n}{\alpha^{2n+1}}\mathrm{P}_n(\hat{\mathbf{r}}\cdot\hat{\mathbf{r}}_0), \quad 0 \le r \le \alpha. \end{aligned} \tag{4.22}$$

4.2 The exterior electric potential

The electric potential u^+ in the exterior of the sphere $|\mathbf{r}| = \alpha$ generated by the dipolar current

$$\mathbf{J}^p(\mathbf{r};\mathbf{r}_0) = \mathbf{Q}\delta(\mathbf{r}-\mathbf{r}_0) \tag{4.23}$$

is provided by the solution of the exterior Dirichlet problem (1.44)–(1.46), where the Dirichlet data are given by the value of u^- on the surface of the sphere $|\mathbf{r}| = \alpha$. These data are

$$\begin{aligned} u^-(\alpha\hat{\mathbf{r}}) &= \frac{1}{4\pi\sigma}\mathbf{Q}\cdot\nabla_{\mathbf{r}_0}\left[\frac{2}{|\alpha\hat{\mathbf{r}}-\mathbf{r}_0|} - \frac{1}{r_0}\right] \\ &+ \frac{1}{\sigma}\mathbf{Q}\cdot\nabla_{\mathbf{r}_0}\sum_{n=1}^{\infty}\sum_{m=-n}^{n}\frac{1}{n(2n+1)}\frac{r_0^n}{\alpha^{n+1}}Y_n^{m*}(\hat{\mathbf{r}}_0)Y_n^m(\hat{\mathbf{r}}). \end{aligned} \tag{4.24}$$

Equation (1.44) and condition (1.46) imply that the solution has the form

$$u^+(\mathbf{r}) = \sum_{n=0}^{\infty}\sum_{m=-n}^{n} B_n^m \frac{1}{r^{n+1}} Y_n^m(\hat{\mathbf{r}}). \tag{4.25}$$

Calculating u^+ as it is given by eq. (4.25) on $|\mathbf{r}| = \alpha$ and equating the resulting expression with the data (4.24) we obtain

$$\begin{aligned} \sum_{n=0}^{\infty}\sum_{m=-n}^{n} B_n^m \frac{1}{\alpha^{n+1}} Y_n^m(\hat{\mathbf{r}}) &= \frac{1}{4\pi\sigma}\mathbf{Q}\cdot\nabla_{\mathbf{r}_0}\left[\frac{2}{|\alpha\hat{\mathbf{r}}-\mathbf{r}_0|} - \frac{1}{r_0}\right] \\ &+ \frac{1}{\sigma}\mathbf{Q}\cdot\nabla_{\mathbf{r}_0}\sum_{n=1}^{\infty}\sum_{m=-n}^{n}\frac{1}{n(2n+1)}\frac{r_0^n}{\alpha^{n+1}}Y_n^{m*}(\hat{\mathbf{r}}_0)Y_n^m(\hat{\mathbf{r}}). \end{aligned} \tag{4.26}$$

Finally, by orthogonality we obtain

$$B_0^0 = \frac{\alpha}{\sqrt{4\pi\sigma}} \frac{\mathbf{Q}\cdot\mathbf{r}_0}{r_0^3} \tag{4.27}$$

and

$$B_n^m = \frac{1}{\sigma n}\mathbf{Q}\cdot\nabla_{\mathbf{r}_0}\, r_0^n\, Y_n^{m*}(\hat{\mathbf{r}}_0) \tag{4.28}$$

for each $n \geq 1$ and $|m| \leq n$. Inserting values (4.27) and (4.28) into eq. (4.25), we obtain the exterior electric potential in the form

$$\begin{aligned} u^+(\mathbf{r}) &= \frac{\mathbf{Q}\cdot\mathbf{r}_0}{4\pi\sigma r_0^3}\frac{\alpha}{r} \\ &+ \frac{1}{\sigma}\mathbf{Q}\cdot\nabla_{\mathbf{r}_0}\sum_{n=1}^{\infty}\sum_{m=-n}^{n}\frac{1}{n}\frac{r_0^n}{r^{n+1}}Y_n^{m*}(\hat{\mathbf{r}}_0)Y_n^m(\hat{\mathbf{r}}), \quad r > \alpha. \end{aligned} \tag{4.29}$$

Note that the appearance of the monopolic term

$$\frac{\mathbf{Q}\cdot\mathbf{r}_0}{4\pi\sigma r_0^3}\frac{\alpha}{r}$$

in expression (4.29) for the exterior potential is due to our choice of having a vanishing interior potential at the center of the sphere. Hence, we see that the value of u^- at the center of the sphere is connected to the position of the dipole.

4.3 Closed expressions of the electric potential

Expressions (4.22) and (4.29) for the electric potential are given in the form of eigenexpansions of the Laplacian in spherical coordinates. In what follows, we will demonstrate how we can manipulate these expressions and obtain closed-form solutions for the corresponding potentials.

First, we show how we can sum up the series

$$f(\rho) = \sum_{n=1}^{\infty}\frac{\rho^n}{n}P_n(\cos\theta) \tag{4.30}$$

for $\rho < 1$. Since the Legendre polynomial is bounded by 1, it follows that the series converges absolutely and almost uniformly. Hence, within the region $\rho < 1$, we can differentiate and integrate the series term by term. Applying the operator $\rho\frac{d}{d\rho}$ on relation (4.30) and differentiating term by term we obtain

$$\rho f'(\rho) = \sum_{n=1}^{\infty} \rho^n P_n(\cos\theta). \tag{4.31}$$

Then using expansion (4.12), we rewrite (4.31) in the form

$$\rho f'(\rho) = \frac{1}{\sqrt{1-2\rho\cos\theta+\rho^2}} - 1 \tag{4.32}$$

which is an ordinary differential equation of the first order that has to be solved with the initial condition

$$f(0) = 0 \tag{4.33}$$

dictated via (4.30). This initial value problem is solved as follows:

$$\begin{aligned} f(\rho) &= \int \frac{d\rho}{\rho\sqrt{1-2\rho\cos\theta+\rho^2}} - \int \frac{d\rho}{\rho} + c \\ &= -\ln\frac{2-2\rho\cos\theta+2\sqrt{1-2\rho\cos\theta+\rho^2}}{\rho} - \ln\rho + c \\ &= -\ln\left(1-\rho\cos\theta+\sqrt{1-2\rho\cos\theta+\rho^2}\right) - \ln 2 + c. \end{aligned} \tag{4.34}$$

Condition (4.33) implies that

$$c = \ln 4 \tag{4.35}$$

therefore, the function $f(\rho)$ is given by

$$f(\rho) = -\ln\frac{1-\rho\cos\theta+\sqrt{1-2\rho\cos\theta+\rho^2}}{2}. \tag{4.36}$$

Next we combine formulae (4.30) and (4.36) to write the series on the right-hand side of eq. (4.22) in the closed form:

$$\begin{aligned} &\sum_{n=1}^{\infty} \frac{n+1}{n}\frac{r_0^n r^n}{\alpha^{2n+1}} P_n(\hat{\mathbf{r}}\cdot\hat{\mathbf{r}}_0) \\ &= \frac{1}{\alpha}\sum_{n=1}^{\infty}\left(\frac{r_0}{\alpha}\right)^n\left(\frac{r}{\alpha}\right)^n P_n(\hat{\mathbf{r}}\cdot\hat{\mathbf{r}}_0) + \frac{1}{\alpha}\sum_{n=1}^{\infty}\frac{1}{n}\left(\frac{r_0}{\alpha}\right)^n\left(\frac{r}{\alpha}\right)^n P_n(\hat{\mathbf{r}}\cdot\hat{\mathbf{r}}_0) \\ &= \frac{1}{\alpha}\sum_{n=0}^{\infty}\rho^n P_n(\hat{\mathbf{r}}\cdot\hat{\mathbf{r}}_0) - \frac{1}{\alpha} + \frac{1}{\alpha}\sum_{n=1}^{\infty}\frac{\rho^n}{n} P_n(\hat{\mathbf{r}}\cdot\hat{\mathbf{r}}_0) \\ &= \frac{1}{\alpha}\frac{1}{\sqrt{1-2\rho(\hat{\mathbf{r}}\cdot\hat{\mathbf{r}}_0)+\rho^2}} - \frac{1}{\alpha} - \frac{1}{\alpha}\ln\frac{1-\rho(\hat{\mathbf{r}}\cdot\hat{\mathbf{r}}_0)+\sqrt{1-2\rho(\hat{\mathbf{r}}\cdot\hat{\mathbf{r}}_0)+\rho^2}}{2} \end{aligned} \tag{4.37}$$

where

$$\rho = \left(\frac{r_0}{\alpha}\right)\left(\frac{r}{\alpha}\right) < 1 \tag{4.38}$$

since both r_0 and r live in the interior of the sphere of radius α. Substituting expression (4.37) into (4.22), we find the following closed-form expression for the interior electric potential:

$$\begin{aligned} u^-(\mathbf{r}) = {} & \frac{1}{4\pi\sigma}\mathbf{Q}\cdot\nabla_{\mathbf{r}_0}\left[\frac{1}{|\mathbf{r}-\mathbf{r}_0|}-\frac{1}{r_0}\right] + \frac{1}{4\pi\sigma\alpha}\mathbf{Q}\cdot\nabla_{\mathbf{r}_0}\frac{1}{\sqrt{1-2\rho(\hat{\mathbf{r}}\cdot\hat{\mathbf{r}}_0)+\rho^2}} \\ & - \frac{1}{4\pi\sigma\alpha}\mathbf{Q}\cdot\nabla_{\mathbf{r}_0}\ln\frac{1-\rho(\hat{\mathbf{r}}\cdot\hat{\mathbf{r}}_0)+\sqrt{1-2\rho(\hat{\mathbf{r}}\cdot\hat{\mathbf{r}}_0)+\rho^2}}{2} \end{aligned} \tag{4.39}$$

with ρ given in eq. (4.38). On the surface $r=\alpha$, where the electroencephalography data are collected, the electric potential takes the form

$$\begin{aligned} u^-(\alpha\hat{\mathbf{r}}) = {} & \frac{1}{4\pi\sigma}\mathbf{Q}\cdot\nabla_{\mathbf{r}_0}\left[\frac{1}{|\alpha\hat{\mathbf{r}}-\mathbf{r}_0|}-\frac{1}{r_0}\right] \\ & + \frac{1}{4\pi\sigma\alpha}\mathbf{Q}\cdot\nabla_{\mathbf{r}_0}\frac{1}{\sqrt{1-2\frac{r_0}{\alpha}(\hat{\mathbf{r}}\cdot\hat{\mathbf{r}}_0)+\left(\frac{r_0}{\alpha}\right)^2}} \\ & - \frac{1}{4\pi\sigma\alpha}\mathbf{Q}\cdot\nabla_{\mathbf{r}_0}\ln\frac{1-\frac{r_0}{\alpha}(\hat{\mathbf{r}}\cdot\hat{\mathbf{r}}_0)+\sqrt{1-2\frac{r_0}{\alpha}(\hat{\mathbf{r}}\cdot\hat{\mathbf{r}}_0)+\left(\frac{r_0}{\alpha}\right)^2}}{2}. \end{aligned} \tag{4.40}$$

The following function of $\mathbf{r}$ plays an important role in the theory of EEG and MEG for the spherical model of the brain [81]:

$$F(\mathbf{r};\mathbf{r}_0) = |\mathbf{r}-\mathbf{r}_0|[r|\mathbf{r}-\mathbf{r}_0| + \mathbf{r}\cdot(\mathbf{r}-\mathbf{r}_0)] \tag{4.41}$$

It depends on the position $\mathbf{r}_0$ of the dipole alone.

In terms of the function F we obtain

$$1-\frac{r_0}{r}\hat{\mathbf{r}}\cdot\hat{\mathbf{r}}_0+\sqrt{1-2\frac{r_0}{r}\hat{\mathbf{r}}\cdot\hat{\mathbf{r}}_0+\left(\frac{r_0}{r}\right)^2} = \frac{F(\mathbf{r};\mathbf{r}_0)}{r^2|\mathbf{r}-\mathbf{r}_0|} \tag{4.42}$$

which on the boundary $r=a$ can be written as

$$1-\frac{r_0}{\alpha}\hat{\mathbf{r}}\cdot\hat{\mathbf{r}}_0+\sqrt{1-2\frac{r_0}{\alpha}\hat{\mathbf{r}}\cdot\hat{\mathbf{r}}_0+\left(\frac{r_0}{\alpha}\right)^2} = \frac{F(\alpha\hat{\mathbf{r}};\mathbf{r}_0)}{\alpha^2|\alpha\hat{\mathbf{r}}-\mathbf{r}_0|}. \tag{4.43}$$

Introducing identity (4.42) in expression (4.39) for the interior electric potential, we obtain

$$u^-(\mathbf{r}) = \frac{1}{4\pi\sigma}\mathbf{Q}\cdot\nabla_{\mathbf{r}_0}\left[\frac{1}{|\mathbf{r}-\mathbf{r}_0|} - \frac{1}{r_0} + \frac{\alpha}{|r_0\mathbf{r}-\alpha^2\hat{\mathbf{r}}_0|} - \frac{1}{\alpha}\ln\frac{F\left(\frac{\alpha^2}{r}\hat{\mathbf{r}};\mathbf{r}_0\right)}{2\left(\frac{\alpha^2}{r}\right)^2\left|\frac{\alpha^2}{r}\hat{\mathbf{r}}-\mathbf{r}_0\right|}\right] \tag{4.44}$$

and the boundary data

$$u^-(\alpha\hat{\mathbf{r}}) = \frac{1}{4\pi\sigma}\mathbf{Q}\cdot\nabla_{\mathbf{r}_0}\left[\frac{2}{|\alpha\hat{\mathbf{r}}-\mathbf{r}_0|} - \frac{1}{r_0} - \frac{1}{\alpha}\ln\frac{F(\alpha\hat{\mathbf{r}};\mathbf{r}_0)}{2\alpha^2|\alpha\hat{\mathbf{r}}-\mathbf{r}_0|}\right]. \tag{4.45}$$

Formulae (4.44) and (4.45) provide the interior and the boundary values of the electric potential, respectively, in closed form.

In what follows, we will implement the same program for the exterior electric potential given by relation (4.39). Utilizing the addition theorem (4.14) and relation (4.36), we obtain

$$\begin{aligned}
&\sum_{n=1}^{\infty}\sum_{m=-n}^{n}\frac{1}{n}\frac{r_0^n}{r^{n+1}}Y_n^{m*}(\hat{\mathbf{r}}_0)Y_n^m(\hat{\mathbf{r}})\\
&= \frac{1}{4\pi}\sum_{n=1}^{\infty}\frac{2n+1}{n}\frac{r_0^n}{r^{n+1}}P_n(\hat{\mathbf{r}}\cdot\hat{\mathbf{r}}_0)\\
&= \frac{1}{2\pi}\left[\frac{1}{|\mathbf{r}-\mathbf{r}_0|} - \frac{1}{r}\right] - \frac{1}{4\pi r}\ln\frac{1-\frac{r_0}{r}\hat{\mathbf{r}}\cdot\hat{\mathbf{r}}_0 + \sqrt{1-2\frac{r_0}{r}\hat{\mathbf{r}}\cdot\hat{\mathbf{r}}_0 + \left(\frac{r_0}{r}\right)^2}}{2}.
\end{aligned} \tag{4.46}$$

Therefore, eq. (4.29) is written as

$$u^+(\mathbf{r}) = \frac{\mathbf{Q}\cdot\mathbf{r}_0}{4\pi\sigma r_0^3}\frac{\alpha}{r} + \frac{1}{4\pi\sigma}\mathbf{Q}\cdot\nabla_{\mathbf{r}_0}\left[\frac{2}{|\mathbf{r}-\mathbf{r}_0|} - \frac{1}{r}\ln\frac{F(\mathbf{r};\mathbf{r}_0)}{2r^2|\mathbf{r}-\mathbf{r}_0|}\right] \tag{4.47}$$

which provides the exterior electric potential in closed form. For $r = \alpha$, eq. (4.47) takes the form

$$\begin{aligned}
u^+(\alpha\hat{\mathbf{r}}) &= \frac{\mathbf{Q}\cdot\mathbf{r}_0}{4\pi\sigma r_0^3} + \frac{1}{4\pi\sigma}\mathbf{Q}\cdot\nabla_{\mathbf{r}_0}\left[\frac{2}{|\alpha\hat{\mathbf{r}}-\mathbf{r}_0|} - \frac{1}{\alpha}\ln\frac{F(\alpha\hat{\mathbf{r}};\mathbf{r}_0)}{2\alpha^2|\alpha\hat{\mathbf{r}}-\mathbf{r}_0|}\right]\\
&= \frac{1}{4\pi\sigma}\mathbf{Q}\cdot\nabla_{\mathbf{r}_0}\left[\frac{2}{|\alpha\hat{\mathbf{r}}-\mathbf{r}_0|} - \frac{1}{r_0} - \frac{1}{\alpha}\ln\frac{F(\alpha\hat{\mathbf{r}};\mathbf{r}_0)}{2\alpha^2|\alpha\hat{\mathbf{r}}-\mathbf{r}_0|}\right]
\end{aligned} \tag{4.48}$$

which coincides with expression (4.45). This is consistent with the fact that the electric potential is continuous across the interface between the conductive brain tissue and the nonconductive air.

4.4 Connection with the method of images

We denote by Ω a fundamental domain, by $\partial\Omega$ its smooth boundary and by Ω^c the exterior domain which is the complement of the set $\Omega \cup \partial\Omega$. By the term *method of images* we refer to any technique that can replace the boundary conditions by certain singularities of the solution in the exterior space Ω^c. Perhaps, the simplest of these cases is demonstrated in the following example.

Example 4.1 Suppose we are seeking a harmonic function in the upper half space $x_3 > 0$, that is, generated by a positive unit monopolic source at the point $\mathbf{r}_0 = (x_{01}, x_{02}, x_{03})$ with $x_{03} > 0$, where this harmonic function vanishes on the plane $x_3 = 0$. The mathematical formulation of this problem is given by

$$\Delta u(\mathbf{r}) = \delta(\mathbf{r} - \mathbf{r}_0), \quad x_3 > 0, \quad x_{03} > 0 \tag{4.49}$$

$$u(x_1, x_2, 0) = 0. \tag{4.50}$$

The solution of this problem has the form

$$u(\mathbf{r}) = -\frac{1}{4\pi}\frac{1}{|\mathbf{r} - \mathbf{r}_0|} + w(\mathbf{r}) \tag{4.51}$$

where the function $w(\mathbf{r})$ is harmonic for $x_3 > 0$ and satisfies the condition that for $x_3 = 0$ has the value $\frac{1}{4\pi}\frac{1}{|\mathbf{r}-\mathbf{r}_0|}$. The regular function w can be calculated via spectral methods, by separating variables for the Laplacian in Cartesian coordinates and using expansions of the corresponding eigenfunctions. However, we can use an alternative approach, which is based on the observation that the function w is the harmonic function, generated by a negative unit monopole located at the symmetric with respect to the plane $x_3 = 0$ point $\mathbf{r}_0^{'}$ of $\mathbf{r}_0$. Then, the solution of eqs. (4.49) and (4.50) is given by the function

$$u(\mathbf{r}) = \frac{1}{4\pi}\left[\frac{1}{\left|\mathbf{r} - \mathbf{r}_0^{'}\right|} - \frac{1}{|\mathbf{r} - \mathbf{r}_0|}\right] \tag{4.52}$$

which holds for every $\mathbf{r}$ with $x_3 > 0$. Because of symmetry we have

$$|\bar{\mathbf{r}} - \mathbf{r}_0| = \left|\bar{\mathbf{r}} - \mathbf{r}_0^{'}\right| \tag{4.53}$$

for every $\bar{\mathbf{r}}$ with $\bar{x}_3 = 0$. Hence, it is obvious that the introduction of an *image* with intensity -1, at the symmetric to $\mathbf{r}_0$ position $\mathbf{r}_0^{'}$, eliminates the need for calculating the boundary condition (4.50), since this condition is automatically satisfied by symmetry.

Although the basic idea about the method of images appears in the monumental work of Green, published individually in 1828 under the title *An Essay on the Application of Mathematical Analysis to the Theories of Electricity and Magnetism*, it was Lord Kelvin who made a complete use of images around the middle of the nineteenth century.

Example 4.2 A more interesting case was analyzed by Kelvin: the problem (4.49), (4.50) when the plane boundary $x_3 = 0$ is replaced by a sphere of radius α having a monopole with intensity +1 in its interior. The harmonic function is specified to vanish on the surface $r = \alpha$ of the sphere. This problem is defined by

$$\Delta u(\mathbf{r}) = \delta(\mathbf{r} - \mathbf{r}_0), \quad r < \alpha, \quad r_0 < \alpha \tag{4.54}$$

$$u(\mathbf{r}) = 0, \quad r = \alpha. \tag{4.55}$$

The image that solves this interior Dirichlet problem is located at the position

$$\mathbf{r}_0' = \frac{\alpha^2}{r_0^2}\mathbf{r}_0 \tag{4.56}$$

with intensity

$$q' = -\frac{\alpha}{r_0}. \tag{4.57}$$

The proof of this result can be found in ref. [72]. The solution of the problem (4.54), (4.55) is written as

$$u(\mathbf{r}) = \frac{1}{4\pi}\left[\frac{\alpha}{r_0}\frac{1}{\left|\mathbf{r} - \frac{\alpha^2}{r_0^2}\mathbf{r}_0\right|} - \frac{1}{|\mathbf{r} - \mathbf{r}_0|}\right]. \tag{4.58}$$

The structure of solution (4.58) motivates the form of the solution of the interior and exterior electric potentials in terms of images that lie on the complement of the fundamental domain. We start with the last term on the right-hand side of eq. (4.44), which we will rewrite in a form that can be recognized in terms of images. If we denote this term by L we have

$$\begin{aligned} L &= -\frac{1}{4\pi\sigma\alpha}\mathbf{Q}\cdot\nabla_{\mathbf{r}_0}\ln\frac{\alpha^2 - \mathbf{r}\cdot\mathbf{r}_0 + |r_0\mathbf{r} - \alpha^2\hat{\mathbf{r}}|}{2\alpha^2} \\ &= -\frac{1}{4\pi\sigma\alpha}\mathbf{Q}\cdot\nabla_{\mathbf{r}_0}\ln\frac{\mathbf{r}\cdot\left(\frac{\alpha^2}{r}\hat{\mathbf{r}} - \mathbf{r}_0\right) + r\left|\frac{\alpha^2}{r}\hat{\mathbf{r}} - \mathbf{r}_0\right|}{2\alpha^2} \\ &= -\frac{1}{4\pi\sigma\alpha}\mathbf{Q}\cdot\frac{-\mathbf{r} + r\nabla_{\mathbf{r}_0}\left|\frac{\alpha^2}{r}\hat{\mathbf{r}} - \mathbf{r}_0\right|}{\mathbf{r}\cdot\left(\frac{\alpha^2}{r}\hat{\mathbf{r}} - \mathbf{r}_0\right) + r\left|\frac{\alpha^2}{r}\hat{\mathbf{r}} - \mathbf{r}_0\right|}. \end{aligned} \tag{4.59}$$

Using

$$\nabla_{\mathbf{r}_0}\left|\frac{\alpha^2}{r}\hat{\mathbf{r}}-\mathbf{r}_0\right| = -\frac{\frac{\alpha^2}{r}\hat{\mathbf{r}}-\mathbf{r}_0}{\left|\frac{\alpha^2}{r}\hat{\mathbf{r}}-\mathbf{r}_0\right|} \tag{4.60}$$

it follows that relation (4.59) can be written as

$$L = \frac{\mathbf{Q}}{4\pi\sigma\alpha}\cdot\frac{r\left(\frac{\alpha^2}{r}\hat{\mathbf{r}}-\mathbf{r}_0\right)+\mathbf{r}\left|\frac{\alpha^2}{r}\hat{\mathbf{r}}-\mathbf{r}_0\right|}{\left|\frac{\alpha^2}{r}\hat{\mathbf{r}}-\mathbf{r}_0\right|\left[\mathbf{r}\cdot\left(\frac{\alpha^2}{r}\hat{\mathbf{r}}-\mathbf{r}_0\right)+r\left|\frac{\alpha^2}{r}\hat{\mathbf{r}}-\mathbf{r}_0\right|\right]}. \tag{4.61}$$

Expression (4.61) of L involves the function F given by eq. (4.41). Hence,

$$L = \frac{\alpha}{4\pi\sigma r}\mathbf{Q}\cdot\frac{\left(\frac{\alpha^2}{r}\hat{\mathbf{r}}-\mathbf{r}_0\right)+\hat{\mathbf{r}}\left|\frac{\alpha^2}{r}\hat{\mathbf{r}}-\mathbf{r}_0\right|}{F\left(\frac{\alpha^2}{r}\hat{\mathbf{r}};\mathbf{r}_0\right)}. \tag{4.62}$$

It is easy to identify the term $\frac{\alpha^2}{r}\hat{\mathbf{r}}$ in formula (4.62) as the Kelvin image of the point $\mathbf{r}$ with respect to the sphere of radius α. Obviously, the image point $\frac{\alpha^2}{r}\hat{\mathbf{r}}$ lies outside the sphere $r = a$.

The function $|\mathbf{r}-\mathbf{r}_0|^{-1}$ can be considered either as potential at point $\mathbf{r}$ due to a monopole at point $\mathbf{r}_0$, or as the potential at point $\mathbf{r}_0$ due to a monopole at point $\mathbf{r}$. Similarly, the vector potential of a dipole at point $\mathbf{r}_0$ is proportional to the field

$$\nabla_{\mathbf{r}_0} = \frac{1}{|\mathbf{r}-\mathbf{r}_0|} = \frac{\mathbf{r}-\mathbf{r}_0}{|\mathbf{r}-\mathbf{r}_0|^3} \tag{4.63}$$

while the vector potential of a dipole at point $\mathbf{r}$ is proportional to the opposite field

$$\nabla_{\mathbf{r}} = \frac{1}{|\mathbf{r}-\mathbf{r}_0|} = -\frac{\mathbf{r}-\mathbf{r}_0}{|\mathbf{r}-\mathbf{r}_0|^3}. \tag{4.64}$$

In other words, the differential operator $\nabla_{\mathbf{r}}+\nabla_{\mathbf{r}_0}$ annihilates the monopolic potential $|\mathbf{r}-\mathbf{r}_0|^{-1}$. By distributing dipoles in the direction $\hat{\mathbf{r}}$ all the way from the Kelvin point $\frac{\alpha^2}{r}$ to infinity, we generated a potential that has the form

$$\begin{aligned}\int_{\frac{\alpha^2}{r}}^{\infty}\left(\nabla_{\mathbf{r}_0}\frac{1}{|t\hat{\mathbf{r}}-\mathbf{r}_0|}\right)dt &= \int_{\frac{\alpha^2}{r}}^{\infty}\frac{t\hat{\mathbf{r}}-\mathbf{r}_0}{|t\hat{\mathbf{r}}-\mathbf{r}_0|^3}dt \\ &= \hat{\mathbf{r}}\int_{\frac{\alpha^2}{r}}^{\infty}\frac{tdt}{|t\hat{\mathbf{r}}-\mathbf{r}_0|^3}-\mathbf{r}_0\int_{\frac{\alpha^2}{r}}^{\infty}\frac{dt}{|t\hat{\mathbf{r}}-\mathbf{r}_0|^3}.\end{aligned} \tag{4.65}$$

By employing the integral formulae

$$\int \frac{xdx}{\left(\sqrt{ax^2+bx+c}\right)^3} = \frac{2(bx+2c)}{(b^2-4ac)\sqrt{ax^2+bx+c}} \tag{4.66}$$

$$\int \frac{dx}{\left(\sqrt{ax^2+bx+c}\right)^3} = \frac{-2(2ax+b)}{(b^2-4ac)\sqrt{ax^2+bx+c}}, \quad a>0 \tag{4.67}$$

we obtain the following:

$$\int_{\frac{\alpha^2}{r}}^{\infty} \frac{tdt}{|t\hat{\mathbf{r}}-\mathbf{r}_0|^3} = \frac{1}{r_0}\frac{\cos\theta}{\sin^2\theta} + \frac{r_0^2 r - r_0\alpha^2\cos\theta}{r_0^2 r\sin^2\theta\left|\frac{\alpha^2}{r}\hat{\mathbf{r}}-\mathbf{r}_0\right|} \tag{4.68}$$

and

$$\int_{\frac{\alpha^2}{r}}^{\infty} \frac{dt}{|t\hat{\mathbf{r}}-\mathbf{r}_0|^3} = \frac{1}{r_0^2\sin^2\theta} + \frac{r_0 r\cos\theta-\alpha^2}{r_0^2 r\sin^2\theta\left|\frac{\alpha^2}{r}\hat{\mathbf{r}}-\mathbf{r}_0\right|} \tag{4.69}$$

where

$$\cos\theta = \hat{\mathbf{r}}\cdot\hat{\mathbf{r}}_0. \tag{4.70}$$

Finally, by inserting expressions (4.68) and (4.69) in (4.65) we find the following:

$$\begin{aligned}
\int_{\frac{\alpha^2}{r}}^{\infty}\left(\nabla_{\mathbf{r}_0}\frac{1}{|t\hat{\mathbf{r}}-\mathbf{r}_0|}\right)dt &= \hat{\mathbf{r}}\left[\frac{1}{r_0}\frac{\cos\theta}{\sin^2\theta} + \frac{r_0 r-\alpha^2\cos\theta}{rr_0\sin^2\theta\left|\frac{\alpha^2}{r}\hat{\mathbf{r}}-\mathbf{r}_0\right|}\right] \\
&\quad -\hat{\mathbf{r}}_0\left[\frac{1}{r_0}\frac{1}{\sin^2\theta} + \frac{r_0 r\cos\theta-\alpha^2}{rr_0\sin^2\theta\left|\frac{\alpha^2}{r}\hat{\mathbf{r}}-\mathbf{r}_0\right|}\right] \\
&= \frac{\hat{\mathbf{r}}\cos\theta-\hat{\mathbf{r}}_0}{r_0\sin^2\theta} + \frac{\hat{\mathbf{r}}(r_0 r-\alpha^2\cos\theta)-\hat{\mathbf{r}}_0(rr_0\cos\theta-\alpha^2)}{rr_0\sin^2\theta\left|\frac{\alpha^2}{r}\hat{\mathbf{r}}-\mathbf{r}_0\right|} \\
&= \frac{\hat{\mathbf{r}}}{r_0\sin^2\theta}\left[\cos\theta + \frac{r_0-\frac{\alpha^2}{r}\cos\theta}{\left|\frac{\alpha^2}{r}\hat{\mathbf{r}}-\mathbf{r}_0\right|}\right] - \frac{\hat{\mathbf{r}}_0}{r_0\sin^2\theta}\left[1+\frac{r_0\cos\theta-\frac{\alpha^2}{r}}{\left|\frac{\alpha^2}{r}\hat{\mathbf{r}}-\mathbf{r}_0\right|}\right] \\
&= \frac{\hat{\mathbf{r}}}{r_0\sin^2\theta}\left[\hat{\mathbf{r}}_0\cdot\hat{\mathbf{r}} - \frac{\hat{\mathbf{r}}_0\cdot\left(\frac{\alpha^2}{r}\hat{\mathbf{r}}-\mathbf{r}_0\right)}{\left|\frac{\alpha^2}{r}\hat{\mathbf{r}}-\mathbf{r}_0\right|}\right] - \frac{\hat{\mathbf{r}}_0}{r_0\sin^2\theta}\left[\hat{\mathbf{r}}_0\cdot\hat{\mathbf{r}} - \frac{\hat{\mathbf{r}}\cdot\left(\frac{\alpha^2}{r}\hat{\mathbf{r}}-\mathbf{r}_0\right)}{\left|\frac{\alpha^2}{r}\hat{\mathbf{r}}-\mathbf{r}_0\right|}\right].
\end{aligned} \tag{4.71}$$

Introducing the symbol

$$\mathbf{R} = \frac{\alpha^2}{r}\hat{\mathbf{r}} - \mathbf{r}_0 \tag{4.72}$$

with

$$R = |\mathbf{R}| \tag{4.73}$$

and using the identities

$$\frac{\left(R - \frac{\alpha^2}{r}\right)\cos\theta + r_0}{r_0 \sin^2\theta} = \frac{R + \frac{\alpha^2}{r}}{R + \frac{\alpha^2}{r} - r_0\cos\theta} \tag{4.74}$$

and

$$\frac{R - \frac{\alpha^2}{r} + r_0\cos\theta}{r_0 \sin^2\theta} = \frac{r_0}{R + \frac{\alpha^2}{r} - r_0\cos\theta} \tag{4.75}$$

we can rewrite relation (4.71) in the form

$$\int_{\frac{\alpha^2}{r}}^{\infty} \left(\nabla_{\mathbf{r}_0} \frac{1}{|t\hat{\mathbf{r}} - \mathbf{r}_0|}\right) dt = \frac{\alpha^2}{r} \frac{\left(\frac{\alpha^2}{r}\hat{\mathbf{r}} - \mathbf{r}_0\right) + \hat{\mathbf{r}}\left|\frac{\alpha^2}{r}\hat{\mathbf{r}} - \mathbf{r}_0\right|}{F\left(\frac{\alpha^2}{r}\hat{\mathbf{r}}; \mathbf{r}_0\right)}. \tag{4.76}$$

Hence, by using expression (4.76) for the contribution of the distributed dipoles along the ray $\hat{\mathbf{r}}$ in formula (4.62), we obtain

$$L = \frac{\mathbf{Q}}{4\pi\sigma\alpha} \cdot \int_{\frac{\alpha^2}{r}}^{\infty} \left(\nabla_{\mathbf{r}_0} \frac{1}{|t\hat{\mathbf{r}} - \mathbf{r}_0|}\right) dt. \tag{4.77}$$

4.5 The inhomogeneous model

In the inhomogeneous spherical model of the head–brain system, the region Ω_c is a homogeneous sphere, of radius c, with conductivity σ_c, and Ω_f, Ω_b, Ω_s are concentric homogeneous shells with outer radii f, b, s, and conductivities σ_f, σ_b, σ_s, respectively. The region exterior to Ω_s is denoted by Ω_e. In each of the regions Ω_i, $i = c, f, b, s, e$, we need to find harmonic functions v_i, $i = c, f, b, s, e$ supported in Ω_i, which satisfy the transmission conditions (2.29), (2.30), (2.32), (2.33), (2.35), (2.36) and (2.38) and the Neumann condition (2.39). Because of eqs. (2.26) and (2.27), the potential v_i, $i = c, f, b, s, e$, is due to a monopolic excitation. They admit the following expansions:

$$v_c(\mathbf{r};\mathbf{r}_0)=\sum_{n=0}^{\infty}\sum_{m=-n}^{n} C_n^m(r_0)r^n Y_n^{m*}(\hat{\mathbf{r}}_0)Y_n^m(\hat{\mathbf{r}}) \tag{4.78}$$

$$v_f(\mathbf{r};\mathbf{r}_0)=\sum_{n=0}^{\infty}\sum_{m=-n}^{n}\left[F_{1n}^m(r_0)r^n+\frac{F_{2n}^m(r_0)}{r^{n+1}}\right]Y_n^{m*}(\hat{\mathbf{r}}_0)Y_n^m(\hat{\mathbf{r}}) \tag{4.79}$$

$$v_b(\mathbf{r};\mathbf{r}_0)=\sum_{n=0}^{\infty}\sum_{m=-n}^{n}\left[B_{1n}^m(r_0)r^n+\frac{B_{2n}^m(r_0)}{r^{n+1}}\right]Y_n^{m*}(\hat{\mathbf{r}}_0)Y_n^m(\hat{\mathbf{r}}) \tag{4.80}$$

$$v_s(\mathbf{r};\mathbf{r}_0)=\sum_{n=0}^{\infty}\sum_{m=-n}^{n}\left[S_{1n}^m(r_0)r^n+\frac{S_{2n}^m(r_0)}{r^{n+1}}\right]Y_n^{m*}(\hat{\mathbf{r}}_0)Y_n^m(\hat{\mathbf{r}}) \tag{4.81}$$

$$v_e(\mathbf{r};\mathbf{r}_0)=\sum_{n=0}^{\infty}\sum_{m=-n}^{n}\frac{E_{2n}^m(r_0)}{r^{n+1}}Y_n^{m*}(\hat{\mathbf{r}}_0)Y_n^m(\hat{\mathbf{r}}). \tag{4.82}$$

The $n=0$ term specifies the arbitrary constant of the potentials and we assume that they are zero; all sums in eqs. (4.78)–(4.82) start with $n=1$.

The eight boundary conditions that determine the eight constants

$$C_n^m=C, F_{1n}^m=F_1, F_{2n}^m=F_2, B_{1n}^m=B_1, B_{2n}^m=B_2, S_{1n}^m=S_1, S_{2n}^m=S_2, E_{2n}^m=E$$

provide eight algebraic equations. We write these eight equations as follows:

On the interface $r=c$

$$\frac{4\pi}{\sigma_c(2n+1)}\frac{r_0^n}{c^{n+1}}+\frac{1}{\sigma_c}Cc^n=F_1c^n+\frac{F_2}{c^{n+1}} \tag{4.83}$$

$$-\frac{n+1}{2n+1}4\pi\frac{r_0^n}{c^{n+2}}+nCc^{n-1}=n\sigma_fF_1c^{n-1}-\frac{n+1}{c^{n+2}}\sigma_fF_2 \tag{4.84}$$

on the interface $r=f$

$$F_1f^n+\frac{F_2}{f^{n+1}}=B_1f^n+\frac{B_2}{f^{n+1}} \tag{4.85}$$

$$\sigma_fnF_1f^{n-1}-\sigma_f(n+1)\frac{F_2}{f^{n+2}}=\sigma_bnB_1f^{n-1}-\sigma_b(n+1)\frac{B_2}{f^{n+2}} \tag{4.86}$$

on the interface $r=b$

$$B_1b^n+\frac{B_2}{b^{n+1}}=S_1b^n+\frac{S_2}{b^{n+1}} \tag{4.87}$$

$$\sigma_bnB_1b^{n-1}-\sigma_b(n+1)\frac{B_2}{b^{n+2}}=\sigma_snS_1b^{n-1}-\sigma_s(n+1)\frac{S_2}{b^{n+2}} \tag{4.88}$$

and on the boundary $r=s$

$$S_1s^n+\frac{S_2}{s^{n+1}}=\frac{E_2}{s^{n+1}} \tag{4.89}$$

$$nS_1s^{n-1} - (n+1)\frac{S_2}{s^{n+2}} = 0. \tag{4.90}$$

We set

$$\tilde{\mathbf{W}}(x) = \begin{bmatrix} x^n & x^{-(n+1)} \\ nx^{n-1} & -(n+1)x^{-(n+2)} \end{bmatrix} \tag{4.91}$$

$$\tilde{\mathbf{W}}(x;\sigma) = \begin{bmatrix} x^n & x^{-(n+1)} \\ \sigma nx^{n-1} & -\sigma(n+1)x^{-(n+2)} \end{bmatrix}. \tag{4.92}$$

Since

$$\det \tilde{\mathbf{W}}(x) = -\frac{2n+1}{x^2} \neq 0 \tag{4.93}$$

and

$$\det \tilde{\mathbf{W}}(x;\sigma) = -\sigma\frac{2n+1}{x^2} \neq 0 \tag{4.94}$$

both matrices are invertible. Then, eqs. (4.83) and (4.84) are written as

$$\begin{pmatrix} F_1 \\ F_2 \end{pmatrix} = \frac{1}{\sigma_c}\tilde{\mathbf{W}}^{-1}(c;\sigma_f)\cdot\tilde{\mathbf{W}}(c;\sigma_c)\begin{pmatrix} C \\ \frac{4\pi r_0^n}{2n+1} \end{pmatrix}. \tag{4.95}$$

Similarly, eqs. (4.85) and (4.86) are written as

$$\begin{pmatrix} B_1 \\ B_2 \end{pmatrix} = \tilde{\mathbf{W}}^{-1}(f;\sigma_b)\cdot\tilde{\mathbf{W}}(f;\sigma_f)\begin{pmatrix} F_1 \\ F_2 \end{pmatrix} \tag{4.96}$$

eqs. (4.87) and (4.88) are written as

$$\begin{pmatrix} S_1 \\ S_2 \end{pmatrix} = \tilde{\mathbf{W}}^{-1}(b;\sigma_s)\cdot\tilde{\mathbf{W}}(b;\sigma_b)\begin{pmatrix} B_1 \\ B_2 \end{pmatrix} \tag{4.97}$$

and eqs. (4.89) and (4.90) are written as

$$\begin{pmatrix} s^{-(n+1)}E_2 \\ 0 \end{pmatrix} = \tilde{\mathbf{W}}(s)\begin{pmatrix} S_1 \\ S_2 \end{pmatrix}. \tag{4.98}$$

Furthermore, we write

$$\tilde{\mathbf{M}}_c = \tilde{\mathbf{W}}^{-1}(c;\sigma_f)\cdot\tilde{\mathbf{W}}(c;\sigma_c) \tag{4.99}$$

$$\tilde{\mathbf{M}}_f = \tilde{\mathbf{W}}^{-1}(f;\sigma_b)\cdot\tilde{\mathbf{W}}(f;\sigma_f) \tag{4.100}$$

$$\tilde{\mathbf{M}}_b = \tilde{\mathbf{W}}^{-1}(b;\sigma_s)\cdot\tilde{\mathbf{W}}(b;\sigma_b) \tag{4.101}$$

$$\tilde{\mathbf{M}}_s = \tilde{\mathbf{W}}(s) \tag{4.102}$$

which lead to

$$\begin{pmatrix} F_1 \\ F_2 \end{pmatrix} = \frac{1}{\sigma_c}\tilde{\mathbf{M}}_c \begin{pmatrix} C \\ \frac{4\pi r_0^n}{2n+1} \end{pmatrix} \tag{4.103}$$

$$\begin{pmatrix} B_1 \\ B_2 \end{pmatrix} = \tilde{\mathbf{M}}_f \begin{pmatrix} F_1 \\ F_2 \end{pmatrix} \tag{4.104}$$

$$\begin{pmatrix} S_1 \\ S_2 \end{pmatrix} = \tilde{\mathbf{M}}_b \begin{pmatrix} B_1 \\ B_2 \end{pmatrix} \tag{4.105}$$

$$\begin{pmatrix} s^{-(n+1)}E_2 \\ 0 \end{pmatrix} = \tilde{\mathbf{M}}_s \begin{pmatrix} S_1 \\ S_2 \end{pmatrix}. \tag{4.106}$$

Then

$$\begin{pmatrix} s^{-(n+1)}E \\ 0 \end{pmatrix} = \tilde{\mathbf{M}}_s \cdot \tilde{\mathbf{M}}_b \cdot \tilde{\mathbf{M}}_f \cdot \tilde{\mathbf{M}}_c \frac{1}{\sigma_c} \begin{pmatrix} C \\ \frac{4\pi r_0^n}{2n+1} \end{pmatrix} = \tilde{\mathbf{M}} \begin{pmatrix} C \\ \frac{4\pi r_0^n}{2n+1} \end{pmatrix}. \tag{4.107}$$

If

$$\tilde{\mathbf{M}} = \begin{bmatrix} m_{11} & m_{12} \\ m_{21} & m_{22} \end{bmatrix} = \tilde{\mathbf{M}}_s \cdot \tilde{\mathbf{M}}_b \cdot \tilde{\mathbf{M}}_f \cdot \tilde{\mathbf{M}}_c \frac{1}{\sigma_c} \tag{4.108}$$

then eq. (4.107) gives

$$m_{11}C + \frac{4\pi r_0^n}{2n+1} m_{12} = s^{-(n+1)}E_2 \tag{4.109}$$

$$m_{21}C + \frac{4\pi r_0^n}{2n+1} m_{22} = 0. \tag{4.110}$$

Hence,

$$C = -\frac{4\pi r_0^n}{2n+1}\frac{m_{22}}{m_{21}} \tag{4.111}$$

and then eq. (4.103) gives F_1, F_2, eq. (4.104) gives B_1, B_2, eq. (4.105) gives S_1, S_2 and eq. (4.106) gives E.

After performing the above calculations and obtain

$$\tilde{\mathbf{W}}^{-1}(x;\sigma) = \frac{x^2}{\sigma(2n+1)} \begin{bmatrix} \sigma(n+1)x^{-(n+2)} & x^{-(n+1)} \\ \sigma n x^{n-1} & -x^n \end{bmatrix} \tag{4.112}$$

and

$$\tilde{\mathbf{W}}^{-1}(x;\sigma_1)\cdot\tilde{\mathbf{W}}(x;\sigma_2)=\frac{1}{2n+1}\begin{bmatrix}(n+1)+n\frac{\sigma_2}{\sigma_1} & (n+1)\left(1-\frac{\sigma_2}{\sigma_1}\right)x^{-(2n+1)}\\ n\left(1-\frac{\sigma_2}{\sigma_1}\right)x^{2n+1} & n+(n+1)\frac{\sigma_2}{\sigma_1}\end{bmatrix}. \tag{4.113}$$

With this notation we find

$$\tilde{\mathbf{M}}_c=\frac{1}{2n+1}\frac{\sigma_c}{\sigma_f}\begin{bmatrix}n+(n+1)\frac{\sigma_f}{\sigma_c} & (n+1)\left(\frac{\sigma_f}{\sigma_c}-1\right)c^{-(2n+1)}\\ n\left(\frac{\sigma_f}{\sigma_c}-1\right)c^{2n+1} & (n+1)+n\frac{\sigma_f}{\sigma_c}\end{bmatrix} \tag{4.114}$$

$$\tilde{\mathbf{M}}_f=\frac{1}{2n+1}\frac{\sigma_f}{\sigma_b}\begin{bmatrix}n+(n+1)\frac{\sigma_b}{\sigma_f} & (n+1)\left(\frac{\sigma_b}{\sigma_f}-1\right)f^{-(2n+1)}\\ n\left(\frac{\sigma_b}{\sigma_f}-1\right)f^{2n+1} & (n+1)+n\frac{\sigma_b}{\sigma_f}\end{bmatrix} \tag{4.115}$$

$$\tilde{\mathbf{M}}_b=\frac{1}{2n+1}\frac{\sigma_b}{\sigma_s}\begin{bmatrix}n+(n+1)\frac{\sigma_s}{\sigma_b} & (n+1)\left(\frac{\sigma_s}{\sigma_b}-1\right)b^{-(2n+1)}\\ n\left(\frac{\sigma_s}{\sigma_b}-1\right)b^{2n+1} & (n+1)+n\frac{\sigma_s}{\sigma_b}\end{bmatrix} \tag{4.116}$$

$$\tilde{\mathbf{M}}_s=\begin{bmatrix}s^n & s^{-(n+1)}\\ ns^{n-1} & -(n+1)s^{-(n+2)}\end{bmatrix}. \tag{4.117}$$

We introduce $\tilde{\mathbf{N}}_c$, $\tilde{\mathbf{N}}_f$, $\tilde{\mathbf{N}}_b$, $\tilde{\mathbf{N}}_s$ via the relations

$$\tilde{\mathbf{N}}_c=(2n+1)\frac{\sigma_f}{\sigma_c}\tilde{\mathbf{M}}_c \tag{4.118}$$

$$\tilde{\mathbf{N}}_f=(2n+1)\frac{\sigma_b}{\sigma_f}\tilde{\mathbf{M}}_f \tag{4.119}$$

$$\tilde{\mathbf{N}}_b=(2n+1)\frac{\sigma_s}{\sigma_b}\tilde{\mathbf{M}}_b \tag{4.120}$$

$$\tilde{\mathbf{N}}_s=\tilde{\mathbf{M}}_s \tag{4.121}$$

and

$$\tilde{\mathbf{N}}=\tilde{\mathbf{N}}_s\cdot\tilde{\mathbf{N}}_b\cdot\tilde{\mathbf{N}}_f\cdot\tilde{\mathbf{N}}_c. \tag{4.122}$$

Then

$$\tilde{\mathbf{M}}=\frac{1}{\sigma_s(2n+1)^3}\tilde{\mathbf{N}}. \tag{4.123}$$

If we set

$$\tilde{\mathbf{N}} = \begin{bmatrix} n_{11} & n_{12} \\ n_{21} & n_{22} \end{bmatrix} \tag{4.124}$$

then

$$\begin{pmatrix} s^{-(n+1)}E \\ 0 \end{pmatrix} = \frac{1}{\sigma_s(2n+1)^3} \begin{bmatrix} n_{11} & n_{12} \\ n_{21} & n_{22} \end{bmatrix} \begin{pmatrix} C \\ \frac{4\pi r_0^n}{2n+1} \end{pmatrix}. \tag{4.125}$$

Consequently

$$C = -\frac{4\pi r_0^n}{2n+1}\frac{n_{22}}{n_{21}} \tag{4.126}$$

and

$$E = \frac{4\pi r_0^n s^{n+1}}{\sigma_s(2n+1)^4}\left(n_{12} - n_{11}\frac{n_{22}}{n_{21}}\right) = -\frac{4\pi r_0^n s^{n+1}}{\sigma_s(2n+1)^4}\frac{\det\tilde{\mathbf{N}}}{n_{21}}. \tag{4.127}$$

Inserting eq. (4.127) in (4.82) we obtain, for $r = s$, the formula

$$v_e(s\hat{\mathbf{r}};\mathbf{r}_0) = -\sum_{n=1}^{\infty}\sum_{m=-n}^{n}\frac{4\pi r_0^n}{\sigma_s(2n+1)^4}\frac{\det\tilde{\mathbf{N}}}{n_{21}}Y_n^{m*}(\hat{\mathbf{r}}_0)Y_n^m(\hat{\mathbf{r}}) \tag{4.128}$$

for the values of the exterior potential on the surface of the head, where EEG is measured.

In view of the finite expansion

$$P_n(\hat{\mathbf{r}}\cdot\hat{\mathbf{r}}_0) = \sum_{m=-n}^{n}\frac{4\pi}{2n+1}Y_n^m(\hat{\mathbf{r}})Y_n^{m*}(\hat{\mathbf{r}}_0) \tag{4.129}$$

expansion (4.128) is written as

$$v_e(s\hat{\mathbf{r}};\mathbf{r}_0) = -\sum_{n=1}^{\infty}\frac{\det\tilde{\mathbf{N}}}{\sigma_s n_{21}(2n+1)^3}r_0^n Y_n^m(\hat{\mathbf{r}})Y_n^{m*}(\hat{\mathbf{r}}_0) \tag{4.130}$$

where n_{21} is the 21-entry of the matrix $\tilde{\mathbf{N}}$. Comparing eq. (4.130) with (3.8) we obtain the coefficients s_n in the form

$$s_n = -\frac{\det\tilde{\mathbf{N}}}{\sigma_s n_{21}(2n+1)^3}, \quad n = 1, 2, \ldots \tag{4.131}$$

4.6 The exterior magnetic field

As we have shown in Chapter 3, the magnetic field $\mathbf{B}$ in the exterior of the brain–head system is both solenoidal and irrotational:

$$\nabla \cdot \mathbf{B} = 0, \quad \nabla \times \mathbf{B} = \mathbf{0}. \tag{4.132}$$

Thus, it can be represented as the gradient of a harmonic function U, which we will call the magnetic potential

$$\mathbf{B} = \mu_0 \nabla U, \quad \mathbf{r} \in \Omega_e. \tag{4.133}$$

Furthermore, because of the identity

$$\hat{\mathbf{r}} \cdot \hat{\mathbf{n}}(\mathbf{r}') \times \frac{\mathbf{r} - \mathbf{r}'}{|\mathbf{r} - \mathbf{r}'|^3} = 0 \tag{4.134}$$

with $\hat{\mathbf{n}}(\mathbf{r}')$ denoting the unit normal on the boundary of the sphere, it follows that for the spherical model of MEG there is no contribution from the interfaces between spherical shells of different conductivities. Consequently, the MEG problems for the homogeneous and the inhomogeneous models are identical. For a single dipole excitation at the point $\mathbf{r}_0$ with moment $\mathbf{Q}$, the radial component of the exterior magnetic field is given by

$$\hat{\mathbf{r}} \cdot \mathbf{B}(\mathbf{r}; \mathbf{r}_0) = -\frac{\mu_0}{4\pi} \frac{\mathbf{Q} \times \mathbf{r}_0 \cdot \hat{\mathbf{r}}}{|\mathbf{r} - \mathbf{r}_0|^3}. \tag{4.135}$$

Using

$$\hat{\mathbf{r}} \cdot \mathbf{B}(\mathbf{r}; \mathbf{r}_0) = \mu_0 \frac{\partial}{\partial r} U(\mathbf{r}; \mathbf{r}_0) \tag{4.136}$$

it follows that

$$\frac{\partial}{\partial r} U(\mathbf{r}; \mathbf{r}_0) = -\frac{1}{4\pi} \frac{\mathbf{Q} \times \mathbf{r}_0 \cdot \hat{\mathbf{r}}}{|\mathbf{r} - \mathbf{r}_0|^3}. \tag{4.137}$$

Integrating along a ray in the direction $\hat{\mathbf{r}}$ from r to infinity relation (4.134) gives

$$\begin{aligned} U(\mathbf{r}; \mathbf{r}_0) &= -\int_r^{+\infty} \frac{\partial}{\partial r'} U(r'\hat{\mathbf{r}}; \mathbf{r}_0)\, dr' = \frac{1}{4\pi} \int_r^{+\infty} \frac{\mathbf{Q} \times \mathbf{r}_0 \cdot \hat{\mathbf{r}}}{|\mathbf{r}' - \mathbf{r}_0|^3} dr' \\ &= \frac{\mathbf{Q} \times \mathbf{r}_0 \cdot \hat{\mathbf{r}}}{4\pi} \int_r^{+\infty} \frac{dt}{|t\hat{\mathbf{r}} - \mathbf{r}_0|^3}. \end{aligned} \tag{4.138}$$

After long but straightforward calculations we can show that

$$\int_r^{+\infty} \frac{dt}{|t\hat{\mathbf{r}} - \mathbf{r}_0|^3} = \frac{r}{F(\mathbf{r}; \mathbf{r}_0)} \tag{4.139}$$

where F is the function

$$F(\mathbf{r};\mathbf{r}_0) = |\mathbf{r}-\mathbf{r}_0|[r|\mathbf{r}-\mathbf{r}_0| + \mathbf{r}\cdot(\mathbf{r}-\mathbf{r}_0)] = |\mathbf{r}-\mathbf{r}_0|^2\mathbf{r}\cdot\left[\hat{\mathbf{r}} + \frac{\mathbf{r}-\mathbf{r}_0}{|\mathbf{r}-\mathbf{r}_0|}\right] \tag{4.140}$$

which also appeared in expressions (4.41)–(4.45) in relation to the image representation of the electric potential.

Consequently,

$$U(\mathbf{r};\mathbf{r}_0) = \frac{\mathbf{Q}\times\mathbf{r}_0\cdot\mathbf{r}}{4\pi F(\mathbf{r};\mathbf{r}_0)} \tag{4.141}$$

and

$$\begin{aligned}\mathbf{B}(\mathbf{r};\mathbf{r}_0) &= \frac{\mu_0}{4\pi}\nabla_{\mathbf{r}}\frac{(\mathbf{Q}\times\mathbf{r}_0\cdot\mathbf{r})}{F(\mathbf{r};\mathbf{r}_0)}\\ &= \frac{\mu_0}{4\pi}\frac{\nabla_{\mathbf{r}}(\mathbf{Q}\times\mathbf{r}_0\cdot\mathbf{r})}{F(\mathbf{r};\mathbf{r}_0)} - \frac{\mu_0}{4\pi}(\mathbf{Q}\times\mathbf{r}_0\cdot\mathbf{r})\frac{\nabla_{\mathbf{r}}F(\mathbf{r};\mathbf{r}_0)}{F^2(\mathbf{r};\mathbf{r}_0)}.\end{aligned} \tag{4.142}$$

Straightforward calculations lead to

$$\nabla_{\mathbf{r}}(\mathbf{Q}\times\mathbf{r}_0\cdot\mathbf{r}) = \mathbf{Q}\times\mathbf{r}_0 \tag{4.143}$$

and

$$\begin{aligned}\nabla_{\mathbf{r}}F(\mathbf{r};\mathbf{r}_0) &= \nabla_{\mathbf{r}}\left(r|\mathbf{r}-\mathbf{r}_0|^2\right) + \nabla_{\mathbf{r}}[|\mathbf{r}-\mathbf{r}_0|\mathbf{r}\cdot(\mathbf{r}-\mathbf{r}_0)]\\ &= \hat{\mathbf{r}}|\mathbf{r}-\mathbf{r}_0|^2 + 2r(\mathbf{r}-\mathbf{r}_0) + \frac{\mathbf{r}-\mathbf{r}_0}{|\mathbf{r}-\mathbf{r}_0|}\mathbf{r}\cdot(\mathbf{r}-\mathbf{r}_0) + |\mathbf{r}-\mathbf{r}_0|(2\mathbf{r}-\mathbf{r}_0)\\ &= |\mathbf{r}-\mathbf{r}_0|(r+|\mathbf{r}-\mathbf{r}_0|)\hat{\mathbf{r}} + \left[|\mathbf{r}-\mathbf{r}_0|^2 + 2r|\mathbf{r}-\mathbf{r}_0| + \mathbf{r}\cdot(\mathbf{r}-\mathbf{r}_0)\right]\frac{\mathbf{r}-\mathbf{r}_0}{|\mathbf{r}-\mathbf{r}_0|}.\end{aligned} \tag{4.144}$$

Hence, we finally obtain

$$\begin{aligned}\mathbf{B}(\mathbf{r};\mathbf{r}_0) = \frac{\mu_0}{4\pi}\frac{\mathbf{Q}\times\mathbf{r}_0}{F(\mathbf{r};\mathbf{r}_0)} - \frac{\mu_0}{4\pi}\frac{\mathbf{Q}\times\mathbf{r}_0\cdot\mathbf{r}}{F^2(\mathbf{r};\mathbf{r}_0)}\Bigg[&|\mathbf{r}-\mathbf{r}_0|(r+|\mathbf{r}-\mathbf{r}_0|)\hat{\mathbf{r}}\\ &+ \left(|\mathbf{r}-\mathbf{r}_0|^2 + 2r|\mathbf{r}-\mathbf{r}_0| + \mathbf{r}\cdot(\mathbf{r}-\mathbf{r}_0)\right)\frac{\mathbf{r}-\mathbf{r}_0}{|\mathbf{r}-\mathbf{r}_0|}\Bigg]\end{aligned} \tag{4.145}$$

for the exterior magnetic field in closed form.

For the case of a distributed current, the magnetic field is given by the integral

$$\mathbf{B}(\mathbf{r}) = \frac{\mu_0}{4\pi}\int_{\Omega_c}\mathbf{B}(\mathbf{r};\mathbf{r}_0)\,dv(\mathbf{r}_0) \tag{4.146}$$

where Ω_c denotes the cerebrum.

Chapter 5
The ellipsoidal model

The ellipsoidal model of the head-brain system was analyzed in Section 3.3. In the present chapter, we will obtain analytic solutions for EEG and MEG in ellipsoidal geometry for both the homogeneous and the 3-shell inhomogeneous problems (just as was done in Chapter 4 for the spherical model). In this way analytical expressions for the coefficients C_{in}^{km} that appear in expressions (3.96)–(3.100) will be derived.

In the present chapter, as well as in all other cases where we will have ellipsoidal coordinates we will denote the magnetic parameter by $\hat{\mu}_0$ in order to distinguish it from the coordinate μ_0 of the ellipsoidal system.

5.1 EEG for the homogeneous model

We consider the reference ellipsoid (3.28) as the boundary S of the head Ω and we assume the simple homogeneous model where the shells surrounding the cerebrum are ignored. Then, for an isolated dipole excitation $\{\mathbf{r}_0, \mathbf{Q}\}$ the interior electric potential u^- solves the Neumann problem

$$\sigma\, \Delta_{\mathbf{r}} u^-(\mathbf{r}, \mathbf{r}_0) = \mathbf{Q} \cdot \nabla_{\mathbf{r}} \delta(\mathbf{r} - \mathbf{r}_0), \quad \mathbf{r} \in \Omega \tag{5.1}$$

$$\frac{\partial}{\partial n(\mathbf{r})} u^-(\mathbf{r}, \mathbf{r}_0) = 0, \quad \mathbf{r} \in S \tag{5.2}$$

where $\mathbf{r} = (\rho, \mu, \nu)$ and $\mathbf{r}_0 = (\rho_0, \mu_0, \nu_0)$. After we obtain the solution of the problems (5.1) and (5.2), we can define the Dirichlet problem,

$$\Delta_{\mathbf{r}} u^+(\mathbf{r}, \mathbf{r}_0) = 0, \quad \mathbf{r} \in \Omega^c \tag{5.3}$$

$$u^+(\mathbf{r}, \mathbf{r}_0) = u^-(\mathbf{r}, \mathbf{r}_0), \quad \mathbf{r} \in S \tag{5.4}$$

$$u^+(\mathbf{r}, \mathbf{r}_0) = O\left(\frac{1}{r^2}\right), \quad r \to +\infty \tag{5.5}$$

for the exterior electric potential u^+. Writing u^- in the form

$$u^-(\mathbf{r}, \mathbf{r}_0) = \frac{1}{4\pi\sigma} \left(\mathbf{Q} \cdot \nabla_{\mathbf{r}_0}\right) \frac{1}{|\mathbf{r} - \mathbf{r}_0|} + w(\mathbf{r}, \mathbf{r}_0) \tag{5.6}$$

where $w(\mathbf{r}, \mathbf{r}_0)$ is an interior harmonic function, we obtain the boundary condition

$$\frac{\partial}{\partial \rho} w(\mathbf{r}, \mathbf{r}_0) = -\frac{1}{4\pi\sigma} \left(\mathbf{Q} \cdot \nabla_{\mathbf{r}_0}\right) \frac{\partial}{\partial \rho} \frac{1}{|\mathbf{r} - \mathbf{r}_0|} \tag{5.7}$$

for $\rho = a_1$.

https://doi.org/10.1515/9783110547535-006

The expansion of the fundamental solution $|\mathbf{r}-\mathbf{r}_0|^{-1}$ in the ellipsoidal system [26] is given by

$$\frac{1}{|\mathbf{r}-\mathbf{r}_0|}=\sum_{n=0}^{\infty}\sum_{m=1}^{2n+1}\frac{4\pi}{2n+1}\frac{1}{\gamma_n^m}\mathbb{E}_n^m(\rho_0,\mu_0,\nu_0)\mathbb{F}_n^m(\rho,\mu,\nu),\quad \rho_0<\rho \tag{5.8}$$

Therefore, if we assume an expansion of w in terms of interior ellipsoidal harmonics in the form

$$w(\rho,\mu,\nu;\rho_0,\mu_0,\nu_0)=\sum_{n=0}^{\infty}\sum_{m=1}^{2n+1}B_n^m(\rho_0,\mu_0,\nu_0)\mathbb{E}_n^m(\rho,\mu,\nu) \tag{5.9}$$

with $h_2\le\rho<a_1$, then the Neumann condition (5.7) takes the form

$$\begin{aligned}&\sum_{n=0}^{\infty}\sum_{m=1}^{2n+1}B_n^m(\rho_0,\mu_0,\nu_0)\left(E_n^m(a_1)\right)'S_n^m(\mu,\nu)\\&=-\frac{1}{4\pi\sigma}\sum_{n=0}^{\infty}\sum_{m=1}^{2n+1}\frac{4\pi}{2n+1}\frac{1}{\gamma_n^m}(\mathbf{Q}\cdot\nabla_{\mathbf{r}_0})\mathbb{E}_n^m(\rho_0,\mu_0,\nu_0)\left(F_n^m(a_1)\right)'S_n^m(\mu,\nu).\end{aligned} \tag{5.10}$$

Utilizing the orthogonality relation (3.47), we resolve the relation (5.10) in its spectral components to obtain,

$$B_n^m(\rho_0,\mu_0,\nu_0)=-\frac{(\mathbf{Q}\cdot\nabla_{\mathbf{r}_0})\mathbb{E}_n^m(\rho_0,\mu_0,\nu_0)}{(2n+1)\sigma\gamma_n^m}\frac{\left(F_n^m(a_1)\right)'}{\left(E_n^m(a_1)\right)'}. \tag{5.11}$$

However,

$$\begin{aligned}\frac{\left(F_n^m(a_1)\right)'}{\left(E_n^m(a_1)\right)'}&=(2n+1)\frac{\left(E_n^m(a_1)\right)'I_n^m(a_1)+E_n^m(a_1)\left(I_n^m(a_1)\right)'}{\left(E_n^m(a_1)\right)'}\\&=(2n+1)\left[I_n^m(a_1)-\frac{1}{a_2a_3E_n^m(a_1)\left(E_n^m(a_1)\right)'}\right].\end{aligned} \tag{5.12}$$

Therefore,

$$B_n^m(\rho_0,\mu_0,\nu_0)=\frac{(\mathbf{Q}\cdot\nabla_{\mathbf{r}_0})\mathbb{E}_n^m(\rho_0,\mu_0,\nu_0)}{\sigma\gamma_n^m}\left[\frac{1}{a_2a_3E_n^m(a_1)\left(E_n^m(a_1)\right)'}-I_n^m(a_1)\right]. \tag{5.13}$$

Inserting, (5.13) in (5.9) and the resulting equation in (5.6), we obtain the interior electric potential

$$u^{-}(\rho,\mu,\nu;\rho_0,\mu_0,\nu_0) = \frac{1}{4\pi\sigma}\left(\mathbf{Q}\cdot\nabla_{\mathbf{r}_0}\right)\left[\frac{1}{|\mathbf{r}-\mathbf{r}_0|}\right.$$
$$+4\pi\sum_{n=1}^{\infty}\sum_{m=1}^{2n+1}\frac{1}{\gamma_n^m}$$
$$\left.\left(\frac{1}{a_2a_3E_n^m(a_1)\left(E_n^m(a_1)\right)'} - I_n^m(a_1)\right)\mathbb{E}_n^m(\rho_0,\mu_0,\nu_0)\mathbb{E}_n^m(\rho,\mu,\nu)\right]. \quad (5.14)$$

Keeping in (5.14) only the terms of degree 1 and 2 we obtain the following Cartesian approximation of the electric potential:

$$u^{-}(\mathbf{r},\mathbf{r}_0) = \frac{3}{4\pi\sigma}\sum_{m=1}^{3}\left[I_1^m(\rho) - I_1^m(a_1) + \frac{1}{a_1a_2a_3}\right]Q_mx_m$$
$$-\frac{5}{4\pi\sigma}\frac{(\Lambda-a_1^2)(\Lambda-a_2^2)(\Lambda-a_3^2)}{(\Lambda-\Lambda')}\left[I_2^1(\rho) - I_2^1(a_1) + \frac{1}{2\Lambda a_1a_2a_3}\right]$$
$$\cdot\left[\sum_{k=1}^{3}\frac{Q_kx_{0k}}{\Lambda-a_k^2}\right]\left[\sum_{m=1}^{3}\frac{x_m^2}{\Lambda-a_m^2}+1\right]$$
$$+\frac{5}{4\pi\sigma}\frac{(\Lambda'-a_1^2)(\Lambda'-a_2^2)(\Lambda'-a_3^2)}{(\Lambda-\Lambda')}\left[I_2^2(\rho) - I_2^2(a_1) + \frac{1}{2\Lambda' a_1a_2a_3}\right]$$
$$\cdot\left[\sum_{k=1}^{3}\frac{Q_kx_{0k}}{\Lambda'-a_k^2}\right]\left[\sum_{m=1}^{3}\frac{x_m^2}{\Lambda'-a_m^2}+1\right]$$
$$+\frac{15}{4\pi\sigma}\left[I_2^3(\rho) - I_2^3(a_1) + \frac{1}{a_1a_2a_3\left(a_1^2+a_2^2\right)}\right](Q_1x_{02}+Q_2x_{01})x_1x_2$$
$$+\frac{15}{4\pi\sigma}\left[I_2^4(\rho) - I_2^4(a_1) + \frac{1}{a_1a_2a_3\left(a_3^2+a_1^2\right)}\right](Q_3x_{01}+Q_1x_{03})x_3x_1$$
$$+\frac{15}{4\pi\sigma}\left[I_2^5(\rho) - I_2^5(a_1) + \frac{1}{a_1a_2a_3\left(a_2^2+a_3^2\right)}\right](Q_2x_{03}+Q_3x_{02})x_2x_3$$
$$+O\left(el^3\right), \rho_0 < \rho < a_1 \quad (5.15)$$

where the symbol $O(el^3)$ indicates ellipsoidal terms in (ρ,μ,ν) of degree greater or equal to 3.

For the exterior electric potential u^+ which solves the Dirichlet problem (5.3)–(5.5), we can use the expansions (5.8) and (5.14) to generate the Dirichlet data.

$$u^{+}(a_1,\mu,\nu;\rho_0,\mu_0,\nu_0) = \frac{1}{\sigma}\sum_{n=1}^{\infty}\sum_{m=1}^{2n+1}\frac{\left(\mathbf{Q}\cdot\nabla_{\mathbf{r}_0}\right)\mathbb{E}_n^m(\rho_0,\mu_0,\nu_0)}{\gamma_n^m a_2a_3\left(E_n^m(a_1)\right)'}S_n^m(\mu,\nu). \quad (5.16)$$

Then, the solution of (5.3)–(5.5) is given by

$$u^+(\rho,\mu,\nu;\rho_0,\mu_0,\nu_0)=\frac{1}{\sigma a_2 a_3}\sum_{n=1}^{\infty}\sum_{m=1}^{2n+1}\frac{(\mathbf{Q}\cdot\nabla_{\mathbf{r}_0})\mathbb{E}_n^m(\rho_0,\mu_0,\nu_0)}{\gamma_n^m(E_n^m(a_1))'F_n^m(a_1)}\mathbb{F}_n^m(\rho,\mu,\nu),\quad \rho>a_1. \tag{5.17}$$

5.2 MEG for the homogeneous model

We recall that the exterior magnetic field for the dipole excitation $\{\mathbf{r}_0,\mathbf{Q}\}$ is given by the Geselowitz representation (1.72) which for the case of the ellipsoid is written as

$$\mathbf{B}(\mathbf{r};\mathbf{r}_0)=\frac{\mu_0}{4\pi}\mathbf{Q}\times\nabla_{\mathbf{r}_0}\frac{1}{|\mathbf{r}-\mathbf{r}_0|}-\frac{\mu_0\sigma}{4\pi}\int_{\rho=a_1}u^-(\mathbf{r}';\mathbf{r}_0)\hat{\mathbf{n}}(\mathbf{r}')\times\nabla_{\mathbf{r}'}\frac{1}{|\mathbf{r}-\mathbf{r}'|}ds(\mathbf{r}') \tag{5.18}$$

where the interior electric potential (5.14), on the boundary $\rho=a_1$, takes the form

$$u^-(a_1,\mu',\nu';\mathbf{r}_0)=\frac{1}{4\pi\sigma}(\mathbf{Q}\cdot\nabla_{\mathbf{r}_0})\frac{1}{|\mathbf{r}'-\mathbf{r}_0|}\bigg|_{\rho=a_1}+\frac{1}{\sigma}\sum_{n=1}^{\infty}\sum_{m=1}^{2n+1}\frac{1}{\gamma_n^m}(\mathbf{Q}\cdot\nabla_{\mathbf{r}_0})\mathbb{E}_n^m(\mathbf{r}_0)$$
$$\left(\frac{1}{a_1a_2a_3(E_n^m(a_1))'}-E_n^m(a_1)I_n^m(a_1)\right)E_n^m(\mu')E_n^m(\nu'). \tag{5.19}$$

Keeping in (5.15) only the dipole and quadrupole terms, we obtain the Cartesian representation

$$\begin{aligned}u^-(a_1,\mu',\nu';\mathbf{r}_0)=&\frac{3}{4\pi\sigma}\sum_{m=1}^{3}\frac{Q_m x'_m}{a_1a_2a_3}\\&-\frac{5}{8\pi\sigma}\frac{(\Lambda-a_1^2)(\Lambda-a_2^2)(\Lambda-a_3^2)}{a_1a_2a_3\Lambda(\Lambda-\Lambda')}\left[\sum_{k=1}^{3}\frac{Q_k x_{0k}}{\Lambda-a_k^2}\right]\left[\sum_{m=1}^{3}\frac{x_m'^2}{\Lambda-a_m^2}+1\right]\\&+\frac{5}{8\pi\sigma}\frac{(\Lambda'-a_1^2)(\Lambda'-a_2^2)(\Lambda'-a_3^2)}{a_1a_2a_3\Lambda'(\Lambda-\Lambda')}\left[\sum_{k=1}^{3}\frac{Q_k x_{0k}}{\Lambda'-a_k^2}\right]\left[\sum_{m=1}^{3}\frac{x_m'^2}{\Lambda'-a_m^2}+1\right]\\&+\frac{15}{4\pi\sigma}\frac{Q_1x_{02}+Q_2x_{01}}{a_1a_2a_3(a_1^2+a_2^2)}x_1'x_2'+\frac{15}{4\pi\sigma}\frac{Q_3x_{01}+Q_1x_{03}}{a_1a_2a_3(a_1^2+a_3^2)}x_1'x_3'\\&+\frac{15}{4\pi\sigma}\frac{Q_2x_{03}+Q_3x_{02}}{a_1a_2a_3(a_2^2+a_3^2)}x_2'x_3'+O(el_3).\end{aligned} \tag{5.20}$$

Similarly, for the ellipsoidal representation,

$$u^-(a_1,\mu',\nu';\mathbf{r}_0)=\sum_{m=1}^{3}J_mE_1^m(\mu')E_1^m(\nu')+\sum_{m=1}^{5}\theta_mE_2^m(\mu')E_2^m(\nu')+O(el_3), \tag{5.21}$$

where

$$J_m = \frac{a_m h_m}{\sigma V h_1 h_2 h_3}(\mathbf{Q}\cdot\hat{\mathbf{x}}_m), \quad m=1,2,3 \tag{5.22}$$

$$\theta_1 = -\frac{5}{6\sigma V(\Lambda-\Lambda')}\left(\mathbf{Q}\otimes\mathbf{r}_0 : \tilde{\Lambda}\right) \tag{5.23}$$

$$\theta_2 = \frac{5}{6\sigma V(\Lambda-\Lambda')}\left(\mathbf{Q}\otimes\mathbf{r}_0 : \tilde{\Lambda}'\right) \tag{5.24}$$

$$\theta_3 = \frac{5a_1a_2a_3}{\sigma V h_1h_2h_3}\frac{\mathbf{Q}\otimes\mathbf{r}_0 : (\hat{\mathbf{x}}_1\otimes\hat{\mathbf{x}}_2+\hat{\mathbf{x}}_2\otimes\hat{\mathbf{x}}_1)}{a_3h_3\left(a_1^2+a_2^2\right)} \tag{5.25}$$

$$\theta_4 = \frac{5a_1a_2a_3}{\sigma V h_1h_2h_3}\frac{\mathbf{Q}\otimes\mathbf{r}_0 : (\hat{\mathbf{x}}_1\otimes\hat{\mathbf{x}}_3+\hat{\mathbf{x}}_3\otimes\hat{\mathbf{x}}_1)}{a_2h_2\left(a_1^2+a_3^2\right)} \tag{5.26}$$

$$\theta_5 = \frac{5a_1a_2a_3}{\sigma V h_1h_2h_3}\frac{\mathbf{Q}\otimes\mathbf{r}_0 : (\hat{\mathbf{x}}_2\otimes\hat{\mathbf{x}}_3+\hat{\mathbf{x}}_3\otimes\hat{\mathbf{x}}_2)}{a_1h_1\left(a_2^2+a_3^2\right)} \tag{5.27}$$

$$V = \frac{4\pi}{3}a_1a_2a_3 \tag{5.28}$$

with V denoting the volume of the ellipsoid $\rho = a_1$; also

$$\tilde{\Lambda} = \sum_{m=1}^{3}\frac{\tilde{\mathbf{x}}_m\otimes\tilde{\mathbf{x}}_m}{\Lambda-a_m^2} \tag{5.29}$$

$$\tilde{\Lambda}' = \sum_{m=1}^{3}\frac{\tilde{\mathbf{x}}_m\otimes\tilde{\mathbf{x}}_m}{\Lambda'-a_m^2}. \tag{5.30}$$

The expression (5.21) provides the factor u^- of the integrand in (5.18). The corresponding expression of the factor $\tilde{\mathbf{n}}(\mathbf{r}')\times\frac{\mathbf{r}-\mathbf{r}'}{|\mathbf{r}-\mathbf{r}'|^3}$ can also be calculated as follows: first, by utilizing the identity

$$\tilde{\mathbf{n}}(\mathbf{r}') = \hat{\mathbf{x}}\rho' = \frac{3V}{4\pi}l_{a_1}(\mu',\nu')\tilde{\mathbf{M}}^{-1}(a_1)\cdot\mathbf{r}' \tag{5.31}$$

where l_{a_1} is the weighting function defined in (3.93) and

$$\tilde{\mathbf{M}}(\rho) = \sum_{m=1}^{3}\left(\rho^2-a_1^2+a_m^2\right)\,\tilde{\mathbf{x}}_m\otimes\tilde{\mathbf{x}}_m. \tag{5.32}$$

This implies

$$\tilde{\mathbf{M}}^{-1}(a_1) = \sum_{m=1}^{3}\frac{\tilde{\mathbf{x}}_m\otimes\tilde{\mathbf{x}}_m}{a_m^2}. \tag{5.33}$$

Second, employing the identity

$$\frac{\mathbf{r}-\mathbf{r}'}{|\mathbf{r}-\mathbf{r}'|^3} = 3\mathbf{r}\cdot\tilde{\mathbf{H}}_1(\rho) + \tilde{\mathbf{F}}(\mathbf{r})\cdot\mathbf{r}' + O(el_2') \tag{5.34}$$

with

$$\tilde{\mathbf{H}}_1(\rho) = \sum_{m=1}^{3} I_1^m(\rho)\tilde{\mathbf{x}}_m \otimes \tilde{\mathbf{x}}_m \tag{5.35}$$

$$\tilde{\mathbf{H}}_2(\rho) = \sum_{\substack{i,j=1\\ i\neq j}}^{3} I_2^{i+j}(\rho)\tilde{\mathbf{x}}_i \otimes \tilde{\mathbf{x}}_j \otimes \tilde{\mathbf{x}}_i \otimes \tilde{\mathbf{x}}_j \tag{5.36}$$

and

$$\tilde{\mathbf{F}}(\mathbf{r}) = -\frac{I\!F_2^1(\mathbf{r})}{\Lambda-\Lambda'}\tilde{\Lambda} + \frac{I\!F_2^2(\mathbf{r})}{\Lambda-\Lambda'}\tilde{\Lambda}' + 15\mathbf{r}\otimes\mathbf{r}:\tilde{\tilde{\mathbf{H}}}_2(\rho). \tag{5.37}$$

Then,

$$\begin{aligned}
\tilde{\mathbf{n}}(\mathbf{r}')\times\nabla_{\mathbf{r}'}\frac{1}{|\mathbf{r}-\mathbf{r}'|}\bigg|_{\rho=a_1} &= \hat{\rho}'\times\frac{\mathbf{r}-\mathbf{r}'}{|\mathbf{r}-\mathbf{r}'|^3}\bigg|_{\rho=a_1}\\
&= a_1a_2a_3 l_{a_1}(\mu',\nu')\Bigg[\frac{1}{3}\sum_{m=1}^{3} a_m^2\left(\hat{\mathbf{x}}_m\cdot\tilde{\mathbf{M}}^{-1}(a_1)\times\tilde{\mathbf{F}}(\mathbf{r})\cdot\hat{\mathbf{x}}_m\right)\\
&+\frac{1}{3h_1h_2h_3}\sum_{m=1}^{3} a_m h_m\left(\hat{\mathbf{x}}_m\cdot\tilde{\mathbf{M}}^{-1}(a_1)\times\tilde{\mathbf{H}}_1(\rho)\cdot\mathbf{r}\right)E_1^m(\mu')E_1^m(\nu')\\
&-\frac{1}{3(\Lambda-\Lambda')}\sum_{m=1}^{3}\frac{a_m^2}{\Lambda-a_m^2}\left(\hat{\mathbf{x}}_m\cdot\tilde{\mathbf{M}}^{-1}(a_1)\times\tilde{\mathbf{F}}(\mathbf{r})\cdot\hat{\mathbf{x}}_m\right)E_2^1(\mu')E_2^1(\nu')\\
&+\frac{1}{3(\Lambda-\Lambda')}\sum_{m=1}^{3}\frac{a_m^2}{\Lambda'-a_m^2}\left(\hat{\mathbf{x}}_m\cdot\tilde{\mathbf{M}}^{-1}(a_1)\times\tilde{\mathbf{F}}(\mathbf{r})\cdot\hat{\mathbf{x}}_m\right)E_2^1(\mu')E_2^1(\nu')\\
&+\frac{1}{h_1^2h_2^2h_3^2}\sum_{\substack{i,j=1\\ i\neq j}}^{3}\frac{a_ia_j}{h_{6-(i+j)}}\left(\hat{\mathbf{x}}_i\cdot\tilde{\mathbf{M}}^{-1}(a_1)\times\tilde{\mathbf{F}}(\mathbf{r})\cdot\hat{\mathbf{x}}_j\right)E_2^{i+j}(\mu')E_2^{i+j}(\nu')\Bigg]\\
&+O(el_3').
\end{aligned} \tag{5.38}$$

The expression (5.38) involves three dipole terms and five quadrupole terms, but as we will show below, the three dipole terms vanish. Indeed,

$$\begin{aligned}
\sum_{m=1}^{3} a_m^2\left(\hat{\mathbf{x}}_m\cdot\tilde{\mathbf{M}}^{-1}(a_1)\times\tilde{\mathbf{F}}(\mathbf{r})\cdot\hat{\mathbf{x}}_m\right) &= \sum_{m=1}^{3}\hat{\mathbf{x}}_m\times\tilde{\mathbf{F}}(\mathbf{r})\cdot\hat{\mathbf{x}}_m\\
&= -\frac{I\!F_2^1(\mathbf{r})}{\Lambda-\Lambda'}\sum_{m=1}^{3}\frac{\tilde{\mathbf{x}}_m\times\tilde{\mathbf{x}}_m}{\Lambda-a_m^2} + \frac{I\!F_2^2(\mathbf{r})}{\Lambda-\Lambda'}\sum_{m=1}^{3}\frac{\tilde{\mathbf{x}}_m\times\tilde{\mathbf{x}}_m}{\Lambda'-a_m^2}
\end{aligned}$$

$$+15\sum_{m=1}^{3}\hat{\mathbf{x}}_m \times \left(\mathbf{r}\otimes\mathbf{r}:\tilde{\tilde{\mathbf{H}}}_2(\rho)\right)\cdot\hat{\mathbf{x}}_m$$
$$=15\sum_{\substack{i,j=1\\ i\neq j}}^{3} x_i x_j I_2^{i+j}(\rho)\hat{\mathbf{x}}_j\times\hat{\mathbf{x}}_j=\mathbf{0}. \tag{5.39}$$

The relation (5.39) is consistent with the laws of magnetostatics and in particular with the nonexistence of magnetic monopoles. Inserting (5.21) and (5.39) in the integral on the right hand side of (5.18) and using orthogonally, we obtain

$$\int_{\rho=a_1} u^-(\mathbf{r}';\mathbf{r}_0)\hat{\mathbf{n}}(\mathbf{r}')\times\nabla_{\mathbf{r}'}\frac{1}{|\mathbf{r}-\mathbf{r}'|}ds(\mathbf{r}')=\sum_{m=1}^{3}J_m\gamma_1^m\boldsymbol{\beta}_m+\sum_{m=1}^{3}\theta_m\gamma_2^m\boldsymbol{\delta}_m+O(el_3) \tag{5.40}$$

where

$$\boldsymbol{\beta}_m=3\frac{a_1a_2a_3}{h_1h_2h_3}\frac{h_m}{a_m}\hat{\mathbf{x}}_m\otimes\mathbf{r}\underset{\cdot}{\times}\tilde{\mathbf{H}}_1(\rho),\quad m=1,2,3 \tag{5.41}$$

$$\boldsymbol{\delta}_1=-\frac{a_1a_2a_3}{3(\Lambda-\Lambda')}\tilde{\Lambda}\underset{\cdot}{\times}\tilde{\mathbf{F}}(\mathbf{r}) \tag{5.42}$$

$$\boldsymbol{\delta}_2=\frac{a_1a_2a_3}{3(\Lambda-\Lambda')}\tilde{\Lambda}\underset{\cdot}{\times}\tilde{\mathbf{F}}(\mathbf{r}) \tag{5.43}$$

$$\boldsymbol{\delta}_3=\frac{a_1a_2a_3}{h_1h_2h_3^2}\left[\frac{a_2}{a_1}\hat{\mathbf{x}}_1\otimes\hat{\mathbf{x}}_2+\frac{a_1}{a_2}\hat{\mathbf{x}}_2\otimes\hat{\mathbf{x}}_1\right]\underset{\cdot}{\times}\tilde{\mathbf{F}}(\mathbf{r}) \tag{5.44}$$

$$\boldsymbol{\delta}_4=\frac{a_1a_2a_3}{h_1h_2^2h_3}\left[\frac{a_3}{a_1}\hat{\mathbf{x}}_1\otimes\hat{\mathbf{x}}_3+\frac{a_1}{a_3}\hat{\mathbf{x}}_3\otimes\hat{\mathbf{x}}_1\right]\underset{\cdot}{\times}\tilde{\mathbf{F}}(\mathbf{r}) \tag{5.45}$$

$$\boldsymbol{\delta}_5=\frac{a_1a_2a_3}{h_1^2h_2h_3}\left[\frac{a_3}{a_2}\hat{\mathbf{x}}_2\otimes\hat{\mathbf{x}}_3+\frac{a_2}{a_3}\hat{\mathbf{x}}_3\otimes\hat{\mathbf{x}}_2\right]\underset{\cdot}{\times}\tilde{\mathbf{F}}(\mathbf{r}). \tag{5.46}$$

But

$$\sum_{m=1}^{3}J_m\gamma_1^\mu\boldsymbol{\beta}_m=\frac{3}{\sigma}\sum_{m=1}^{3}(\mathbf{Q}\cdot\hat{\mathbf{x}}_m)\,\hat{\mathbf{x}}_m\otimes\mathbf{r}\underset{\cdot}{\times}\tilde{\mathbf{H}}_1(\rho)=\frac{3}{\sigma}\mathbf{Q}\otimes\mathbf{r}\underset{\cdot}{\times}\tilde{\mathbf{H}}_1(\rho). \tag{5.47}$$

After extended manipulations involving the ellipsoidal harmonics, which are given in [27], we arrive at the expression

$$\mathbf{B}(\mathbf{r}';\mathbf{r}_0)=\frac{\hat{\mu}_0}{4\pi}\frac{\mathbb{F}_2^1(\rho,\mu,\nu)}{\Lambda-\Lambda'}\sum_{i=1}^{3}\frac{d_i}{\Lambda-a_i^2}\hat{\mathbf{x}}_i-\frac{\hat{\mu}_0}{4\pi}\frac{\mathbb{F}_2^2(\rho,\mu,\nu)}{\Lambda-\Lambda'}\sum_{i=1}^{3}\frac{d_i}{\Lambda'-a_i^2}\hat{\mathbf{x}}_i$$
$$-\frac{15\hat{\mu}_0}{4\pi}\sum_{\substack{i,j=1\\ i\neq j}}^{3}d_ix_ix_jI_2^{i+j}(\rho)\hat{\mathbf{x}}_j+O(\rho^{-4}) \tag{5.48}$$

where the dipole dependent vector $\mathbf{d}$ is given by

$$\mathbf{d} = \frac{a_2^2 Q_2 x_{03} - a_3^2 Q_3 x_{02}}{a_2^2 + a_3^2} \hat{\mathbf{x}}_1 + \frac{a_3^2 Q_3 x_{01} - a_1^2 Q_1 x_{03}}{a_1^2 + a_3^2} \hat{\mathbf{x}}_2 + \frac{a_1^2 Q_1 x_{02} - a_2^2 Q_2 x_{01}}{a_1^2 + a_2^2} \hat{\mathbf{x}}_3. \tag{5.49}$$

Very elaborate manipulations are needed to calculate the octapolic term $n = 2$ for the field $\mathbf{B}$. The complete analysis can be found in reference [35].

5.3 The complete magnetic field

In this section, we demonstrate an analytic technique to obtain the general expression of $\mathbf{B}$ in ellipsoidal geometry. The idea is as follows: we know that in order to obtain the magnetic potential U for the spherical model of the brain we need to integrate the radial component $\hat{\mathbf{r}} \cdot \mathbf{B}(\mathbf{r})$ along a ray from the exterior observation point $\mathbf{r}$ to infinity in the direction $\hat{\mathbf{r}}$. This ray is the intersection of the cone with the direction $\hat{\mathbf{r}}$ as its generating line and the azimouthal plane that contains $\hat{\mathbf{r}}$. This ray is the intersection of the two coordinate surfaces which specify the orientation of points in space.

For the ellipsoidal system we can follow the corresponding curve in space specified by the intersection of the hyperboloid of one sheet and the hyperboloid of two sheets that pass through the point (ρ, μ, ν), that is, the $\mu =$ constant and the $\nu =$ constant coordinate surfaces, (since the coordinates μ, ν are those that specify the spatial orientation in terms of ellipsoidal coordinates). This particular curve is defined by

$$\mathbf{C}(\rho) = \frac{1}{h_1 h_2 h_3} \Big[h_1 \rho \mu_0 \nu_0 \hat{\mathbf{x}}_1 + h_2 \sqrt{\rho^2 - h_3^2} \sqrt{\mu_0^2 - h_3^2} \sqrt{h_3^2 - \nu_0^2} \hat{\mathbf{x}}_2 + h_3 \sqrt{\rho^2 - h_2^2} \sqrt{h_2^2 - \mu_0^2} \sqrt{h_2^2 - \nu_0^2} \hat{\mathbf{x}}_3 \Big] \tag{5.50}$$

where $\rho \in [\rho_0, +\infty)$ is the parameter of the curve and $\mathbf{r}_0 = (\rho_0, \mu_0, \nu_0)$ is the point from where the ellipsoidal *ray* emanates. If the point where the magnetic potential is evaluated belongs to the surface $\rho = a_1$, then the parameterization of (5.50) is $\rho \in [a_1, +\infty)$. We write the electric potential on the ellipsoid $\rho = a_1$ as

$$u(a_1, \mu', \nu'; \mathbf{r}_0) = \sum_{n=1}^{\infty} \sum_{m=1}^{2n+1} A_n^m(\mathbf{r}_0) E_n^m(\mu') E_n^m(\nu') \tag{5.51}$$

where

$$A_n^m(\mathbf{r}_0) = \frac{\mathbf{Q} \cdot \nabla_{\mathbf{r}_0} \mathbb{E}_n^m(\mathbf{r}_0)}{\sigma a_2 a_3 \gamma_n^m (E_n^m(a_1))'}. \tag{5.52}$$

Also

$$\frac{\mathbf{r}-\mathbf{r}'}{|\mathbf{r}-\mathbf{r}'|^3}\bigg|_{\rho'=a_1} = -\nabla_{\mathbf{r}}\frac{1}{|\mathbf{r}-\mathbf{r}'|}\bigg|_{\rho'=a_1} = -\sum_{n=0}^{\infty}\sum_{m=1}^{2n+1}\frac{4\pi}{2n+1}\frac{1}{\gamma_n^m}(\nabla_{\mathbf{r}}I\!F_n^m(\mathbf{r}))I\!E_n^m(\mathbf{r}')\bigg|_{\rho'=a_1}$$

$$= -\sum_{n=0}^{\infty}\sum_{m=1}^{2n+1}\mathbf{B}_n^m(\mathbf{r})E_n^m(\mu')E_n^m(\nu') \tag{5.53}$$

with

$$\mathbf{B}_n^m(\mathbf{r}) = \frac{4\pi}{(2n+1)\gamma_n^m}\left(\nabla_{\mathbf{r}}I\!F_n^m(\mathbf{r})\right)E_n^m(a_1) \tag{5.54}$$

and

$$\hat{\rho}' = a_1a_2a_3l_{a_1}(\mu',\nu')\sum_{i=1}^{3}\frac{x_i'}{a_i^2}\bigg|_{\rho'=a_1}\hat{\mathbf{x}}_i = \frac{a_1a_2a_3}{h_1h_2h_3}\sum_{i=1}^{3}\frac{h_i}{a_i}\,l_{a_1}(\mu',\nu')E_1^i(\mu')E_1^i(\nu')\hat{\mathbf{x}}_i. \tag{5.55}$$

Therefore,

$$-\int_{\rho'=a_1} u^-(\mathbf{r}';\mathbf{r}_0)\hat{\rho}'\times\frac{\mathbf{r}-\mathbf{r}'}{|\mathbf{r}-\mathbf{r}'|^3}ds(\mathbf{r}') = \sum_{n=0}^{\infty}\sum_{m=1}^{2n+1}\sum_{k=1}^{\infty}\sum_{\lambda=1}^{2k+1}A_k^\lambda(\mathbf{r}_0)\mathbf{C}_{nk}^{m\lambda}\times\mathrm{B}_n^m(\mathbf{r}) \tag{5.56}$$

where B_n^m are given in (4.97) and

$$\mathbf{C}_{nk}^{m\lambda} = \frac{a_1a_2a_3}{h_1h_2h_3}\sum_{i=1}^{3}\frac{h_i}{a_i}$$

$$\int_{\rho'=a_1} E_n^m(\mu')E_n^m(\nu')E_k^\lambda(\mu')E_k^\lambda(\nu')E_1^i(\mu')E_1^i(\nu')l_{a_1}(\mu',\nu')ds(\mu',\nu')\hat{\mathbf{x}}_i. \tag{5.57}$$

Expanding the first term on the right hand side of (5.18), we obtain

$$\mathbf{Q}\times\frac{\mathbf{r}-\mathbf{r}_0}{|\mathbf{r}-\mathbf{r}_0|^3} = -\mathbf{Q}\times\nabla_{\mathbf{r}}\frac{1}{|\mathbf{r}-\mathbf{r}_0|} = -\sum_{n=0}^{\infty}\sum_{m=1}^{2n+1}\frac{4\pi}{2n+1}\frac{1}{\gamma_n^m}I\!E_n^m(\mathbf{r}_0)\mathbf{Q}\times\nabla_{\mathbf{r}}I\!F_n^m(\mathbf{r}). \tag{5.58}$$

Then, in view of (5.56) and (5.58), formula (5.18) becomes

$$\mathbf{B}(\mathbf{r};\mathbf{r}_0) = \hat{\mu}_0\sum_{n=1}^{\infty}\sum_{m=1}^{2n+1}\frac{\nabla_{\mathbf{r}}I\!F_n^m(\mathbf{r})}{(2n+1)\gamma_n^m}\times\Bigg[\mathbf{Q}I\!E_n^m(\mathbf{r}_0)$$

$$-\frac{1}{a_2a_3}\sum_{k=1}^{\infty}\sum_{\lambda=1}^{2k+1}\mathbf{C}_{nk}^{m\lambda}\frac{E_n^m(a_1)}{(E_k^\lambda(a_1))'}\frac{\mathbf{Q}\cdot\nabla_{\mathbf{r}_0}I\!E_k^\lambda(\mathbf{r}_0)}{\gamma_k^\lambda}\Bigg] \tag{5.59}$$

where $\mathbf{C}_{nk}^{m\lambda}$ are given in (5.57). In the expansion (5.59), we have used the fact that the $n=0$ term vanishes. Indeed, if we denote the $n=0$ term by $\mathbf{B}_0(\mathbf{r};\mathbf{r}_0)$, we have

$$\mathbf{B}_0(\mathbf{r};\mathbf{r}_0)=\hat{\mu}_0\frac{\nabla_{\mathbf{r}}I_0^1(\rho)}{\gamma_0^1}\times\left[\mathbf{Q}-\frac{1}{a_2a_3}\sum_{k=1}^{\infty}\sum_{\lambda=1}^{2k+1}\mathbf{C}_{0k}^{1\lambda}\frac{1}{(E_k^\lambda(a_1))'}\frac{\mathbf{Q}\cdot\nabla_{\mathbf{r}_0}\mathbb{E}_k^\lambda(\mathbf{r}_0)}{\gamma_k^\lambda}\right] \tag{5.60}$$

where

$$\begin{aligned}\mathbf{C}_{0k}^{1\lambda}&=\frac{a_1a_2a_3}{h_1h_2h_3}\sum_{i=1}^{3}\frac{h_i}{a_i}\hat{\mathbf{x}}_i\int_{\rho'=a_1}E_k^\lambda(\mu')E_k^\lambda(\nu')E_1^i(\mu')E_1^i(\nu')l_{a_1}(\mu',\nu')ds'\\&=\frac{a_1a_2a_3}{h_1h_2h_3}\sum_{i=1}^{3}\frac{h_i}{a_i}\gamma_1^i\delta_{k1}\delta_{i\lambda}\hat{\mathbf{x}}_i=\frac{4\pi}{3}a_1a_2a_3h_1h_2h_3\sum_{i=1}^{3}\frac{\delta_{k1}\delta_{i\lambda}}{a_ih_i}\hat{\mathbf{x}}_i\\&=\frac{4\pi}{3}a_1a_2a_3h_1h_2h_3\frac{\delta_{k1}}{a_\lambda h_\lambda}\hat{\mathbf{x}}_\lambda.\end{aligned} \tag{5.61}$$

Inserting (5.61) in (5.60), we find

$$\begin{aligned}\mathbf{B}_0(\mathbf{r};\mathbf{r}_0)&=\frac{\hat{\mu}_0}{4\pi}\nabla_{\mathbf{r}}I_0^1(\mathbf{r})\times\left[\mathbf{Q}-\sum_{\lambda=1}^{3}\frac{h_\lambda}{h_1h_2h_3}\mathbf{Q}\cdot\nabla_{\mathbf{r}_0}\left(\frac{h_1h_2h_3}{h_\lambda}x_{0\lambda}\right)\hat{\mathbf{x}}_\lambda\right]\\&=\frac{\hat{\mu}_0}{4\pi}\nabla_{\mathbf{r}}I_0^1(\mathbf{r})\times\left[\mathbf{Q}-\sum_{\lambda=1}^{3}\mathbf{Q}\cdot\hat{\mathbf{x}}_\lambda\otimes\hat{\mathbf{x}}_\lambda\right]=\mathbf{0}.\end{aligned} \tag{5.62}$$

The above calculations verify that the dipole term contribution from the source $\{\mathbf{r}_0,\mathbf{Q}\}$ and from the interface at $\rho=a_1$ cancel each other. Therefore, the exterior magnetic field starts with the quadrupolic contribution

$$\nabla_{\mathbf{r}}IF_1^i(\mathbf{r})=O\left(\frac{1}{r^3}\right),\quad r\to\infty. \tag{5.63}$$

The term $\mathbf{C}_{1k}^{m\lambda}$ with $k\geq3$ has to vanish by orthogonality, since S_k^λ should have the same degree as $S_1^m\cdot S_1^i$. Also, note that the surface ellipsoidal harmonic

$$S_n^m(\mu,\nu)S_1^i(\mu,\nu)=E_n^m(\mu)E_n^m(\nu)E_1^i(\mu)E_1^i(\nu) \tag{5.64}$$

lives in the subspace generated by the surface harmonics of degree less or equal to $n+1$. So, by orthogonality

$$\mathbf{C}_{nk}^{m\lambda}=\mathbf{0},\quad k\geq n+2. \tag{5.65}$$

Hence, in the expansion (5.59) the $k-$ summation goes from 1 to $n+1$. In fact, introducing the source dependent dyadic

$$\tilde{\mathbf{D}}_n^m(\mathbf{r}_0) = \frac{1}{(2n+1)\gamma_n^m}$$

$$\left[-\mathbb{E}_n^m(\mathbf{r}_0)\tilde{\mathbf{I}} + \frac{1}{a_2 a_3} \sum_{k=1}^{n+1} \sum_{\lambda=1}^{2k+1} \frac{1}{\gamma_k^\lambda} \frac{E_n^m(a_1)}{\left(E_k^\lambda(a_1)\right)'} \left(\nabla_{\mathbf{r}_0} \mathbb{E}_k^\lambda(\mathbf{r}_0)\right) \otimes \mathbf{C}_{nk}^{m\lambda} \right] \tag{5.66}$$

we can write the complete exterior magnetic field as

$$\mathbf{B}(\mathbf{r};\mathbf{r}_0) = \hat{\mu}_0 \sum_{n=1}^{\infty} \sum_{m=1}^{2n+1} \mathbf{Q} \cdot \tilde{\mathbf{D}}_n^m(\mathbf{r}_0) \times \nabla_{\mathbf{r}} \mathbb{F}_n^m(\mathbf{r}). \tag{5.67}$$

In view of the differential representation

$$\mathbf{B}(\mathbf{r};\mathbf{r}_0) = \hat{\mu}_0 \nabla_{\mathbf{r}} U(\mathbf{r};\mathbf{r}_0) \tag{5.68}$$

an integration along the ellipsoidal coordinate curve (5.50) gives

$$\begin{aligned} U(\mathbf{r};\mathbf{r}_0) &= -\int_{\rho}^{+\infty} \frac{\partial}{\partial \rho'} U(\rho',\mu,\nu;\mathbf{r}_0) d\rho' \\ &= -\frac{1}{\hat{\mu}_0} \int_{\rho}^{+\infty} h_{\rho'} \hat{\rho}' \cdot \nabla_{\rho'} \left(\mu_0 U(\rho',\mu,\nu;\mathbf{r}_0)\right) d\rho' \\ &= -\frac{1}{\hat{\mu}_0} \int_{\rho}^{+\infty} h_{\rho'} \hat{\rho}' \cdot \mathbf{B}(\rho',\mu,\nu;\mathbf{r}_0) d\rho' \end{aligned} \tag{5.69}$$

with $\rho \geq a_1$ and the contour of integration is $\mathbf{C}(\rho)$.

Some further calculations lead to the formula

$$\begin{aligned} &\hat{\rho}' \cdot \left(\mathbf{Q} \cdot \tilde{\mathbf{D}}_n^m\right) \times \nabla_{\rho'} \mathbb{F}_n^m(\rho',\mu,\nu;\mathbf{r}_0) \\ &= F_n^m(\rho') \left[\frac{\hat{\mathbf{v}}}{h_\mu} \frac{\partial}{\partial \mu} - \frac{\hat{\boldsymbol{\mu}}}{h_\nu} \frac{\partial}{\partial \nu} \right] E_n^m(\mu) E_n^m(\nu) \cdot \left(\mathbf{Q} \cdot \tilde{\mathbf{D}}_n^m(\mathbf{r}_0)\right). \end{aligned} \tag{5.70}$$

Utilizing formulae (3.52) and (3.53), we can perform a series of calculations that lead to

$$\left[\frac{\hat{\mathbf{v}}}{h_\mu} \frac{\partial}{\partial \mu} - \frac{\hat{\boldsymbol{\mu}}}{h_\nu} \frac{\partial}{\partial \nu} \right] E_n^m(\mu) E_n^m(\nu) = \sum_{i=1}^{3} \frac{E_1^i(\rho')}{\sqrt{\rho'^2 - \mu^2}\sqrt{\rho'^2 - \nu^2}} f_{ni}^m(\mu,\nu) \hat{\mathbf{x}}_i \tag{5.71}$$

with

$$f_{ni}^m(\mu,\nu) = \frac{E_1^2(\mu)E_1^2(\nu)E_1^3(\mu)E_1^3(\nu)}{h_1 h_2 h_3 (\mu^2 - \nu^2)} h_i E_1^i(\mu) E_1^i(\nu)$$

$$\cdot \left[\frac{\nu(E_n^m(\mu))'E_n^m(\nu)}{\nu^2 - a_1^2 + a_i^2} - \frac{\mu E_n^m(\mu)(E_n^m(\nu))'}{\mu^2 - a_1^2 + a_i^2}\right]$$

$$= \frac{h_i E_2^5(\mu) E_2^5(\nu)}{h_1 h_2 h_3 (\mu^2 - \nu^2)} \left[E_1^i(\mu)(E_1^i(\nu))'(E_n^m(\mu))'E_n^m(\nu)\right.$$

$$\left. - (E_1^i(\mu))'E_1^i(\nu) E_n^m(\mu)(E_n^m(\nu))'\right]. \tag{5.72}$$

Hence, (5.69) reads

$$U(\mathbf{r};\mathbf{r}_0) = -\sum_{n=1}^{\infty}\sum_{m=1}^{2n+1}\sum_{i=1}^{3} f_{ni}^m(\mu,\nu)\left(\mathbf{Q}\cdot\tilde{\mathbf{D}}_n^m(\mathbf{r}_0)\cdot\hat{\mathbf{x}}_i\right)\int_{\rho}^{+\infty}\frac{F_n^m(\rho')E_1^i(\rho')}{\sqrt{\rho'^2 - h_3^2}\sqrt{\rho'^2 - h_2^2}}\,d\rho' \tag{5.73}$$

where the coefficients $f_{ni}^m(\mu,\nu)$ depend only on the orientation point (μ,ν) of the point $\mathbf{r}$, while the dependence on the distance ρ enters via the integral factor. The quantity $\mathbf{Q}\cdot\tilde{\mathbf{D}}_n^m\cdot\hat{\mathbf{x}}_i$ depends solely on the dipole source $\{\mathbf{r}_0,\mathbf{Q}\}$. The expression (5.73) provides the magnetic potential in a separable form.

5.4 EEG for the 3-shell model

In the present section, we follow the steps that are described in Section 4.5, for the solution of the spherical model, in the case where the 3-shell model is based on the system of confocal ellipsoids.

In order to keep the presentation as closely related as possible to the spherical geometry, we will use the same notation as in the spherical case, introducing a cup "^" on the top of the corresponding coefficients. For the notation of the ellipsoidal harmonics, we refer to Section 3.3. We write (ρ,μ,ν) and (ρ_0,μ_0,ν_0) for the ellipsoidal coordinates of the observation point and the location of the monopole, respectively. Furthermore, we assume that the electric potentials $\hat{u}_i$ due to a dipolic excitation are obtained from the corresponding electric potentials $\hat{\upsilon}_i$ due to a monopolic excitation, via the action of the source depended directional derivative

$$\frac{1}{4\pi}\mathbf{Q}\cdot\nabla_{\mathbf{r}_0}$$

as in the expressions (2.26) and (2.27).

Then, in view of (5.8), we obtain

$$\hat{u}_c(\mathbf{r};\mathbf{r}_0) = \frac{1}{4\pi\sigma_c}(\mathbf{Q}\cdot\nabla_{\mathbf{r}_0})\left[\frac{1}{|\mathbf{r},\mathbf{r}_0|} + \hat{\upsilon}_c(\mathbf{r};\mathbf{r}_0)\right] \tag{5.74}$$

$$\hat{u}_s(\mathbf{r};\mathbf{r}_0) = \frac{1}{4\pi}(\mathbf{Q}\cdot\nabla_{\mathbf{r}_0})\hat{\upsilon}_i(\mathbf{r};\mathbf{r}_0) \tag{5.75}$$

for $j=f,b,s,e$ where the monopolic potentials admit the following expansions:

$$\hat{\upsilon}_c(\mathbf{r};\mathbf{r}_0)=\sum_{n=1}^{\infty}\sum_{m=1}^{2n+1}\hat{C}_{1n}^m(\rho_0)E_n^m(\rho)S_n^m(\mu_0,\nu_0)S_n^m(\mu,\nu) \tag{5.76}$$

$$\hat{\upsilon}_f(\mathbf{r};\mathbf{r}_0)=\sum_{n=1}^{\infty}\sum_{m=1}^{2n+1}\left[\hat{F}_{1n}^m(\rho_0)E_n^m(\rho)+\hat{F}_{2n}^m(\rho_0)F_n^m(\rho)\right]S_n^m(\mu_0,\nu_0)S_n^m(\mu,\nu) \tag{5.77}$$

$$\hat{\upsilon}_b(\mathbf{r};\mathbf{r}_0)=\sum_{n=1}^{\infty}\sum_{m=1}^{2n+1}\left[\hat{B}_{1n}^m(\rho_0)E_n^m(\rho)+\hat{B}_{2n}^m(\rho_0)F_n^m(\rho)\right]S_n^m(\mu_0,\nu_0)S_n^m(\mu,\nu) \tag{5.78}$$

$$\hat{\upsilon}_s(\mathbf{r},\mathbf{r}_0)=\sum_{n=1}^{\infty}\sum_{m=1}^{2n+1}\left[\hat{S}_{1n}^m(\rho_0)E_n^m(\rho)+\hat{S}_{2n}^m(\rho_0)F_n^m(\rho)\right]S_n^m(\mu_0,\nu_0)S_n^m(\mu,\nu) \tag{5.79}$$

$$\hat{\upsilon}_e(\mathbf{r},\mathbf{r}_0)=\sum_{n=1}^{\infty}\sum_{m=1}^{2n+1}\hat{E}_{2n}^m(\rho_0)F_n^m(\rho)S_n^m(\mu_0,\nu_0)S_n^m(\mu,\nu). \tag{5.80}$$

Comparing the expansions (4.78)–(4.82) for the sphere with the corresponding expansions (5.76)–(5.80) for the ellipsoid we see that all we need to do is to replace:

- the interior radial function r^n with the interior Lamé function $E_n^m(\rho)$
- the exterior radial function $r^{-(n+1)}$ with the exterior Lamé function $F_n^m(\rho)$
- the spherical surface harmonic $Y_n^m(\hat{\mathbf{r}})$ with the ellipsoidal surface harmonic $S_n^m(\mu,\nu)$

and insert the hat "^" on the top of every coefficient of the ellipsoidal expansions.

In the ellipsoidal system we have

$$\frac{\partial}{\partial n}=\hat{\mathbf{r}}\cdot\nabla_{\mathbf{r}}=\frac{1}{h_\rho}\frac{\partial}{\partial\rho} \tag{5.81}$$

and since in the relative transmission conditions the operator $\frac{\partial}{\partial n}$ appears in both sides of the equations, the factors $1/h_\rho$ are eliminated; thus we only need to replace $\frac{\partial}{\partial r}$ by $\frac{\partial}{\partial\rho}$. Using (5.74) and (5.75) and the ellipsoidal addition theorem

$$P_n(\hat{\mathbf{r}}\cdot\hat{\mathbf{r}}_0)=\sum_{m=1}^{2n+1}\frac{4\pi}{2n+1}\frac{1}{\gamma_n^m}S_n^m(\mu_0,\nu_0)S_n^m(\mu,\nu) \tag{5.82}$$

where γ_n^m are the normalizing constants of the surface ellipsoidal harmonics, we obtain the conditions below.

On the interface $\rho=c_1$ we have the conditions

$$\frac{1}{\sigma_c}\left[\frac{4\pi}{\gamma_n^m}\frac{F_n^m(c_1)}{2n+1}E_n^m(\rho_0)+E_n^m(c_1)\hat{C}_{1n}^m(\rho_0)\right]$$
$$=E_n^m(c_1)\hat{F}_{1n}^m(\rho_0)+F_n^m(c_1)\hat{F}_{2n}^m(\rho_0) \tag{5.83}$$

and

$$\frac{1}{\sigma_f}\left[\frac{4\pi}{\gamma_n^m}\frac{F_n^{m\prime}(c_1)}{2n+1}E_n^m(\rho_0)+E_n^{m\prime}(c_1)\hat{C}_{1n}^m(\rho_0)\right]$$
$$=E_n^{m\prime}(c_1)\hat{F}_{1n}^m(\rho_0)+F_n^{m\prime}(c_1)\hat{F}_{2n}^m(\rho_0). \tag{5.84}$$

On the interface $\rho=f_1$ we have

$$E_n^m(f_1)\hat{F}_{1n}^m(\rho_0)+F_n^m(f_1)\hat{F}_{2n}^m(\rho_0)=E_n^m(f_1)\hat{B}_{1n}^m(\rho_0)+F_n^m(f_1)\hat{B}_{2n}^m(\rho_0) \tag{5.85}$$

and

$$\sigma_f\left[E_n^{m\prime}(f_1)\hat{F}_{1n}^m(\rho_0)+F_n^{m\prime}(f_1)\hat{F}_{2n}^m(\rho_0)\right]=\sigma_b\left[E_n^{m\prime}(f_1)\hat{B}_{1n}^m(\rho_0)+F_n^{m\prime}(f_1)\hat{B}_{2n}^m(\rho_0)\right]. \tag{5.86}$$

On the interface $\rho=b_1$ we have

$$E_n^m(b_1)\hat{B}_{1n}^m(\rho_0)+F_n^m(b_1)\hat{B}_{2n}^m(\rho_0)=E_n^m(b_1)\hat{S}_{1n}^m(\rho_0)+F_n^m(b_1)\hat{S}_{2n}^m(\rho_0) \tag{5.87}$$

and

$$\sigma_b\left[E_n^{m\prime}(b_1)\hat{B}_{1n}^m(\rho_0)+F_n^{m\prime}(b_1)\hat{B}_{2n}^m(\rho_0)\right]=\sigma_s\left[E_n^{m\prime}(b_1)\hat{S}_{1n}^m(\rho_0)+F_n^{m\prime}(b_1)\hat{S}_{2n}^m(\rho_0)\right]. \tag{5.88}$$

Finally, on the interface $\rho=s_1$ we obtain the transmission condition

$$E_n^m(s_1)\hat{S}_{1n}^m(\rho_0)+F_n^m(s_1)\hat{S}_{2n}^m(\rho_0)=F_n^m(s_1)\hat{E}_{2n}^m(\rho_0) \tag{5.89}$$

and the Neumann condition

$$E_n^{m\prime}(s_1)\hat{S}_{1n}^m(\rho_0)+F_n^{m\prime}(s_1)\hat{S}_{2n}^m(\rho_0)=0. \tag{5.90}$$

We rewrite the conditions (5.83)–(5.90) in matrix form on each of the interfaces $\rho=c_1,f_1,b_1,s_1$ as follows:

we define the Wronskian matrix

$$\hat{\mathbf{W}}_n^m(x)=\begin{bmatrix}E_n^m(x) & F_n^m(x)\\ E_n^{m\prime}(x) & F_n^{m\prime}(x)\end{bmatrix} \tag{5.91}$$

which is invertible since the Lamé functions are linearly independent. Its inverse is given by

$$\left(\hat{\mathbf{W}}_n^m(x)\right)^{-1}=\frac{\sqrt{x^2-h_3^2}\sqrt{x^2-h_2^2}}{2n+1}\begin{bmatrix}-F_n^{m\prime}(x) & F_n^m(x)\\ E_n^{m\prime}(x) & -E_n^m(x)\end{bmatrix}. \tag{5.92}$$

We also define the parametric Wronskian matrix

$$\hat{\mathbf{W}}_n^m\left(x;\frac{\sigma_1}{\sigma_2}\right)=\begin{bmatrix} E_n^m(x) & F_n^m(x)\\ \frac{\sigma_1}{\sigma_2}E_n^{m\prime}(x) & \frac{\sigma_1}{\sigma_2}F_n^{m\prime}(x)\end{bmatrix}. \tag{5.93}$$

Furthermore, for simplicity of notation we suppress for the moment the indices n and m, setting $\hat{C}_1=\hat{C}_{1n}^m$, $\hat{F}_1=\hat{F}_{1n}^m$, $\hat{F}_2=\hat{F}_{2n}^m$, $\hat{B}_1=\hat{B}_{1n}^m$, $\hat{B}_2=\hat{B}_{2n}^m$, $\hat{S}_1=\hat{S}_{1n}^m$, $\hat{S}_2=\hat{S}_{2n}^m$, $\hat{E}_2=\hat{E}_{2n}^m$, $\hat{\mathbf{W}}(x)=\hat{\mathbf{W}}_n^m(x)$, $\hat{\mathbf{W}}(x;\frac{\sigma_1}{\sigma_2})=\hat{\mathbf{W}}_n^m(x;\frac{\sigma_1}{\sigma_2})$.

Then, the conditions (5.83)–(5.90) takee the form

$$\hat{\mathbf{W}}(c_1)\begin{bmatrix}\hat{F}_1(\rho_0)\\ \hat{F}_2(\rho_0)\end{bmatrix}=\frac{1}{\sigma_c}\hat{\mathbf{W}}\left(c_1;\frac{\sigma_c}{\sigma_f}\right)\begin{bmatrix}\hat{C}_1(\rho_0)\\ \frac{4\pi}{2n+1}\frac{E_n^m(\rho_0)}{\gamma_n^m}\end{bmatrix} \tag{5.94}$$

$$\hat{\mathbf{W}}(f_1)\begin{bmatrix}\hat{B}_1(\rho_0)\\ \hat{B}_2(\rho_0)\end{bmatrix}=\hat{\mathbf{W}}\left(f_1;\frac{\sigma_f}{\sigma_b}\right)\begin{bmatrix}\hat{F}_1(\rho_0)\\ \hat{F}_2(\rho_0)\end{bmatrix} \tag{5.95}$$

$$\hat{\mathbf{W}}(b_1)\begin{bmatrix}\hat{S}_1(\rho_0)\\ \hat{S}_2(\rho_0)\end{bmatrix}=\hat{\mathbf{W}}\left(b_1;\frac{\sigma_b}{\sigma_s}\right)\begin{bmatrix}\hat{B}_1(\rho_0)\\ \hat{B}_2(\rho_0)\end{bmatrix} \tag{5.96}$$

and

$$\hat{\mathbf{W}}(s_1)\begin{bmatrix}\hat{S}_1(\rho_0)\\ \hat{S}_2(\rho_0)\end{bmatrix}=\begin{bmatrix}0\\ F(s_1)\hat{E}_2(\rho_0)\end{bmatrix}. \tag{5.97}$$

The above notation leads to the expressions

$$\begin{bmatrix}\hat{F}_1\\ \hat{F}_2\end{bmatrix}=\frac{1}{\sigma_c}\hat{\mathbf{M}}_c(n,m)\begin{bmatrix}\hat{C}_1(\rho_0)\\ \frac{4\pi}{2n+1}\frac{E_n^m(\rho_0)}{\gamma_n^m}\end{bmatrix} \tag{5.98}$$

$$\begin{bmatrix}\hat{B}_1\\ \hat{B}_2\end{bmatrix}=\hat{\mathbf{M}}_f(n,m)\begin{bmatrix}\hat{F}_1\\ \hat{F}_2\end{bmatrix} \tag{5.99}$$

$$\begin{bmatrix}\hat{S}_1\\ \hat{S}_2\end{bmatrix}=\hat{\mathbf{M}}_b(n,m)\begin{bmatrix}\hat{B}_1\\ \hat{B}_2\end{bmatrix} \tag{5.100}$$

$$\begin{bmatrix}F_n^m(s_1)\hat{E}_2\\ 0\end{bmatrix}=\hat{\mathbf{M}}_s(n,m)\begin{bmatrix}\hat{S}_1\\ \hat{S}_2\end{bmatrix} \tag{5.101}$$

where

$$\hat{\mathbf{M}}_c(n,m) = \frac{c_2c_3}{2n+1}\begin{bmatrix} -E_n^m(c_1)F_n^{m\prime}(c_1) + \frac{\sigma_c}{\sigma_f}E_n^{m\prime}(c_1)F_n^m(c_1) & -\left(1-\frac{\sigma_c}{\sigma_f}\right)F_n^m(c_1)F_n^{m\prime}(c_1) \\ \left(1-\frac{\sigma_c}{\sigma_f}\right)E_n^m(c_1)E_n^{m\prime}(c_1) & E_n^{m\prime}(c_1)F_n^m(c_1) - \frac{\sigma_c}{\sigma_f}E_n^m(c_1)F_n^{m\prime}(c_1) \end{bmatrix} \tag{5.102}$$

$$\hat{\mathbf{M}}_f(n,m) = \frac{f_2f_3}{2n+1}\begin{bmatrix} -E_n^m(f_1)F_n^{m\prime}(f_1) + \frac{\sigma_f}{\sigma_b}E_n^{m\prime}(f_1)F_n^m(f_1) & -\left(1-\frac{\sigma_f}{\sigma_b}\right)F_n^m(f_1)F_n^{m\prime}(f_1) \\ \left(1-\frac{\sigma_f}{\sigma_b}\right)E_n^m(f_1)E_n^{m\prime}(f_1) & E_n^{m\prime}(f_1)F_n^m(f_1) - \frac{\sigma_f}{\sigma_b}E_n^m(f_1)F_n^{m\prime}(f_1) \end{bmatrix} \tag{5.103}$$

$$\hat{\mathbf{M}}_b(n,m) = \frac{b_2b_3}{2n+1}\begin{bmatrix} -E_n^m(b_1)F_n^{m\prime}(b_1) + \frac{\sigma_b}{\sigma_s}E_n^{m\prime}(b_1)F_n^m(b_1) & -\left(1-\frac{\sigma_b}{\sigma_s}\right)F_n^m(b_1)F_n^{m\prime}(b_1) \\ \left(1-\frac{\sigma_b}{\sigma_s}\right)E_n^m(b_1)E_n^{m\prime}(b_1) & E_n^{m\prime}(b_1)F_n^m(b_1) - \frac{\sigma_b}{\sigma_s}E_n^m(b_1)F_n^{m\prime}(b_1) \end{bmatrix} \tag{5.104}$$

$$\hat{\mathbf{M}}_s(n,m) = \begin{bmatrix} E_n^m(s_1) & F_n^m(s_1) \\ E_n^{m\prime}(s_1) & F_n^{m\prime}(s_1) \end{bmatrix} \tag{5.105}$$

and

$$\begin{bmatrix} \hat{m}_{11}(n,m) & \hat{m}_{12}(n,m) \\ \hat{m}_{21}(n,m) & \hat{m}_{22}(n,m) \end{bmatrix} = \hat{\mathbf{M}}(n,m) = \frac{1}{\sigma_c}\hat{\mathbf{M}}_s \cdot \hat{\mathbf{M}}_b \cdot \hat{\mathbf{M}}_f \cdot \hat{\mathbf{M}}_c. \tag{5.106}$$

In eq. (5.101), we substitute successively, the vector $[\hat{S}_1, \hat{S}_2]^{\mathrm{T}}$ from (5.100), the vector $[\hat{B}_1, \hat{B}_2]^{\mathrm{T}}$ from (5.99) and the vector $[\hat{F}_1, \hat{F}_2]^{\mathrm{T}}$ from (5.98), to obtain the vector equation

$$\begin{bmatrix} F_n^m(s_1)\hat{E}_2 \\ 0 \end{bmatrix} = \hat{\mathbf{M}}\}(n,m)\begin{bmatrix} \hat{C}_1(\rho_0) \\ \frac{4\pi}{2n+1}\frac{E_n^m(\rho_0)}{\gamma_n^m} \end{bmatrix} \tag{5.107}$$

where $\hat{\mathbf{M}}$ is given in (5.106). The above gives rise to two scalar equations:

$$F_n^m(s_1)\hat{E}_2 = \hat{m}_{11}(n,m)\hat{C}_1(\rho_0) + \hat{m}_{12}(n,m)\frac{4\pi}{2n+1}\frac{E_n^m(\rho_0)}{\gamma_n^m} \tag{5.108}$$

$$0 = \hat{m}_{21}(n,m)\hat{C}_1(\rho_0) + \hat{m}_{22}(n,m)\frac{4\pi}{2n+1}\frac{E_n^m(\rho_0)}{\gamma_n^m}. \tag{5.109}$$

Hence, the $\hat{C}_{1n}^m(\rho_0)$ coefficients are given by

$$\hat{C}_{1n}^m(\rho_0) = -\frac{4\pi}{2n+1}\frac{E_n^m(\rho_0)}{\gamma_n^m}\frac{\hat{m}_{22}(n,m)}{\hat{m}_{21}(n,m)}. \tag{5.110}$$

Substituting (5.110) in (5.98) we obtain the coefficients $\hat{F}_{1n}^m(\rho_0)$ and $\hat{F}_{2n}^m(\rho_0)$ in the form

$$\begin{bmatrix} \hat{F}_{1n}^m(\rho_0) \\ \hat{F}_{2n}^m(\rho_0) \end{bmatrix} = \frac{4\pi}{2n+1}\frac{E_n^m(\rho_0)}{\sigma_c\gamma_n^m}\hat{\mathbf{M}}_c(n,m)\begin{bmatrix} -\frac{\hat{m}_{22}(n,m)}{\hat{m}_{21}(n,m)} \\ 1 \end{bmatrix}. \tag{5.111}$$

Next, we can insert (5.111) in (5.99) to obtain the coefficients $\hat{B}_{1n}^m(\rho_0)$ and $\hat{B}_{2n}^m(\rho_0)$ and finally eq. (5.100) gives the coefficients $\hat{S}_{1n}^m(\rho_0)$ and $\hat{S}_{2n}^m(\rho_0)$. The coefficients $\hat{E}_{2n}^m(\rho_0)$ are obtained from eq. (5.108):

$$\begin{aligned} \hat{E}_{2n}^m(\rho_0) &= -\frac{4\pi}{2n+1}\frac{E_n^m(\rho_0)}{\gamma_n^m F_n^m(s_1)}\left(\frac{\hat{m}_{11}(n,m)\hat{m}_{22}(n,m)}{\hat{m}_{21}(n,m)} - \hat{m}_{12}(n,m)\right) \\ &= -\frac{4\pi}{2n+1}\frac{E_n^m(\rho_0)}{F_n^m(s_1)}\frac{\det\hat{\mathbf{M}}(n,m)}{\gamma_n^m\hat{m}_{21}(n,m)}. \end{aligned} \tag{5.112}$$

Therefore, the values of the electric potential on the boundary $\partial\Omega_s$ is given by

$$\begin{aligned} \hat{\upsilon}_s(\mathbf{r};\mathbf{r}_0)|_{\partial\Omega_s} &= \hat{\upsilon}_s(s_1,\mu,\nu;\rho_0,\mu_0,\nu_0) \\ &= -\sum_{n=1}^{\infty}\sum_{m=1}^{2n+1}\frac{4\pi}{2n+1}\frac{\det\hat{\mathbf{M}}(n,m)}{\gamma_n^m\hat{m}_{21}(n,m)}\mathbb{E}_n^m(\mathbf{r}_0)S_n^m(\mu,\nu). \end{aligned} \tag{5.113}$$

Using the reduction

$$E_n^m(x) \underset{e\to s}{\longrightarrow} x^n \tag{5.114}$$

$$F_n^m(x) \underset{e\to s}{\longrightarrow} \frac{1}{x^{n+1}} \tag{5.115}$$

$$E_n^{m\prime}(x) \underset{e\to s}{\longrightarrow} nx^{n-1} \tag{5.116}$$

$$F_n^{m\prime}(x) \underset{e\to s}{\longrightarrow} -\frac{n+1}{x^{n+2}} \tag{5.117}$$

where the rotation $e \to s$ means: ‘as the ellipsoid degenerates to the sphere’, we see that all the matrices $\hat{\mathbf{M}}_c$, $\hat{\mathbf{M}}_f$, $\hat{\mathbf{M}}_b$, $\hat{\mathbf{M}}_s$ reduce to the corresponding matrices $\mathbf{M}_c$, $\mathbf{M}_f$, $\mathbf{M}_b$, $\mathbf{M}_s$ for the sphere. The same is true for the RHS of (5.98) and (4.103):

$$\begin{bmatrix} \hat{F}_1 \\ \hat{F}_2 \end{bmatrix} \underset{e \to s}{\longrightarrow} \begin{bmatrix} F_1 \\ F_2 \end{bmatrix}. \tag{5.118}$$

Therefore, the forward problem for the ellipsoid is reduced to the forward problem for the sphere. In particular, the solution for the ellipsoid follows from the matrix

$$\hat{\mathbf{M}}(n,m) = \begin{bmatrix} \hat{m}_{11}(n,m) & \hat{m}_{12}(n,m) \\ \hat{m}_{21}(n,m) & \hat{m}_{22}(n,m) \end{bmatrix} = \frac{1}{\sigma_c} \hat{\mathbf{M}}_s \cdot \hat{\mathbf{M}}_b \cdot \hat{\mathbf{M}}_f \cdot \hat{\mathbf{M}}_c \tag{5.119}$$

Then we can obtain the ratio $\hat{m}_{22}/\hat{m}_{21}$ appearing in (5.111).

Finally, having the coefficients $\hat{C}_{1n}^m, \hat{F}_{1n}^m, \hat{F}_{2n}^m, \hat{B}_{1n}^m, \hat{B}_{2n}^m, \hat{S}_{1n}^m, \hat{S}_{2n}^m$, we obtain the monopolic potentials $\hat{v}_c, \hat{v}_f, \hat{v}_b, \hat{v}_s$ and from these we obtain the dipolic potentials $\hat{u}_c, \hat{u}_f, \hat{u}_b, \hat{u}_s$.

5.5 Direct solution of the EEG problem

We can also solve the EEG problem in ellipsoidal geometry for the 3-shell model via a direct solution of the 8×8 algebraic system that is generated by the transmission and boundary conditions.

In fact, using the expansion (5.8) we express the electric potential in Ω_c as

$$\begin{aligned} u_e(\rho,\mu,\nu;\rho_0,\mu_0,\nu_0) &= \frac{1}{4\pi\sigma_c} \mathbf{Q} \cdot \nabla_{\mathbf{r}_0} \frac{1}{|\mathbf{r}-\mathbf{r}_0|} + \sum_{n=1}^{\infty} \sum_{m=1}^{2n+1} t_n^m(\rho_0,\mu_0,\nu_0) \mathbb{E}_n^m(\rho,\mu,\nu) \\ &= \sum_{n=1}^{\infty} \sum_{m=1}^{2n+1} \left[t_n^m(\rho_0,\mu_0,\nu_0) + \frac{1}{\sigma_c \gamma_n^m} \mathbf{Q} \cdot \nabla_{\mathbf{r}_0} \mathbb{E}_n^m(\rho_0,\mu_0,\nu_0) I_n^m(\rho) \right] \mathbb{E}_n^m(\rho,\mu,\nu), \quad \rho < c_1 \end{aligned} \tag{5.120}$$

where the $n = 0$ term is annihilated by the operator $\mathbf{Q} \cdot \nabla_{\mathbf{r}_0}$.

Furthermore, the electric potentials within the shells take the form

$$\begin{aligned} u_f(\rho,\mu,\nu;\rho_0,\mu_0,\nu_0) = \sum_{n=1}^{\infty} \sum_{m=1}^{2n+1} & [r_n^m(\rho_0,\mu_0,\nu_0) \mathbb{E}_n^m(\rho,\mu,\nu) \\ & + q_n^m(\rho_0,\mu_0,\nu_0) \mathbb{F}_n^m(\rho,\mu,\nu)], \quad c_1 < \rho < f_1 \end{aligned} \tag{5.121}$$

$$\begin{aligned} u_b(\rho,\mu,\nu;\rho_0,\mu_0,\nu_0) = \sum_{n=1}^{\infty} \sum_{m=1}^{2n+1} & [k_n^m(\rho_0,\mu_0,\nu_0) \mathbb{E}_n^m(\rho,\mu,\nu) \\ & + m_n^m(\rho_0,\mu_0,\nu_0) \mathbb{F}_n^m(\rho,\mu,\nu)], \quad f_1 < \rho < b_1 \end{aligned} \tag{5.122}$$

$$\begin{aligned} u_s(\rho,\mu,\nu;\rho_0,\mu_0,\nu_0) = \sum_{n=1}^{\infty} \sum_{m=1}^{2n+1} & [g_n^m(\rho_0,\mu_0,\nu_0) \mathbb{E}_n^m(\rho,\mu,\nu) \\ & + h_n^m(\rho_0,\mu_0,\nu_0) \mathbb{F}_n^m(\rho,\mu,\nu)], \quad b_1 < \rho < s_1 \end{aligned} \tag{5.123}$$

Finally, the exterior electric potential has the expansion

$$u_e(\rho,\mu,\nu;\rho_0,\mu_0,\nu_0) = \sum_{n=1}^{\infty}\sum_{m=1}^{2m+1} f_n^m(\rho_0,\mu_0,\nu_0)\mathbb{F}_n^m(\rho,\mu,\nu), \quad \rho > s_1. \tag{5.124}$$

Introducing (5.76)–(5.80) in the transmition conditions

$$u_c(\rho,\mu,\nu;\mathbf{r}_0) = u_f(\rho,\mu,\nu;\mathbf{r}_0), \quad \rho = c_1 \tag{5.125}$$

$$\sigma_c \frac{\partial}{\partial n} u_c(\rho,\mu,\nu;\mathbf{r}_0) = \sigma_f \frac{\partial}{\partial n} u_f(\rho,\mu,\nu;\mathbf{r}_0), \quad \rho = c_1 \tag{5.126}$$

$$u_f(\rho,\mu,\nu;\mathbf{r}_0) = u_b(\rho,\mu,\nu;\mathbf{r}_0), \quad \rho = f_1 \tag{5.127}$$

$$\sigma_f \frac{\partial}{\partial n} u_f(\rho,\mu,\nu;\mathbf{r}_0) = \sigma_b \frac{\partial}{\partial n} u_b(\rho,\mu,\nu;\mathbf{r}_0), \quad \rho = f_1 \tag{5.128}$$

$$u_b(\rho,\mu,\nu;\mathbf{r}_0) = u_s(\rho,\mu,\nu;\mathbf{r}_0), \quad \rho = b_1 \tag{5.129}$$

$$\sigma_b \frac{\partial}{\partial n} u_b(\rho,\mu,\nu;\mathbf{r}_0) = \sigma_s \frac{\partial}{\partial n} u_s(\rho,\mu,\nu;\mathbf{r}_0), \quad \rho = b_1 \tag{5.130}$$

$$u_s(\rho,\mu,\nu;\mathbf{r}_0) = u_e(\rho,\mu,\nu;\mathbf{r}_0), \quad \rho = s_1 \tag{5.131}$$

as well as in the boundary condition

$$\frac{\partial}{\partial n} u_s(\rho,\mu,\nu;\mathbf{r}_0) = 0, \quad \rho = s_1 \tag{5.132}$$

and utilizing the orthogonality conditions of the surface harmonics $S_n^m(\mu,\nu)$ we end up with a linear system of order 8 for the 8 unknown coefficients $t_n^m(\mathbf{r}_0)$, $r_n^m(\mathbf{r}_0)$, $q_n^m(\mathbf{r}_0)$, $k_n^m(\mathbf{r}_0)$, $m_n^m(\mathbf{r}_0)$, $g_n^m(\mathbf{r}_0)$, $h_n^m(\mathbf{r}_0)$, $f_n^m(\mathbf{r}_0)$.

Setting

$$D_n^m(\mathbf{r}_0) = \frac{1}{\gamma_n^m}(\mathbf{Q}\cdot\nabla_{\mathbf{r}_0})\mathbb{E}_n^m(\mathbf{r}_0) \tag{5.133}$$

we can write this 8×8 system in the following form:

$$\left[t_n^m + \frac{1}{\sigma_c} D_n^m(\mathbf{r}_0) I_n^m(c_1)\right] E_n^m(c_1) = r_n^m E_n^m(c_1) + q_n^m F_n^m(c_1) \tag{5.134}$$

$$\begin{aligned}\sigma_c &\left[\frac{1}{\sigma_c} D_n^m(\mathbf{r}_0)\left(I_n^m(c_1)\right)' E_n^m(c_1) + \left(t_n^m + \frac{1}{\sigma_c} D_n^m(\mathbf{r}_0) I_n^m(c_1)\right)\left(E_n^m(c_1)\right)'\right] \\ &= \sigma_f\left[r_n^m\left(E_n^m(c_1)\right)' + q_n^m\left(F_n^m(c_1)\right)'\right]\end{aligned} \tag{5.135}$$

$$r_n^m E_n^m(f_1) + q_n^m F_n^m(f_1) = k_n^m E_n^m(f_1) + m_n^m F_n^m(f_1) \tag{5.136}$$

$$\sigma_f\left[r_n^m\left(E_n^m(f_1)\right)' + q_n^m\left(F_n^m(f_1)\right)'\right] = \sigma_b\left[k_n^m\left(E_n^m(f_1)\right)' + m_n^m\left(F_n^m(f_1)\right)'\right] \tag{5.137}$$

$$k_n^m E_n^m(b_1) + m_n^m F_n^m(b_1) = g_n^m E_n^m(b_1) + h_n^m F_n^m(b_1) \tag{5.138}$$

$$\sigma_b \left[k_n^m \left(E_n^m(b_1)\right)' + m_n^m \left(F_n^m(b_1)\right)'\right] = \sigma_s \left[g_n^m \left(E_n^m(b_1)\right)' + h_n^m \left(F_n^m(b_1)\right)'\right] \tag{5.139}$$

$$g_n^m E_n^m(s_1) + h_n^m F_n^m(s_1) = f_n^m F_n^m(s_1) \tag{5.140}$$

$$g_n^m \left(E_n^m(s_1)\right)' + h_n^m \left(F_n^m(s_1)\right) = 0. \tag{5.141}$$

Solving the system (5.134)–(5.141) in the way described in Section 5.4 or in some other suitable manner, we arrive at the potentials

$$\begin{aligned} u_c(\rho,\mu,\nu;\mathbf{r}_0) &= u_f(c_1,\mu,\nu;\mathbf{r}_0) \\ &+ \frac{1}{\sigma_c}\sum_{n=1}^{\infty}\sum_{m=1}^{2n+1} D_n^m(\mathbf{r}_0) I_n^m(\rho,c_1) \mathbb{E}_n^m(\rho,\mu,\nu), \quad \rho\in[h_2,c_1) \end{aligned} \tag{5.142}$$

$$\begin{aligned} u_f(\rho,\mu,\nu;\mathbf{r}_0) &= u_b(f_1,\mu,\nu;\mathbf{r}_0) \\ &+ \frac{1}{\sigma_f}\sum_{n=1}^{\infty}\sum_{m=1}^{2n+1} \frac{G_{2,n}^m}{G_{3,n}^m} D_n^m(\mathbf{r}_0) I_n^m(\rho,f_1) \mathbb{E}_n^m(\rho,\mu,\nu), \quad c_1<\rho<f_1 \end{aligned} \tag{5.143}$$

$$\begin{aligned} u_b(\rho,\mu,\nu;\mathbf{r}_0) &= u_s(b_1,\mu,\nu;\mathbf{r}_0) \\ &+ \frac{1}{\sigma_b}\sum_{n=1}^{\infty}\sum_{m=1}^{2n+1} \frac{G_{1,n}^m}{G_{3,n}^m} D_n^m(\mathbf{r}_0) I_n^m(\rho,b_1) \mathbb{E}_n^m(\rho,\mu,\nu), \quad f_1<\rho<b_1 \end{aligned} \tag{5.144}$$

$$\begin{aligned} u_s(\rho,\mu,\nu;\mathbf{r}_0) &= u_e(s_1,\mu,\nu;\mathbf{r}_0) \\ &+ \sum_{n=1}^{\infty}\sum_{m=1}^{2n+1} \frac{1}{G_{3,n}^m} D_n^m(\mathbf{r}_0) I_n^m(\rho,s_1) \mathbb{E}_n^m(\rho,\mu,\nu), \quad b_1<\rho<s_1 \end{aligned} \tag{5.145}$$

and

$$u_e(\rho,\mu,\nu;\mathbf{r}_0) = \sum_{n=1}^{\infty}\sum_{m=1}^{2n+1} \frac{1}{S_n^m}\frac{1}{G_{3,n}^m} D_n^m(\mathbf{r}_0) \frac{I_n^m(\rho)}{I_n^m(s_1)} \mathbb{E}_n^m(\rho,\mu,\nu), \quad \rho>s_1. \tag{5.146}$$

In the above expressions, the various constants are defined as follows:

$$C_n^m = E_n^m(c_1)\left(E_n^m(c_1)\right)' c_2 c_3 \tag{5.147}$$

$$\hat{F}_n^m = E_n^m(f_1)\left(E_n^m(f_1)\right)' f_2 f_3 \tag{5.148}$$

$$B_n^m = E_n^m(b_1)\left(E_n^m(b_1)\right)' b_2 b_3 \tag{5.149}$$

$$S_n^m = E_n^m(s_1)\left(E_n^m(s_1)\right)' s_2 s_3 \tag{5.150}$$

with the notation $\hat{F}_n^m$ introduced in order to be distinguished from the Lamé function F_n^m

$$I_n^m(x,y) = I_n^m(x) - I_n^m(y) = \int_x^y \frac{dt}{\left(E_n^m(t)\right)^2 \sqrt{t^2 - h_3^2}\sqrt{t^2 - h_2^2}}. \tag{5.151}$$

The constants $G_{i,n}^m$, $i = 1,2,3$, are defined as follows:

$$G_{1,n}^m = \sigma_b + (\sigma_b - \sigma_s)\left(I_n^m(b_1,s_1) + \frac{1}{S_n^m} - \frac{1}{B_n^m}\right)B_n^m \tag{5.152}$$

$$\begin{aligned} G_{2,n}^m &= \sigma_f + (\sigma_f - \sigma_b)\left(I_n^m(f_1,s_1) + \frac{1}{S_n^m} - \frac{1}{\hat{F}_n^m}\right)\hat{F}_n^m \\ &+ (\sigma_b - \sigma_s)\left(I_n^m(b_1,s_1) + \frac{1}{S_n^m} - \frac{1}{B_n^m}\right)B_n^m \\ &+ \frac{(\sigma_f - \sigma_b)(\sigma_b - \sigma_s)}{\sigma_b} I_n^m(f_1,b_1)\left(I_n^m(b_1,s_1) + \frac{1}{S_n^m} - \frac{1}{B_n^m}\right)B_n^m\hat{F}_n^m \end{aligned} \tag{5.153}$$

$$\begin{aligned} G_{3,n}^m &= \sigma_c + (\sigma_c - \sigma_f)\left(I_n^m(c_1,s_1) + \frac{1}{S_n^m} - \frac{1}{C_n^m}\right)C_n^m \\ &+ (\sigma_f - \sigma_b)\left(I_n^m(f_1,s_1) + \frac{1}{S_n^m} - \frac{1}{\hat{F}_n^m}\right)\hat{F}_n^m \\ &+ (\sigma_b - \sigma_s)\left(I_n^m(b_1,s_1) + \frac{1}{S_n^m} - \frac{1}{B_n^m}\right)B_n^m \\ &+ \frac{(\sigma_c - \sigma_f)(\sigma_f - \sigma_b)}{\sigma_f} I_n^m(c_1,f_1)\left(I_n^m(f_1,s_1) + \frac{1}{S_n^m} - \frac{1}{\hat{F}_n^m}\right)\hat{F}_n^m C_n^m \\ &+ \frac{(\sigma_c - \sigma_f)(\sigma_b - \sigma_s)}{\sigma_b} I_n^m(c_1,b_1)\left(I_n^m(b_1,s_1) + \frac{1}{S_n^m} - \frac{1}{B_n^m}\right)B_n^m C_n^m \\ &+ \frac{(\sigma_f - \sigma_b)(\sigma_b - \sigma_s)}{\sigma_b} I_n^m(f_1,b_1)\left(I_n^m(b_1,s_1) + \frac{1}{S_n^m} - \frac{1}{B_n^m}\right)B_n^m\hat{F}_n^m \\ &+ \frac{(\sigma_c - \sigma_f)(\sigma_f - \sigma_b)(\sigma_b - \sigma_s)}{\sigma_f\sigma_b} I_n^m(c_1,f_1)\left(I_n^m(f_1,b_1) - \frac{1}{\hat{F}_n^m}\right) \\ &\left(I_n^m(b_1,s_1) + \frac{1}{S_n^m} - \frac{1}{B_n^m}\right)B_n^m\hat{F}_n^m C_n^m. \end{aligned} \tag{5.154}$$

The parenthesis on the right hand side of eq. (5.152) represents the expression

$$\left(I_n^m(b_1,s_1) + \frac{1}{S_n^m} - \frac{1}{B_n^m}\right) = \frac{1}{2n+1}\left[\frac{(F_n^m(b_1))'}{(E_n^m(b_1))'} - \frac{(F_n^m(s_1))'}{(E_n^m(s_1))'}\right] \tag{5.155}$$

and similarly for the corresponding parentheses in the formulae (5.153) and (5.154).

Note the effect of the different conductivities in the expressions (5.152)–(5.154) besides the leading terms of the conductivity σ_b in $G^m_{1,n}$, σ_f in $G^m_{2,n}$ and σ_c in $G^m_{3,n}$, all other terms vanish when the conductivities become equal. In fact, since the conductivities of the cerebrum, the cerebrospinal fluid and the scalp are close to each other, while the conductivity of the skull is rather different, we can assume the following relations already mentioned at the end of Chapter 1:

$$\sigma_c \approx \frac{1}{3}\sigma_f \approx \frac{1}{12}\sigma_b \approx \sigma_s. \tag{5.156}$$

Thus, the only safe assumption is the vanishing of the terms that are proportional to $\sigma_c - \sigma_s$ but this difference does not appear in any term since the conductivity differences concern only neighboring shells.

The terms of degree 1 and 2 of the exterior electric potential in Cartesian form are given by

$$\begin{aligned} u_e(\mathbf{r};\mathbf{r}_0) = {} & \frac{1}{V_s}\sum_{m=1}^{3}\frac{Q_m}{G^m_{3,1}}\frac{I^m_n(\rho)}{I^m_n(s_1)}x_m \\ & - \frac{5}{6V_s(\Lambda_s - \Lambda'_s)}\sum_{m=1}^{3} Q_m x_{0m}\left[\frac{1}{G^1_{3,2}}\frac{I^1_2(\rho)}{I^1_2(s_1)}\frac{IE^1_2(\mathbf{r})}{\Lambda_s(\Lambda_s - s_m^2)}\right. \\ & \left. - \frac{1}{G^2_{3,2}}\frac{I^2_2(\rho)}{I^2_2(s_1)}\frac{IE^2_2(\mathbf{r})}{\Lambda'_s(\Lambda'_s - s_m^2)}\right] \\ & + \frac{5}{V_s}\sum_{\substack{i,j=1\\ i\neq j}}^{3}\frac{Q_i x_{0j}}{G^{6-i-j}_{3,2}\left(s_i^2 + s_j^2\right)}\frac{I^{i+j}_2(\rho)}{I^{i+j}_2(s_1)}x_i x_j + O\left(el^3\right) \end{aligned} \tag{5.157}$$

where

$$V_s = \frac{4\pi}{3}s_1 s_2 s_3 \tag{5.158}$$

and V_s is the volume of the ellipsoidal $\rho = s_1$; also

$$\left.\begin{matrix}\Lambda_s\\ \Lambda'_s\end{matrix}\right\} = s_1^2 - \frac{1}{3}\left[h_2^2 + h_3^2 \pm \sqrt{h_1^4 + h_2^2 h_3^2}\right] \tag{5.159}$$

are the roots of the equation

$$\sum_{m=1}^{3}\frac{1}{\Lambda_s - s_m^2} = 0. \tag{5.160}$$

5.6 MEG for the 3-shell model

The exterior magnetic field for the 3-shell model with an isolated dipole excitation located at $\mathbf{r}_0$ with dipole moment $\mathbf{Q}$ is obtained from (2.55) in the form

$$\begin{aligned}\mathbf{B}(\mathbf{r}) = {} & \frac{\hat{\mu}_0}{4\pi}\mathbf{Q}\times\nabla_{\mathbf{r}_0}\frac{1}{|\mathbf{r}-\mathbf{r}_0|}\\ & -\frac{\hat{\mu}_0}{4\pi}\int_{S_c}(\sigma_c u_c(\boldsymbol{\rho},\mathbf{r}_0)-\sigma_f u_f(\boldsymbol{\rho},\mathbf{r}_0))\mathbf{D}(\boldsymbol{\rho},\mathbf{r})ds(\boldsymbol{\rho})\\ & -\frac{\hat{\mu}_0}{4\pi}\int_{S_f}(\sigma_f u_f(\boldsymbol{\rho},\mathbf{r}_0)-\sigma_b u_b(\boldsymbol{\rho},\mathbf{r}_0))\mathbf{D}(\boldsymbol{\rho},\mathbf{r})ds(\boldsymbol{\rho})\\ & -\frac{\hat{\mu}_0}{4\pi}\int_{S_b}(\sigma_b u_b(\boldsymbol{\rho},\mathbf{r}_0)-\sigma_s u_s(\boldsymbol{\rho},\mathbf{r}_0))\mathbf{D}(\boldsymbol{\rho},\mathbf{r})ds(\boldsymbol{\rho})\\ & -\frac{\hat{\mu}_0}{4\pi}\int_{S_s}\sigma_s u_s(\boldsymbol{\rho},\mathbf{r}_0)\mathbf{D}(\boldsymbol{\rho},\mathbf{r})ds(\boldsymbol{\rho}) \qquad (5.161)\end{aligned}$$

where

$$\mathbf{D}(\boldsymbol{\rho},\mathbf{r}) = \hat{\mathbf{n}}(\boldsymbol{\rho})\times\nabla_{\boldsymbol{\rho}}\frac{1}{|\mathbf{r}-\boldsymbol{\rho}|} \qquad (5.162)$$

is the dipole field generated at the surface point $\boldsymbol{\rho}$ with the normal to the surface unit moment $\hat{\mathbf{n}}(\boldsymbol{\rho})$.

On the interfaces, we have the following continuity conditions:

$$u_c(\boldsymbol{\rho},\mathbf{r}_0) = u_f(\boldsymbol{\rho},\mathbf{r}_0), \quad \boldsymbol{\rho}\in S_c \qquad (5.163)$$

$$u_f(\boldsymbol{\rho},\mathbf{r}_0) = u_b(\boldsymbol{\rho},\mathbf{r}_0), \quad \boldsymbol{\rho}\in S_b \qquad (5.164)$$

$$u_b(\boldsymbol{\rho},\mathbf{r}_0) = u_s(\boldsymbol{\rho},\mathbf{r}_0), \quad \boldsymbol{\rho}\in S_s \qquad (5.165)$$

which we can insert in the expression (5.161) to rewrite this expression in the form

$$\begin{aligned}\mathbf{B}(\mathbf{r}) = {} & \frac{\hat{\mu}_0}{4\pi}\mathbf{Q}\times\frac{\mathbf{r}-\mathbf{r}_0}{|\mathbf{r}-\mathbf{r}_0|^3}+\frac{\hat{\mu}_0}{4\pi}(\sigma_f-\sigma_c)\mathbf{I}_c(\mathbf{r})\\ & +\frac{\hat{\mu}_0}{4\pi}(\sigma_b-\sigma_f)\mathbf{I}_f(\mathbf{r})+\frac{\hat{\mu}_0}{4\pi}(\sigma_s-\sigma_b)\mathbf{I}_b(\mathbf{r})-\frac{\hat{\mu}_0}{4\pi}\sigma_s\mathbf{I}_s(\mathbf{r}) \qquad (5.166)\end{aligned}$$

where

$$\mathbf{I}_i(\mathbf{r}) = \int_{S_i} u_i(\boldsymbol{\rho}', \mathbf{r}_0)\hat{\mathbf{n}}(\boldsymbol{\rho}') \times \frac{\mathbf{r} - \boldsymbol{\rho}'}{|\mathbf{r} - \boldsymbol{\rho}'|^3} ds(\boldsymbol{\rho}') \tag{5.167}$$

for $i = c, f, b, s$.

The explicit calculations for the above expressions are very complicated and for this reason we restrict our interest to the terms of order 1 and 2, that is, to the dipole and quadrupole terms, of the expansion of **B** in ellipsoidal harmonics.

Long and tedious calculations lead to the expansion

$$\begin{aligned}\hat{\mathbf{n}}(\boldsymbol{\rho}') \times \nabla_{\boldsymbol{\rho}'} \left.\frac{1}{|\mathbf{r} - \boldsymbol{\rho}'|}\right|_{\rho' = c_1} &= \hat{\mathbf{n}}(\boldsymbol{\rho}') \times \left.\frac{\mathbf{r} - \boldsymbol{\rho}'}{|\mathbf{r} - \boldsymbol{\rho}'|^3}\right|_{\rho' = c_1} \\ &= l_{c_1}(\mu', \nu') \left[\sum_{m=1}^{3} \boldsymbol{\beta}_m^c(\mathbf{r}) S_1^m(\mu', \nu') + \sum_{m=1}^{5} \boldsymbol{\delta}_m^c(\mathbf{r}) S_2^m(\mu', \nu') \right] + O(el'_3)\end{aligned} \tag{5.168}$$

where we recall that the notation $O(el_3)$ represents ellipsoidal terms of order greater or equal to 3 and $\rho' = c_1$ defines the surface S_c. The coefficients in (5.168) are given by the following formulae:

$$\boldsymbol{\beta}_m^c = 3\frac{c_1 c_2 c_3}{h_1 h_2 h_3} \frac{h_m}{c_m} (\hat{\mathbf{x}}_m \otimes \mathbf{r}) \underset{\cdot}{\times} \tilde{\mathbf{H}}_1(\rho), \quad m = 1, 2, 3 \tag{5.169}$$

$$\boldsymbol{\delta}_1^c = -\frac{c_1 c_2 c_3}{3(\Lambda_c - \Lambda'_c)} \tilde{\Lambda}_c \underset{\cdot}{\times} \tilde{\mathbf{F}}_c(\mathbf{r}) \tag{5.170}$$

$$\boldsymbol{\delta}_2^c = \frac{c_1 c_2 c_3}{3(\Lambda_c - \Lambda'_c)} \tilde{\Lambda}'_c \underset{\cdot}{\times} \tilde{\mathbf{F}}_c(\mathbf{r}) \tag{5.171}$$

$$\boldsymbol{\delta}_3^c = \frac{c_1 c_2 c_3}{h_1 h_2 h_3^2} \left[\frac{c_2}{c_1} \hat{\mathbf{x}}_1 \otimes \hat{\mathbf{x}}_2 + \frac{c_1}{c_2} \hat{\mathbf{x}}_2 \otimes \hat{\mathbf{x}}_1 \right] \underset{\cdot}{\times} \tilde{\mathbf{F}}_c(\mathbf{r}) \tag{5.172}$$

$$\boldsymbol{\delta}_4^c = \frac{c_1 c_2 c_3}{h_1 h_2^2 h_3} \left[\frac{c_3}{c_1} \hat{\mathbf{x}}_1 \otimes \hat{\mathbf{x}}_3 + \frac{c_1}{c_3} \hat{\mathbf{x}}_3 \otimes \hat{\mathbf{x}}_1 \right] \underset{\cdot}{\times} \tilde{\mathbf{F}}_c(\mathbf{r}) \tag{5.173}$$

$$\boldsymbol{\delta}_5^c = \frac{c_1 c_2 c_3}{h_1^2 h_2 h_3} \left[\frac{c_3}{c_2} \hat{\mathbf{x}}_2 \otimes \hat{\mathbf{x}}_3 + \frac{c_2}{c_3} \hat{\mathbf{x}}_3 \otimes \hat{\mathbf{x}}_2 \right] \underset{\cdot}{\times} \tilde{\mathbf{F}}_c(\mathbf{r}) \tag{5.174}$$

where the cross-dot product is taken in the sense

$$(\boldsymbol{a} \otimes \mathbf{b}) \underset{\cdot}{\times} (\mathbf{c} \otimes \mathbf{d}) = (\boldsymbol{a} \times \mathbf{c})(\mathbf{b} \cdot \mathbf{d}) \tag{5.175}$$

and the related polyadics are defined below:

$$\tilde{\Lambda}_c = \sum_{m=1}^{3} \frac{\hat{\mathbf{x}}_m \otimes \hat{\mathbf{x}}_m}{\Lambda_c - c_m^2} \tag{5.176}$$

$$\tilde{\Lambda'}_c = \sum_{m=1}^{3} \frac{\hat{\mathbf{x}}_m \otimes \hat{\mathbf{x}}_m}{\Lambda'_c - c_m^2} \tag{5.177}$$

$$\tilde{\mathbf{F}}_c(\mathbf{r}) = -\frac{I\!F_2^1(\mathbf{r})}{\Lambda_c - \Lambda'_c}\tilde{\Lambda}_c + \frac{I\!F_2^2(\mathbf{r})}{\Lambda_c - \Lambda'_c}\tilde{\Lambda'}_c + 15\mathbf{r} \otimes \mathbf{r} : \tilde{\tilde{\mathbf{H}}}_2(\rho) \tag{5.178}$$

$$\tilde{\mathbf{H}}_1(\rho) = \sum_{m=1}^{3} I_1^m(\rho)\hat{\mathbf{x}}_m \otimes \hat{\mathbf{x}}_m \tag{5.179}$$

$$\tilde{\tilde{\mathbf{H}}}_2(\rho) = \sum_{\substack{i,j=1\\ i\neq j}}^{3} I_2^{i+j}(\rho)\hat{\mathbf{x}}_i \otimes \hat{\mathbf{x}}_j \otimes \hat{\mathbf{x}}_i \otimes \hat{\mathbf{x}}_j. \tag{5.180}$$

As it was mentioned in Section 4.1, the constants

$$\left.\begin{matrix}\Lambda_c\\ \Lambda'_c\end{matrix}\right\} = c_1^2 - \frac{h_2^2 + h_3^2}{3} \pm \frac{2}{3}\sqrt{h_1^4 + h_2^2 h_3^2} \tag{5.181}$$

are the roots of the equation

$$\sum_{m=1}^{3} \frac{1}{\Lambda_c - c_m^2} = 0 \tag{5.182}$$

and correspond to the ellipsoid S_c. Furthermore, because of confocality the constants $\Lambda_f, \Lambda'_f, \Lambda_b, \Lambda'_b, \Lambda_s, \Lambda'_s$ corresponding to the surfaces S_f, S_b, S_c, respectively, satisfy the relations

$$\Lambda_c - \Lambda'_c = \Lambda_f - \Lambda'_f = \Lambda_b - \Lambda'_b = \Lambda_s - \Lambda'_s \tag{5.183}$$

$$\Lambda_c - c_m^2 = \Lambda_f - f_m^2 = \Lambda_b - b_m^2 = \Lambda_s - s_m^2 \tag{5.184}$$

$$\Lambda'_c - c_m^2 = \Lambda'_f - f_m^2 = \Lambda'_b - b_m^2 = \Lambda'_s - s_m^2 \tag{5.185}$$

for $m = 1, 2, 3$.

Because of the relations (5.183)–(5.185), the polyadics $\tilde{\Lambda}_f, \tilde{\Lambda'}_f, \tilde{\mathbf{F}}_f$ for the surface S_f, the polyadics $\tilde{\Lambda}_b, \tilde{\Lambda'}_b, \tilde{\mathbf{F}}_b$ for the surface S_b and $\tilde{\Lambda}_s, \tilde{\Lambda'}_s, \tilde{\mathbf{F}}_s$ for the surface S_s satisfy the following degeneracies:

$$\tilde{\Lambda}_c = \tilde{\Lambda}_f = \tilde{\Lambda}_b = \tilde{\Lambda}_s = \tilde{\Lambda} \tag{5.186}$$

$$\tilde{\Lambda'}_c = \tilde{\Lambda'}_f = \tilde{\Lambda'}_b = \tilde{\Lambda'}_s = \tilde{\Lambda'} \tag{5.187}$$

$$\tilde{\mathbf{F}}_c = \tilde{\mathbf{F}}_f = \tilde{\mathbf{F}}_b = \tilde{\mathbf{F}}_s = \tilde{\mathbf{F}} \tag{5.188}$$

where for simplicity of notation we have dropped the subindices and use the symbols $\tilde{\Lambda}, \tilde{\Lambda}', \tilde{\mathbf{F}}$.

Keeping only ellipsoidal harmonics of degree $n = 1$ and $n = 2$, it is possible to obtain the following expansion, derived analytically in [27]

$$u_c(c_1, \mu', \nu') = \sum_{m=1}^{3} J_m^c E_1^m(\mu') E_1^m(\nu') + \sum_{m=1}^{5} \Theta_m^c E_2^m(\mu') E_2^m(\nu') + O(el'_3) \tag{5.189}$$

where

$$J_m^c = \frac{3h_m}{4\pi h_1 h_2 h_3} A_1^m \mathbf{Q} \cdot \hat{\mathbf{x}}_m, \quad m = 1, 2, 3 \tag{5.190}$$

$$\Theta_1^c = -\frac{5}{4\pi(\Lambda_c - \Lambda'_c)} A_2^1 \mathbf{Q} \otimes \mathbf{r}_0 : \tilde{\Lambda} \tag{5.191}$$

$$\Theta_2^c = \frac{5}{4\pi(\Lambda_c - \Lambda'_c)} A_2^2 \mathbf{Q} \otimes \mathbf{r}_0 : \tilde{\Lambda}' \tag{5.192}$$

$$\Theta_{i+j}^c = \frac{5 h_i h_j}{4\pi (h_1 h_2 h_3)^2} A_2^{i+j} \mathbf{Q} \cdot \left(x_{0i} \hat{\mathbf{x}}_j + x_{0j} \hat{\mathbf{x}}_i \right) \tag{5.193}$$

with $i, j \in \{1, 2, 3\}$ and $i \neq j$.

Furthermore,

$$\begin{aligned} A_n^m = \frac{E_n^m(c_1)}{G_{3,n}^m} \Bigg[& I_n^m(c_1, s_1) + \frac{1}{S_n^m} + \frac{\sigma_f - \sigma_b}{\sigma_f} I_n^m(c_1, f_1) \left(I_n^m(f_1, s_1) + \frac{1}{S_n^m} - \frac{1}{\hat{F}_n^m} \right) \hat{F}_n^m \\ & + \frac{\sigma_b - \sigma_s}{\sigma_b} I_n^m(c_1, b_1) \left(I_n^m(b_1, s_1) + \frac{1}{S_n^m} - \frac{1}{B_n^m} \right) B_n^m \\ & + \frac{(\sigma_f - \sigma_b)(\sigma_b - \sigma_s)}{\sigma_b \sigma_f} I_n^m(c_1, f_1) \left(I_n^m(b_1, s_1) + \frac{1}{S_n^m} - \frac{1}{B_n^m} \right) \\ & \cdot \left(I_n^m(f_1, b_1) - \frac{1}{F_n^m} \right) B_n^m \hat{F}_n^m \Bigg] \end{aligned} \tag{5.194}$$

for $n = 1, 2$ and $m = 1, 2, ..., 2n+1$, where the definitions of $I_n^m, \hat{F}_n^m, B_n^m, S_n^m$ are given in (5.148)–(5.151) and $G_{3,n}^m$ is given in (5.154). Inserting in (5.167), $u_c(c_1, \mu', \nu')$ and the expansion for the dipole field, and using orthogonality, we obtain

$$\mathbf{I}_c(\mathbf{r}) = \sum_{m=1}^{3} J_m^c \gamma_1^m \boldsymbol{\beta}_m^c(\mathbf{r}) + \sum_{m=1}^{5} \Theta_m^c \gamma_2^m \boldsymbol{\delta}_m^c(\mathbf{r}) + O(el_3). \tag{5.195}$$

In expanded form, this equation becomes

$$\begin{aligned}\mathbf{I}_c(\mathbf{r}) &= \sum_{m=1}^{3} \frac{3c_1c_2c_3}{c_m} A_1^m (\mathbf{Q}\cdot\hat{\mathbf{x}}_m)(\hat{\mathbf{x}}_m \otimes \mathbf{r}) \underset{\cdot}{\times} \mathbf{H}_1(\rho) \\ &- \frac{2(\Lambda_c - c_1^2)(\Lambda_c - c_2^2)(\Lambda_c - c_3^2)}{3(\Lambda_c - \Lambda'_c)} c_1c_2c_3A_2^1 \mathbf{Q}\otimes\mathbf{r}_0 : \tilde{\Lambda}\otimes\tilde{\Lambda} \underset{\cdot}{\times} \tilde{\mathbf{F}}(\mathbf{r}) \\ &+ \frac{2(\Lambda'_c - c_1^2)(\Lambda'_c - c_2^2)(\Lambda'_c - c_3^2)}{3(\Lambda_c - \Lambda'_c)} c_1c_2c_3A_2^2 \mathbf{Q}\otimes\mathbf{r}_0 : \tilde{\Lambda}'\otimes\tilde{\Lambda}' \underset{\cdot}{\times} \tilde{\mathbf{F}}(\mathbf{r}) \\ &+ \mathbf{Q}\otimes\mathbf{r}_0 : \left[\sum_{\substack{i,j=1\\ i\neq j}}^{3} \frac{(\hat{\mathbf{x}}_i\otimes\hat{\mathbf{x}}_j + \hat{\mathbf{x}}_j\otimes\hat{\mathbf{x}}_i)\otimes\left(c_i^2\hat{\mathbf{x}}_j\otimes\hat{\mathbf{x}}_i + c_j^2\hat{\mathbf{x}}_i\otimes\hat{\mathbf{x}}_j\right)}{c_ic_j} c_1c_2c_3A_2^{i+j} \right] \\ &+ O(el_3).\end{aligned} \tag{5.196}$$

The previous calculations summarize the steps that lead to the expression (5.196) for $\mathbf{I}_c(\mathbf{r})$. Following similar arguments we can evaluate the integrals $\mathbf{I}_f(\mathbf{r})$, $\mathbf{I}_b(\mathbf{r})$ and $\mathbf{I}_s(\mathbf{r})$. After these four integrals are evaluated and substituted in the expression (5.166), we then perform some further calculations and reductions that can be found analytically in reference [27]. In this way, we finally arrive at the expression

$$\begin{aligned}\mathbf{B}(\mathbf{r}) &= \frac{\hat{\mu}_0}{4\pi}(\mathbf{d}_c + \mathbf{d}_{2f} - \mathbf{d}_{2c} + \mathbf{d}_{1b} - \mathbf{d}_{1f} + \mathbf{d}_{0s} - \mathbf{d}_{0b}) \\ &\cdot \left(\frac{I\!F_2^1(\rho,\mu,\nu)}{\Lambda_c - \Lambda'_c}\tilde{\Lambda} - \frac{I\!F_2^2(\rho,\mu,\nu)}{\Lambda_c - \Lambda'_c}\tilde{\Lambda}' - 15\sum_{i,j=1}^{3} I_2^{i+j}(\rho)x_ix_j\hat{\mathbf{x}}_i\otimes\hat{\mathbf{x}}_j \right) + O(el_3)\end{aligned} \tag{5.197}$$

where

$$\mathbf{d}_c = \left(\mathbf{Q}\cdot\tilde{\mathbf{M}}(c_1)\times\mathbf{r}_0\right)\cdot\tilde{\mathbf{N}}(c_1) \tag{5.198}$$

$$\mathbf{d}_{2c} = \left(\mathbf{Q}\cdot\tilde{\mathbf{M}}(c_1)\times\mathbf{r}_0\right)\cdot\tilde{\mathbf{N}}_2(c_1) \tag{5.199}$$

$$\mathbf{d}_{if} = \left(\mathbf{Q}\cdot\tilde{\mathbf{M}}(f_1)\times\mathbf{r}_0\right)\cdot\tilde{\mathbf{N}}_i(f_1), \quad i = 1,2 \tag{5.200}$$

$$\mathbf{d}_{ib} = \left(\mathbf{Q}\cdot\tilde{\mathbf{M}}(b_1)\times\mathbf{r}_0\right)\cdot\tilde{\mathbf{N}}_i(b_1), \quad i = 0,1 \tag{5.201}$$

$$\mathbf{d}_{0s} = \left(\mathbf{Q}\cdot\tilde{\mathbf{M}}(s_1)\times\mathbf{r}_0\right)\cdot\tilde{\mathbf{N}}_0(s_1) \tag{5.202}$$

$$\tilde{\mathbf{M}}(c_1) = \sum_{i=1}^{3} c_i^2\hat{\mathbf{x}}_i\otimes\hat{\mathbf{x}}_i \tag{5.203}$$

and similarly with $\tilde{\mathbf{M}}(f_1)$, $\tilde{\mathbf{M}}(b_1)$ and $\tilde{\mathbf{M}}(s_1)$.

Furthermore,

$$\tilde{\mathbf{N}}(c_1) = \sum_{i=1}^{3} \frac{\hat{\mathbf{x}}_i \otimes \hat{\mathbf{x}}_i}{c_1^2 + c_2^2 + c_3^2 - c_i^2} \tag{5.204}$$

$$\tilde{\mathbf{N}}_2(c_1) = \sum_{i=1}^{3} \frac{G_{2,2}^{6-i}}{G_{3,2}^{6-i}} \frac{\hat{\mathbf{x}}_i \otimes \hat{\mathbf{x}}_i}{c_1^2 + c_2^2 + c_3^2 - c_i^2} \tag{5.205}$$

$$\tilde{\mathbf{N}}_1(c_1) = \sum_{i=1}^{3} \frac{G_{1,2}^{6-i}}{G_{3,2}^{6-i}} \frac{\hat{\mathbf{x}}_i \otimes \hat{\mathbf{x}}_i}{c_1^2 + c_2^2 + c_3^2 - c_i^2} \tag{5.206}$$

$$\tilde{\mathbf{N}}_0(c_1) = \sum_{i=1}^{3} \frac{\sigma_s}{G_{3,2}^{6-i}} \frac{\hat{\mathbf{x}}_i \otimes \hat{\mathbf{x}}_i}{c_1^2 + c_2^2 + c_3^2 - c_i^2} \tag{5.207}$$

and similarly for the corresponding $\tilde{\mathbf{N}}$'s when the c's are replaced by the f's, the b's and the s's.

If we insert the values of $G_{1,n}^m, G_{2,n}^m, G_{3,n}^m$ in the expressions (3.148)–(3.151) we obtain the values of the coefficients C_{in}^{km} appearing in the expression (3.96)–(3.99).

5.7 Problems

Problem 1 Solve the Dirichlet problem (5.3)–(5.5) to obtain equation (5.17).

Problem 2 Define the Cartesian expression (5.15) starting with the expression (5.14).

Problem 3 Prove the asymptotic expansion (5.34)–(5.37).

Problem 4 Prove the expansion (5.40)–(5.46).

Problem 5 Starting from the asymptotic condition

$$u_e(\mathbf{r}, \mathbf{r}_0) = O\left(\frac{1}{r^2}\right), \quad r \to +\infty$$

prove that all potentials u_c, u_f, u_b, u_s, u_e have ellipsoidal expansions with leading terms a dipole field (that is they start with $n = 1$).

Problem 6 Provide the details that lead to the expansions (5.57) and (5.58).

Problem 7 Assuming the representation $\mathbf{B}(\mathbf{r}) = \nabla \times \mathbf{A}(\mathbf{r})$ for the solenoidal field $\mathbf{B}(\mathbf{r})$, prove that

$$\frac{\mu_0}{4\pi}\mathbf{A}(\mathbf{r}) = \frac{\mathbf{Q}}{|\mathbf{r}-\mathbf{r}_0|} - \sigma \int_{\rho'=a_1} u^-(\mathbf{r}') \frac{\hat{\rho}'}{|\mathbf{r}-\mathbf{r}'|} ds(\mathbf{r}')$$

where $\mathbf{A}$ has the ellipsoidal expansion.

$$\mathbf{A}(\mathbf{r}) = -\hat{\mu}_0 \sum_{n=1}^{\infty} \sum_{m=1}^{2n+1} \left(\mathbf{Q} \cdot \tilde{\mathbf{D}}_n^m(r_0)\right) \mathbb{F}_n^m(\mathbf{r}).$$

Problem 8 Prove that the vector field $\mathbf{A}$ defined in Problem 7 is solenoidal, that is, $\nabla \cdot \mathbf{A} = 0$.

Problem 9 Prove the identity (5.70).

Problem 10 Prove the form (5.72) for the orientation depended coefficients $f_{ni}^m(\mu, \nu)$.

Problem 11 Verify that the electric potentials u_c, u_f, u_b, u_s, u_e given by (5.142), (5.143), (5.144), (5.145), (5.146), respectively, together with notations (5.133) and (5.147)–(5.154) satisfy the transmition conditions (5.125)–(5.131) and the boundary condition (5.132).

Problem 12 Prove the following identifies:

i) $$\frac{(\Lambda - a_1^2)(\Lambda - a_2^2)(\Lambda - a_3^2)}{(\Lambda - a_i^2)^2} = (\Lambda - a_i^2) - \sum_{m=1}^{3} (\Lambda - a_m^2), \quad i = 1, 2, 3.$$

ii) The same as in i) with Λ replaced by Λ'.

iii) $$\frac{(\Lambda - a_1^2)(\Lambda - a_2^2)(\Lambda - a_3^2)}{3(\Lambda - \Lambda')} \tilde{\Lambda} \otimes \tilde{\Lambda} - \frac{(\Lambda' - a_1^2)(\Lambda' - a_2^2)(\Lambda' - a_3^2)}{3(\Lambda - \Lambda')} \tilde{\Lambda}' \otimes \tilde{\Lambda}'$$
$$= \frac{1}{3}\mathbf{I} \otimes \mathbf{I} - \sum_{i=1}^{3} \hat{\mathbf{x}}_i \otimes \hat{\mathbf{x}}_i \otimes \hat{\mathbf{x}}_i \otimes \hat{\mathbf{x}}_i.$$

iv) $$\frac{(\hat{\mathbf{x}}_i \otimes \hat{\mathbf{x}}_j + \hat{\mathbf{x}}_j \otimes \hat{\mathbf{x}}x_i) \otimes \left(a_j^2 \hat{\mathbf{x}}_i \otimes \hat{\mathbf{x}}_j + a_i^2 \hat{\mathbf{x}}_j \otimes \hat{\mathbf{x}}_i\right)}{a_i^2 + a_j^2}$$
$$= \hat{\mathbf{x}}_i \otimes \hat{\mathbf{x}}_j \otimes \hat{\mathbf{x}}_i \otimes \hat{\mathbf{x}}_j + \hat{\mathbf{x}}_j \otimes \hat{\mathbf{x}}_i \otimes \hat{\mathbf{x}}_j \otimes \hat{\mathbf{x}}_i$$
$$+ \frac{\left(a_i^2 \hat{\mathbf{x}}_i \otimes \hat{\mathbf{x}}_j - a_j^2 \hat{\mathbf{x}}_j \otimes \hat{\mathbf{x}}_i\right) \otimes (\hat{\mathbf{x}}_j \otimes \hat{\mathbf{x}}_i - \hat{\mathbf{x}}_i \otimes \hat{\mathbf{x}}_j)}{a_i^2 + a_j^2}$$

for $i \neq j$.

Problem 13 Starting from the integral term $\mathbf{I}_c(\mathbf{r})$ as given by (5.167), utilize the relations given in the text to provide the details that lead to the expression (5.195).

Problem 14 Establish the expansion (5.196).

Problem 15 Write the expansions (5.195) and (5.196) for the integrals $\mathbf{I}_f(\mathbf{r})$, $\mathbf{I}_b(\mathbf{r})$, $\mathbf{I}_s(\mathbf{r})$.

Chapter 6
Determination of localized current via EEG or MEG data

In this chapter and the following chapter, we discuss algorithms that invert the recorded electroencephalographic (EEG) and magnetoencephalographic (MEG) data for the purpose of identifying the localized neuronal current that has generated these data.

We consider the spherical and the ellipsoidal models of the brain–head system, and data that are obtained from EEG or MEG. The general idea is that we expand the data in the geometrically appropriate system of complete orthogonal eigenfunctions and compare the coefficients of this expansion with the knowing structure of the corresponding coefficients obtained from the solution of the related forward problem. This algorithm leads to a set of nonlinear, in general, equations for the structure of the unknown current.

It is important to mention here the fundamental work of Albanese and Monk [1], which proved that it is impossible to identify completely a current within a conducting medium when the support of the current is three-dimensional. On the other hand, they demonstrated that it is possible to identify the support of a current that is a subset of two-, one- or zero-dimensional space. Indeed, as we have shown in [28] if we assume that the current is localized in a small sphere we can find the center of this sphere but it is impossible to calculate its radius. The inversion procedure identifies the sphere as an equivalent dipole located at its center. For this reason, we propose two different inversion techniques. The first concerns the case where the current consists of a number of isolated dipoles. The second assumes a continuously distributed current within the whole brain tissue, which, as we have demonstrated, is not uniquely identifiable. In this case, we propose an optimization method that seeks the current source that has minimum L^2- norm, frequently referred to as minimum energy-norm.

Cases where the support of the current has low dimensionality and lies on a curve, or on a surface patch, are reported in refs. [30, 31, 33, 34, 38]. Furthermore, refs. [39, 41] discuss the conditions under which we can discriminate among single, double or multiple dipolic excitations, while refs. [36, 40] concern the influence of the boundary perturbations of the geometrical model on the inversion algorithms.

The inversion methods that identify localized dipoles are presented in this chapter while the continuously distributed currents are discussed in Chapter 7.

https://doi.org/10.1515/9783110547535-007

6.1 EEG for isolated dipoles in a sphere

We consider the three-shell spherical model with radii

$$0 < c < f < b < s < +\infty$$

where $r = c$ is the boundary of the cerebrum, and $r = f, b, s$ are the external boundaries of the domains of the cerebrospinal fluid, bone and skin, respectively. Furthermore, we assume that we are looking for an equivalent dipole located at $\mathbf{r}_0$ with moment $\mathbf{Q}(\mathbf{r}_0)$.

For the inverse EEG problem, the data are the values of the electric potential u_e on the boundary of the skin $r = s$. Given the source-dependent directional derivative operator

$$D_e(\mathbf{r_0}) = \mathbf{Q}(\mathbf{r_0}) \cdot \nabla_{\mathbf{r_0}} \tag{6.1}$$

and formulae (4.82), we expand the above EEG data in the form

$$u_e(s\hat{\mathbf{r}}; \mathbf{r}_0) = D_e(\mathbf{r_0}) v_e(s\,\hat{\mathbf{r}}; \mathbf{r}_0) = \sum_{n=1}^{\infty} \sum_{m=-n}^{n} A_n^m(\mathbf{r_0}) Y_n^m(\hat{\mathbf{r}}) \tag{6.2}$$

with the known coefficients $A_n^m(\mathbf{r_0})$ representing the quantities

$$A_n^m(\mathbf{r_0}) = -\frac{1}{2n+1} \frac{\det \mathbf{M}(n)}{m_{21}(n)} \left(D_e(\mathbf{r_0}) r_0^n Y_n^{m*}(\hat{\mathbf{r}}_0) \right) \tag{6.3}$$

where $\mathbf{M}$ is the matrix given in eq. (4.107). The matrix $\mathbf{M}$ involves the conductivities $\sigma_c, \sigma_f, \sigma_b, \sigma_s$ and the radii c, f, b, s; therefore, it is known from the physics and the geometry of the model. The normalized complex spherical harmonics are defined by

$$Y_n^m\left(\hat{\mathbf{r}}\right) = \sqrt{\frac{2n+1}{4\pi} \frac{(n-|m|)!}{(n+|m|)!}} P_n^{|m|}(\cos\theta) e^{im\varphi} \tag{6.4}$$

where $n = 0, 1, 2, \ldots$ and $|m| \leq n$, and P_n^m the Legendre functions of the first kind. Since we are seeking the six scalars which are the coordinates of the vectors $\mathbf{Q}$ and $\mathbf{r}_0$ we are going to utilize only six of the known coefficients A_n^m in order to obtain a 6×6, nonlinear, algebraic system. Using Cartesian coordinates we have

$$\mathbf{Q} = Q_1 \hat{\mathbf{x}}_1 + Q_2 \hat{\mathbf{x}}_2 + Q_3 \hat{\mathbf{x}}_2 \tag{6.5}$$

$$\mathbf{r}_0 = x_{01} \hat{\mathbf{x}}_1 + x_{02} \hat{\mathbf{x}}_2 + x_{03} \hat{\mathbf{x}}_3 \tag{6.6}$$

and

$$D_e(\mathbf{r_0}) = \mathbf{Q}(\mathbf{r_0}) \cdot \nabla_{\mathbf{r_0}} = Q_1 \frac{\partial}{\partial x_{01}} + Q_2 \frac{\partial}{\partial x_{02}} + Q_3 \frac{\partial}{\partial x_{03}}. \tag{6.7}$$

Let us use the six known coefficients $A_1^{-1}, A_1^0, A_1^1, A_2^{-1}, A_2^0, A_2^{-1}$. Then we find the following equations:

$$D_e(\mathbf{r_0}) r_0 Y_1^{-1*}(\hat{\mathbf{r}}_0) = \mathbf{Q} \cdot \nabla_{\mathbf{r_0}} \sqrt{\frac{3}{8\pi}} r_0 \sin\theta_0 e^{i\varphi_0}$$

$$= \sqrt{\frac{3}{8\pi}} \mathbf{Q} \cdot \nabla_{\mathbf{r_0}} (x_{01} + i x_{02}) = \sqrt{\frac{3}{8\pi}} (Q_1 + iQ_2) \tag{6.8}$$

$$D_e(\mathbf{r_0}) r_0 Y_1^{0*}(\hat{\mathbf{r}}_0) = \mathbf{Q} \cdot \nabla_{\mathbf{r_0}} \sqrt{\frac{3}{4\pi}} r_0 \cos\theta_0 = \sqrt{\frac{3}{4\pi}} \mathbf{Q} \cdot \nabla_{\mathbf{r_0}} x_{03} = \sqrt{\frac{3}{4\pi}} Q_3 \tag{6.9}$$

$$D_e(\mathbf{r_0}) r_0 Y_1^{1*}(\hat{\mathbf{r}}_0) = \mathbf{Q} \cdot \nabla_{\mathbf{r_0}} \sqrt{\frac{3}{8\pi}} r_0 \sin\theta_0 e^{-i\varphi_0}$$

$$= \sqrt{\frac{3}{8\pi}} \mathbf{Q} \cdot \nabla_{\mathbf{r_0}} (x_{01} - i x_{02}) = \sqrt{\frac{3}{8\pi}} (Q_1 - iQ_2) \tag{6.10}$$

$$D_e(\mathbf{r_0}) r_0^2 Y_2^{-1*}(\hat{\mathbf{r}}_0) = \mathbf{Q} \cdot \nabla_{\mathbf{r_0}} \sqrt{\frac{15}{8\pi}} r_0^2 \sin\theta_0 \cos\theta_0 e^{i\varphi_0} = \sqrt{\frac{15}{8\pi}} \mathbf{Q} \cdot \nabla_{\mathbf{r_0}} (x_{01}x_{03} + i x_{01}x_{02})$$

$$= \sqrt{\frac{15}{8\pi}} [Q_1(ix_{02} + x_{03}) + iQ_2 x_{01} + Q_3 x_{01}] \tag{6.11}$$

$$D_e(\mathbf{r_0}) r_0^2 Y_2^{0*}(\hat{\mathbf{r}}_0) = \mathbf{Q} \cdot \nabla_{\mathbf{r_0}} \sqrt{\frac{5}{16\pi}} r_0^2 (3\cos^2\theta_0 - 1)$$

$$= \sqrt{\frac{5}{16\pi}} \mathbf{Q} \cdot \nabla_{\mathbf{r_0}} (-x_{01}^2 - x_{02}^2 + 2x_{03}^2) = \sqrt{\frac{5}{4\pi}} [-Q_1 x_{01} - Q_2 x_{02} + 2Q_3 x_{03}] \tag{6.12}$$

$$D_e(\mathbf{r_0}) r_0^2 Y_2^{1*}(\hat{\mathbf{r}}_0) = \mathbf{Q} \cdot \nabla_{\mathbf{r_0}} \sqrt{\frac{15}{8\pi}} r_0^2 \sin\theta_0 \cos\theta_0 e^{-i\varphi_0}$$

$$= \sqrt{\frac{15}{8\pi}} \mathbf{Q} \cdot \nabla_{\mathbf{r_0}} (x_{01}x_{03} - i x_{01}x_{02}) = \sqrt{\frac{15}{8\pi}} [Q_1(-ix_{02} + x_{03}) - iQ_2 x_{01} + Q_3 x_{01}]. \tag{6.13}$$

From eqs. (6.3), (6.8)–(6.10) we obtain the moment $\mathbf{Q}$ in the form

$$\mathbf{Q} = \frac{1}{2T_1} \left(A_1^1 + A_1^{-1}, iA_1^1 - iA_1^{-1}, \sqrt{2} A_1^0 \right) \tag{6.14}$$

where

$$T_1 = -\sqrt{\frac{1}{24\pi} \frac{\det \mathbf{M}(1)}{m_{21}(1)}}. \tag{6.15}$$

With the three components Q_1, Q_2, Q_3 known, eqs. (6.3), (6.11)–(6.13) imply the following 3×3 linear system for the determination of the components x_{01}, x_{02}, x_{03}:

$$Q_3 x_{01} + Q_1 x_{03} = \frac{1}{2T_2}\left(A_2^1 + A_2^{-1}\right) \tag{6.16}$$

$$Q_2 x_{01} + Q_1 x_{02} = \frac{i}{2T_2}\left(A_2^1 - A_2^{-1}\right) \tag{6.17}$$

$$Q_1 x_{01} + Q_2 x_{02} - 2Q_3 x_{03} = -\frac{\sqrt{6}}{2T_2} A_2^0 \tag{6.18}$$

where

$$T_2 = -\sqrt{\frac{3}{40\pi}\frac{\det \mathbf{M}(2)}{m_{21}(2)}}. \tag{6.19}$$

In other for the system (6.16)–(6.18) to have a unique solution we need the condition

$$Q_1 Q_2^2 \neq 2Q_1 Q_3^2 + Q_1^3. \tag{6.20}$$

For example, condition (6.20) is not satisfied if $\mathbf{Q}$ lies on the x_2x_3-plane, since then $Q_1 = 0$. Another example concerns the case where $Q_3 = 0$, then we should have $Q_1 \neq 0$ and $|Q_1| \neq |Q_2|$. Similarly, if $Q_2 = 0$, then $Q_1 \neq 0$ is enough.

However, these conditions can be avoided if we rotate the Cartesian system with respect to the direction of $\mathbf{Q}$. We observe that the coefficients A_1^{-1}, A_1^0, A_1^1 provide the moment $\mathbf{Q}$ in a rather simple way, while the location $\mathbf{r}_0$ can be obtained from any three of the coefficients $A_2^{-2}, A_2^{-1}, A_2^0, A_2^1, A_2^2$. A partial check that there is indeed a single dipole, as opposed to more than one dipoles, is to confirm that our solution is compatible with the form of the coefficients A_2^{-2} and A_2^2 as well. If this condition is not satisfied, then we can conclude that there is more than one point source synchronously activated.

If we generalize the previous algorithm to a number of localized dipoles

$$(\mathbf{r}_i, \mathbf{Q}(\mathbf{r}_i)), \quad i = 1, 2, \ldots, N$$

then we need to find $6N$ quantities and therefore we should utilize $6N$ of the known coefficients A_n^m. In this case, the algebraic system of order $6N \times 6N$ is strongly nonlinear since for $n \geq 3$ the polynomials in x_{i1}, x_{i2}, x_{i3}, we obtain after the application of the operator $D(\mathbf{r}_i)$, have degrees $n - 1 \geq 2$.

6.2 MEG for isolated dipoles in a sphere

As we have shown, the different spherical shells are not visible in MEG because of the boundary relation (4.134). Recalling representation (4.136) and relation (4.137), we obtain the formula

$$\frac{\partial}{\partial r}U(\mathbf{r};\mathbf{r}_0) = -\frac{1}{4\pi}\frac{\mathbf{Q}\times\mathbf{r}_0\cdot\hat{\mathbf{r}}_0}{|\mathbf{r}-\mathbf{r}_0|^3} = \frac{1}{4\pi r}\mathbf{Q}\times\mathbf{r}_0\cdot\nabla_{\mathbf{r}}\frac{1}{|\mathbf{r}-\mathbf{r}_0|}$$

$$= \frac{1}{4\pi}\hat{\mathbf{r}}\cdot\mathbf{Q}\times\nabla_{\mathbf{r}_0}\frac{1}{|\mathbf{r}-\mathbf{r}_0|}, \qquad r > s. \tag{6.21}$$

Integrating alone a ray from the exterior point $\mathbf{r}$ to infinity, we arrive at the following representation of the magnetic potential U:

$$U(\mathbf{r};\mathbf{r}_0) = -\int_r^{+\infty}\frac{\partial}{\partial t}U(t\hat{\mathbf{r}};\mathbf{r}_0)dt = \frac{1}{4\pi}\int_r^{+\infty}\frac{\mathbf{Q}\times\mathbf{r}_0\cdot\hat{\mathbf{r}}}{|t\hat{\mathbf{r}}-\mathbf{r}_0|^3}dt. \tag{6.22}$$

Using the identity

$$\frac{\mathbf{Q}\times\mathbf{r}_0\cdot\hat{\mathbf{r}}}{|\mathbf{r}-\mathbf{r}_0|^3} = \frac{1}{r}\mathbf{Q}\times\mathbf{r}_0\cdot\nabla_{\mathbf{r}_0}\frac{1}{|\mathbf{r}-\mathbf{r}_0|} \tag{6.23}$$

and the Laplace expansion

$$\frac{1}{|\mathbf{r}-\mathbf{r}_0|} = \sum_{n=0}^{\infty}\frac{r_0^n}{r^{n+1}}P_n(\hat{\mathbf{r}}\cdot\hat{\mathbf{r}}_0) = \sum_{n=0}^{\infty}\sum_{m=-n}^{n}\frac{4\pi}{2n+1}\frac{r_0^n}{r^{n+1}}Y_n^m(\hat{\mathbf{r}})Y_n^{m*}(\hat{\mathbf{r}}_0) \tag{6.24}$$

we obtain the expansion

$$U(\mathbf{r};\mathbf{r}_0) = \mathbf{Q}\times\mathbf{r}_0\cdot\nabla_{\mathbf{r}_0}\sum_{n=0}^{\infty}\sum_{m=-n}^{n}\frac{1}{2n+1}r_0^n\left(\int_r^{+\infty}\frac{dt}{t^{n+2}}\right)Y_n^m(\hat{\mathbf{r}})Y_n^{m*}(\hat{\mathbf{r}}_0)$$

$$= \mathbf{Q}\times\mathbf{r}_0\cdot\nabla_{\mathbf{r}_0}\sum_{n=1}^{\infty}\sum_{m=-n}^{n}\frac{1}{2n+1}\frac{1}{n+1}\frac{r_0^n}{r^{n+1}}Y_n^m(\hat{\mathbf{r}})Y_n^{m*}(\hat{\mathbf{r}}_0). \tag{6.25}$$

Comparing the expansions of the electric potentials (6.2) and (6.3) with the corresponding expansion (6.25) for the magnetic potential we see that they are very similar; the main difference in the directional derivative is in the direction of $\mathbf{Q}$ for the electric potential and in the direction $\mathbf{Q}\times\mathbf{r}_0$, that is, normal to $\mathbf{Q}$ and $\mathbf{r}_0$, for the magnetic potential. The more complicated form of the electric coefficients A_n^m is due to the fact that the three shells affect the electric but not the magnetic potential.

The magnetoencephalography data concern recordings of the magnetic flux $\mathbf{B}$ outside the head. Hence, by taking the integral of the radial part $\hat{\mathbf{r}}\cdot\mathbf{B}$ along the line described in formulae (6.21) and (6.22) we obtain the function U, which we expand as follows:

$$U(\mathbf{r};\mathbf{r}_0) = \sum_{n=1}^{\infty}\sum_{m=-n}^{n}\mathbf{B}_n^m(\mathbf{r}_0)\frac{1}{r^{n+1}}Y_n^m(\hat{\mathbf{r}}) \tag{6.26}$$

where the known coefficients $\mathbf{B}_n^m(\mathbf{r}_0)$ have the form

$$\mathbf{B}_n^m(\mathbf{r}_0) = \frac{1}{2n+1}\frac{1}{n+1}\mathbf{Q}\times\mathbf{r}_0\cdot\nabla_{\mathbf{r}_0} r_0^n Y_n^{m*}(\hat{\mathbf{r}}_0). \tag{6.27}$$

The goal is to find the pair $(\mathbf{r}_0, \mathbf{Q})$ from the knowledge of some of the coefficients $\mathbf{B}_n^m(\mathbf{r}_0)$. Defining the source operator

$$D_m(\mathbf{r}_0) = \mathbf{Q}\times\mathbf{r}_0\cdot\nabla_{\mathbf{r}_0} \tag{6.28}$$

and utilizing the same six coefficients as in the EEG case we obtain the following equations:

$$\begin{aligned} D_m(\mathbf{r}_0) r_0 Y_1^{-1*}(\hat{\mathbf{r}}_0) &= \sqrt{\frac{3}{8\pi}}\mathbf{Q}\times\mathbf{r}_0\cdot\nabla_{\mathbf{r}_0}(x_{01}+ix_{02}) \\ &= \sqrt{\frac{3}{8\pi}}[(Q_2x_{03}-Q_3x_{02})+i(Q_3x_{01}-Q_1x_{03})] \end{aligned} \tag{6.29}$$

$$D_m(\mathbf{r}_0) r_0 Y_1^{0*}(\hat{\mathbf{r}}_0) = \sqrt{\frac{3}{4\pi}}\mathbf{Q}\times\mathbf{r}_0\cdot\nabla_{\mathbf{r}_0}x_{03} = \sqrt{\frac{3}{4\pi}}(Q_1x_{02}-Q_2x_{01}) \tag{6.30}$$

$$D_m(\mathbf{r}_0) r_0 Y_1^{1*}(\hat{\mathbf{r}}_0) = \sqrt{\frac{3}{8\pi}}[(Q_2x_{03}-Q_3x_{02})-i(Q_3x_{01}-Q_1x_{03})] \tag{6.31}$$

$$\begin{aligned} D_m(\mathbf{r}_0) r_0^2 Y_2^{-1*}(\hat{\mathbf{r}}_0) &= \sqrt{\frac{15}{8\pi}}\mathbf{Q}\times\mathbf{r}_0\cdot\nabla_{\mathbf{r}_0}(x_{01}x_{03}+ix_{01}x_{02}) \\ &= \sqrt{\frac{15}{8\pi}}[(Q_2x_{03}-Q_3x_{02})(x_{03}+ix_{02}) \\ &\quad + i(Q_3x_{01}-Q_1x_{03})x_{01}+(Q_1x_{02}-Q_2x_{01})x_{01}] \end{aligned} \tag{6.32}$$

$$\begin{aligned} D_m(\mathbf{r}_0) r_0^2 Y_2^{0*}(\hat{\mathbf{r}}_0) &= \sqrt{\frac{5}{16\pi}}\mathbf{Q}\times\mathbf{r}_0\cdot\nabla_{\mathbf{r}_0}(-x_{01}^2-x_{02}^2+2x_{03}^2) \\ &= \sqrt{\frac{5}{4\pi}}[-(Q_2x_{03}-Q_3x_{02})x_{01}-(Q_3x_{01}-Q_1x_{03})x_{02} \\ &\quad + 2(Q_1x_{02}-Q_2x_{01})x_{03}] \end{aligned} \tag{6.33}$$

$$\begin{aligned} &D_m(\mathbf{r}_0) r_0^2 Y_2^{1*}(\hat{\mathbf{r}}_0) \\ &= \sqrt{\frac{15}{8\pi}}[(Q_2x_{03}-Q_3x_{02})(x_{03}-ix_{02})-i(Q_3x_{01}-Q_1x_{03})x_{01}+(Q_1x_{02}-Q_2x_{01})x_{01}]. \end{aligned} \tag{6.34}$$

If we set

$$\begin{aligned} \mathbf{f} = (f_1, f_2, f_3) &= \mathbf{Q}\times\mathbf{r}_0 \\ &= (Q_2x_{03}-Q_3x_{02})\hat{\mathbf{x}}_1+(Q_3x_{01}-Q_1x_{03})\hat{\mathbf{x}}_2+(Q_1x_{02}-Q_2x_{01})\hat{\mathbf{x}}_3 \end{aligned} \tag{6.35}$$

we rewrite expressions (6.29)–(6.34) as follows:

$$B_1^{-1} = \frac{1}{4}\sqrt{\frac{1}{6\pi}}(f_1 + if_2) \tag{6.36}$$

$$B_1^0 = \frac{1}{4}\sqrt{\frac{1}{3\pi}}f_3 \tag{6.37}$$

$$B_1^1 = \frac{1}{4}\sqrt{\frac{1}{6\pi}}(f_1 - if_2) \tag{6.38}$$

$$B_2^{-1} = \frac{1}{2}\sqrt{\frac{1}{30\pi}}(f_1 x_{03} + f_3 x_{01} + if_1 x_{02} + if_2 x_{01}) \tag{6.39}$$

$$B_2^0 = \frac{1}{6}\sqrt{\frac{1}{5\pi}}(-f_1 x_{01} - f_2 x_{02} + 2f_3 x_{03}) \tag{6.40}$$

$$B_2^1 = \frac{1}{2}\sqrt{\frac{1}{30\pi}}(f_1 x_{03} + f_3 x_{01} - if_1 x_{02} - if_2 x_{01}). \tag{6.41}$$

Adding and subtracting relations (6.36), (6.38) and (6.39), (6.41) pairwise we write systems (6.36)–(6.41) as follows:

$$B_1^{-1} + B_1^1 = \frac{1}{2}\sqrt{\frac{1}{6\pi}}f_1 \tag{6.42}$$

$$B_1^{-1} - B_1^1 = \frac{i}{2}\sqrt{\frac{1}{6\pi}}f_2 \tag{6.43}$$

$$B_1^0 = \frac{1}{4}\sqrt{\frac{1}{3\pi}}f_3 \tag{6.44}$$

$$B_2^{-1} + B_2^1 = \sqrt{\frac{1}{30\pi}}(f_1 x_{03} + f_3 x_{01}) \tag{6.45}$$

$$B_2^{-1} - B_2^1 = i\sqrt{\frac{1}{30\pi}}(f_1 x_{02} + f_2 x_{01}) \tag{6.46}$$

$$B_2^0 = \frac{1}{6}\sqrt{\frac{1}{5\pi}}(-f_1 x_{01} - f_2 x_{02} + 2f_3 x_{03}). \tag{6.47}$$

Expressions (6.42)–(6.44) give

$$f_1 = 2\sqrt{6\pi}(B_1^1 + B_1^{-1}) \tag{6.48}$$

$$f_2 = i2\sqrt{6\pi}(B_1^1 - B_1^{-1}) \tag{6.49}$$

$$f_3 = 4\sqrt{3\pi}B_1^0. \tag{6.50}$$

After inserting these values in eqs. (6.45)–(6.47) we obtain

$$B_2^{-1}+B_2^1=\sqrt{\frac{4}{5}}(B_1^1+B_1^{-1})x_{03}+\sqrt{\frac{8}{5}}B_1^0x_{01} \tag{6.51}$$

$$B_2^{-1}-B_2^1=i\sqrt{\frac{4}{5}}(B_1^1+B_1^{-1})x_{02}+i\sqrt{\frac{4}{5}}(B_1^1-B_1^{-1})x_{01} \tag{6.52}$$

$$B_2^0=-\sqrt{\frac{2}{15}}(B_1^1+B_1^{-1})x_{01}-\sqrt{\frac{2}{15}}(B_1^1-B_1^{-1})x_{02}+\sqrt{\frac{16}{15}}B_1^0x_{03}. \tag{6.53}$$

Equations (6.51)–(6.53) are now a linear 3×3 system for the unknowns x_{01}, x_{02}, x_{03}. The conditions for the existence of a unique solution for eqs. (6.51)–(6.53) are

$$B_1^1+B_1^{-1}\neq 0 \tag{6.54}$$

and

$$\left(B_1^0\right)^2+B_1^1B_1^{-1}\neq 0. \tag{6.55}$$

Let ρ_1, ρ_2, ρ_3 be the solutions of the system (6.51)–(6.53). Then, eqs. (6.48)–(6.50) provide the following linear 3×3 system:

$$Q_2\rho_3-Q_3\rho_2=f_1 \tag{6.56}$$

$$Q_3\rho_1-Q_1\rho_3=f_2 \tag{6.57}$$

$$Q_1\rho_2-Q_2\rho_1=f_3 \tag{6.58}$$

for the determination of the components Q_1, Q_2, Q_3, where f_1, f_2, f_3 assume the known values given in eqs. (6.48)–(6.50). The determinant of systems (6.56)–(6.58) vanishes since, in the product $\mathbf{Q}\times\mathbf{r}_0$, the component of $\mathbf{Q}$ along the direction of $\mathbf{r}_0$ vanishes; therefore, it is not recoverable. Hence, we can only determine the two components of $\mathbf{Q}$, which are perpendicular to $\mathbf{r}_0$. In this connection, we introduce a Cartesian system $\hat{\mathbf{y}}_1, \hat{\mathbf{y}}_2, \hat{\mathbf{r}}_0$ centered at the point $\mathbf{r}_0$, which is already known with the third axis along the direction $\hat{\mathbf{r}}_0$. Analyzing $\mathbf{Q}$ in this system, we obtain

$$\mathbf{Q}=q_1\hat{\mathbf{y}}_1+q_2\hat{\mathbf{y}}_2+q_3\hat{\mathbf{r}}_0 \tag{6.59}$$

and then

$$\mathbf{Q}\times\mathbf{r}_0=q_1\hat{\mathbf{y}}_1\times\mathbf{r}_0+q_2\hat{\mathbf{y}}_2\times\mathbf{r}_0=q_2\hat{\mathbf{y}}_1-q_1\hat{\mathbf{y}}_2. \tag{6.60}$$

Thus,

$$q_2\hat{\mathbf{y}}_1-q_1\hat{\mathbf{y}}_2=f_1\hat{\mathbf{x}}_1+f_2\hat{\mathbf{x}}_2+f_3\hat{\mathbf{x}}_3. \tag{6.61}$$

Consequently,

$$q_1 = -f_1(\hat{\mathbf{y}}_2 \cdot \hat{\mathbf{x}}_1) - f_2(\hat{\mathbf{y}}_2 \cdot \hat{\mathbf{x}}_2) - f_3(\hat{\mathbf{y}}_2 \cdot \hat{\mathbf{x}}_3) \tag{6.62}$$

and

$$q_2 = f_1(\hat{\mathbf{y}}_1 \cdot \hat{\mathbf{x}}_1) + f_2(\hat{\mathbf{y}}_1 \cdot \hat{\mathbf{x}}_2) + f_3(\hat{\mathbf{y}}_1 \cdot \hat{\mathbf{x}}_3) \tag{6.63}$$

where $(\hat{\mathbf{y}}_i \cdot \hat{\mathbf{x}}_j)$, $i = 1, 2$ and $j = 1, 2, 3$, devotes the component of $\hat{\mathbf{y}}_i$ along the direction $\hat{\mathbf{x}}_j$.

Finally, we mention that for more than one localized dipoles we follow the steps described at the end of the previous section.

6.3 EEG for isolated dipoles in an ellipsoid

For the case of the three-shell model in ellipsoidal geometry we use expansion (5.113) to write the boundary values of the electric potential on the skin in the form

$$\begin{aligned} \hat{u}_s(\mathbf{r};\mathbf{r}_0)|_{\partial\Omega_s} &= \hat{u}_s(s_1, \mu, \nu; \rho_0, \mu_0, \nu_0) \\ &= -\sum_{n=1}^{\infty}\sum_{m=1}^{2n+1} \frac{1}{2n+1} \frac{\det \mathbf{M}(n,m)}{\gamma_n^m \hat{m}_{21}(n,m)} (\mathbf{Q} \cdot \nabla_{\mathbf{r}_0} \mathbb{E}_n^m(\mathbf{r}_0)) S_n^m(\mu, \nu). \end{aligned} \tag{6.64}$$

If we are given the potential $\hat{u}_s$ on $\rho = s_1$, then we expand it in the surface ellipsoidal harmonics $S_n^m(\mu, \nu)$ and obtain

$$\hat{u}_s(\mathbf{r};\mathbf{r}_0)|_{\partial\Omega_s} = \sum_{n=1}^{\infty}\sum_{m=1}^{2n+1} C_n^m(\mathbf{r}_0)\; S_n^m(\mu, \nu) \tag{6.65}$$

where the known coefficients $C_n^m(\mathbf{r}_0)$ have the form

$$C_n^m(\mathbf{r}_0) = -\frac{1}{2n+1} \frac{\det \mathbf{M}(n,m)}{\gamma_n^m \hat{m}_{21}(n,m)} (\mathbf{Q} \cdot \nabla_{\mathbf{r}_0} \mathbb{E}_n^m(\mathbf{r}_0)) = L_n^m (\mathbf{Q} \cdot \nabla_{\mathbf{r}_0} \mathbb{E}_n^m(\mathbf{r}_0)) \tag{6.66}$$

with the quantities

$$L_n^m = -\frac{1}{2n+1} \frac{\det \hat{\underset{\sim}{\mathbf{M}}}(n,m)}{\gamma_n^m \hat{m}_{21}(n,m)} \tag{6.67}$$

as known constants.

The inverse problem we want to solve consists of finding the dipole $(\mathbf{r}_0, \mathbf{Q})$ from six coefficients C_n^m.

Indeed, guided by the steps followed for the sphere, we consider the coefficients $C_1^1, C_1^2, C_1^3, C_2^3, C_2^4, C_2^5$. Using the Cartesian forms of the related ellipsoidal harmonics, we obtain the following equations:

$$D(\mathbf{r}_0)\mathbb{E}_1^1(\mathbf{r}_0) = \mathbf{Q}\cdot\nabla_{\mathbf{r_0}} h_2h_3x_{01} = h_2h_3Q_1 \tag{6.68}$$

$$D(\mathbf{r}_0)\mathbb{E}_1^2(\mathbf{r}_0) = \mathbf{Q}\cdot\nabla_{\mathbf{r_0}} h_1h_3x_{02} = h_1h_3Q_2 \tag{6.69}$$

$$D(\mathbf{r}_0)\mathbb{E}_1^3(\mathbf{r}_0) = \mathbf{Q}\cdot\nabla_{\mathbf{r_0}} h_1h_2x_{03} = h_1h_2Q_3 \tag{6.70}$$

$$D(\mathbf{r}_0)\mathbb{E}_2^3(\mathbf{r}_0) = \mathbf{Q}\cdot\nabla_{\mathbf{r_0}} h_1h_2h_3^2x_{01}x_{02} = h_1h_2h_3^2(Q_1x_{02}+Q_2x_{01}) \tag{6.71}$$

$$D(\mathbf{r}_0)\mathbb{E}_2^4(\mathbf{r}_0) = \mathbf{Q}\cdot\nabla_{\mathbf{r_0}} h_1h_2^2h_3x_{01}x_{03} = h_1h_2^2h_3(Q_1x_{03}+Q_3x_{01}) \tag{6.72}$$

$$D(\mathbf{r}_0)\mathbb{E}_2^5(\mathbf{r}_0) = \mathbf{Q}\cdot\nabla_{\mathbf{r_0}} h_1^2h_2h_3x_{02}x_{03} = h_1^2h_2h_3(Q_2x_{03}+Q_3x_{02}). \tag{6.73}$$

From eqs. (6.68)–(6.70) and (6.66) we obtain the moment

$$\mathbf{Q} = \frac{1}{h_1h_2h_3}\left(h_1\frac{C_1^1}{L_1^1}, h_2\frac{C_1^2}{L_1^2}, h_3\frac{C_1^3}{L_1^3}\right). \tag{6.74}$$

Then we substitute Q_1, Q_2, Q_3 in the expressions for C_2^3, C_2^4, C_2^5 and obtain the system

$$h_2\frac{C_1^2}{L_1^2}x_{01} + h_1\frac{C_1^1}{L_1^1}x_{02} = \frac{1}{h_3}\frac{C_2^3}{L_2^3} \tag{6.75}$$

$$h_3\frac{C_1^3}{L_1^3}x_{01} + h_1\frac{C_1^1}{L_1^1}x_{03} = \frac{1}{h_2}\frac{C_2^4}{L_2^4} \tag{6.76}$$

$$h_3\frac{C_1^3}{L_1^3}x_{02} + h_2\frac{C_1^2}{L_1^2}x_{03} = \frac{1}{h_1}\frac{C_2^5}{L_2^5} \tag{6.77}$$

which has a unique solution that specifies $\mathbf{r}_0$ as long as

$$C_1^1C_1^2C_1^3 \neq 0. \tag{6.78}$$

As in the case of the sphere multiple dipoles demand more coefficients of expansion (6.65).

6.4 MEG for isolated dipoles in an ellipsoid

The expansion for the magnetic flux in the exterior space takes the form

$$\frac{1}{\hat{\mu}_0}\mathbf{B}(\mathbf{r};\mathbf{r}_0) = \sum_{n=1}^{\infty}\sum_{m=1}^{2n+1}\frac{1}{(2n+1)\gamma_n^m}\left[\mathbf{Q}\times\nabla_{\mathbf{r}_0}\mathbb{E}_n^m(\mathbf{r}_0) - \mathbf{Q}\cdot\nabla_{\mathbf{r}_0}\sum_{\substack{l=1\\ l\neq m}}^{2n+1}\mathbf{H}_n^{nl}\mathbb{E}_n^l(\mathbf{r}_0)\right]\mathbb{F}_n^m(\mathbf{r}). \tag{6.79}$$

On the other hand, if we expand the data **B** we obtain

$$\frac{1}{\mu_0}\mathbf{B}(\mathbf{r};\mathbf{r}_0)=\sum_{n=1}^{\infty}\sum_{m=1}^{2n+1}\mathbf{D}_n^m(\mathbf{r}_0)\mathbb{F}_n^m(\mathbf{r}) \tag{6.80}$$

so that the known coefficients $\mathbf{D}_n^m(\mathbf{r}_0)$ have the form

$$\mathbf{D}_n^m(\mathbf{r}_0)=\frac{1}{(2n+1)\gamma_n^m}\left[\mathbf{Q}\times\nabla_{\mathbf{r}_0}\mathbb{E}_n^m(\mathbf{r}_0)-\mathbf{Q}\cdot\nabla_{\mathbf{r}_0}\sum_{\substack{l=1\\l\neq m}}^{2n+1}\mathbf{H}_n^{nl}\mathbb{E}_n^l(\mathbf{r}_0)\right]. \tag{6.81}$$

We recall that the coefficients $\mathbf{H}_n^{nl}$ depend only on the geometrical parameters c_1, f_1, b_1, s_1 and the physical parameters $\sigma_c, \sigma_f, \sigma_b, \sigma_s$. Utilizing the coefficients $\mathbf{D}_1^1, \mathbf{D}_1^2, \mathbf{D}_1^3, \mathbf{D}_2^3, \mathbf{D}_2^4, \mathbf{D}_2^5$ we obtain the following equations:

$$\begin{aligned}\mathbf{D}_1^1(\mathbf{r}_0)&=\frac{h_1h_2h_3}{3\gamma_1^1}\left[\frac{1}{h_1}\mathbf{Q}\times\nabla_{\mathbf{r}_0}x_{01}-\frac{1}{h_2}\mathbf{Q}\cdot\nabla_{\mathbf{r}_0}x_{02}-\frac{1}{h_3}\mathbf{Q}\cdot\nabla_{\mathbf{r}_0}x_{03}\right]\\&=\frac{h_1h_2h_3}{3\gamma_1^1}\left[\frac{Q_3}{h_1}\hat{\mathbf{x}}_2-\frac{Q_2}{h_1}\hat{\mathbf{x}}_3-\frac{Q_2}{h_2}\mathbf{H}_1^{12}-\frac{Q_3}{h_3}\mathbf{H}_1^{13}\right]\end{aligned} \tag{6.82}$$

$$\begin{aligned}\mathbf{D}_1^2(\mathbf{r}_0)&=\frac{h_1h_2h_3}{3\gamma_1^2}\left[\frac{1}{h_2}\mathbf{Q}\times\nabla_{\mathbf{r}_0}x_{02}-\frac{1}{h_1}\mathbf{Q}\cdot\nabla_{\mathbf{r}_0}x_{01}-\frac{1}{h_3}\mathbf{Q}\cdot\nabla_{\mathbf{r}_0}x_{03}\right]\\&=\frac{h_1h_2h_3}{3\gamma_1^2}\left[\frac{Q_1}{h_2}\hat{\mathbf{x}}_3-\frac{Q_3}{h_2}\hat{\mathbf{x}}_1-\frac{Q_1}{h_1}\mathbf{H}_1^{11}-\frac{Q_3}{h_3}\mathbf{H}_1^{13}\right]\end{aligned} \tag{6.83}$$

$$\begin{aligned}\mathbf{D}_1^3(\mathbf{r}_0)&=\frac{h_1h_2h_3}{3\gamma_1^3}\left[\frac{1}{h_3}\mathbf{Q}\times\nabla_{\mathbf{r}_0}x_{03}-\frac{1}{h_1}\mathbf{Q}\cdot\nabla_{\mathbf{r}_0}x_{01}-\frac{1}{h_2}\mathbf{Q}\cdot\nabla_{\mathbf{r}_0}x_{02}\right]\\&=\frac{h_1h_2h_3}{3\gamma_1^3}\left[\frac{Q_2}{h_3}\hat{\mathbf{x}}_1-\frac{Q_1}{h_3}\hat{\mathbf{x}}_2-\frac{Q_1}{h_1}\mathbf{H}_1^{11}-\frac{Q_2}{h_2}\mathbf{H}_1^{12}\right]\end{aligned} \tag{6.84}$$

$$5\gamma_2^3\mathbf{D}_2^3(\mathbf{r}_0)=\mathbf{Q}\times\nabla_{\mathbf{r}_0}\mathbb{E}_2^3(\mathbf{r}_0)-\mathbf{Q}\cdot\nabla_{\mathbf{r}_0}\left[\sum_{i=1}^{5}\mathbf{H}_2^{2i}\mathbb{E}_2^i(\mathbf{r}_0)-\mathbf{H}_2^{23}\mathbb{E}_2^3(\mathbf{r}_0)\right] \tag{6.85}$$

$$5\gamma_2^4\mathbf{D}_2^4(\mathbf{r}_0)=\mathbf{Q}\times\nabla_{\mathbf{r}_0}\mathbb{E}_2^4(\mathbf{r}_0)-\mathbf{Q}\cdot\nabla_{\mathbf{r}_0}\left[\sum_{i=1}^{5}\mathbf{H}_2^{2i}\mathbb{E}_2^i(\mathbf{r}_0)-\mathbf{H}_2^{24}\mathbb{E}_2^4(\mathbf{r}_0)\right] \tag{6.86}$$

$$5\gamma_2^5\mathbf{D}_2^5(\mathbf{r}_0)=\mathbf{Q}\times\nabla_{\mathbf{r}_0}\mathbb{E}_2^5(\mathbf{r}_0)-\mathbf{Q}\cdot\nabla_{\mathbf{r}_0}\left[\sum_{i=1}^{5}\mathbf{H}_2^{2i}\mathbb{E}_2^i(\mathbf{r}_0)-\mathbf{H}_2^{25}\mathbb{E}_2^5(\mathbf{r}_0)\right]. \tag{6.87}$$

Subtracting eq. (6.86) from (6.85), eq. (6.87) from (6.86) and eq. (6.85) from (6.87) we obtain the following equations:

$$\begin{aligned}5\left(\gamma_2^3\mathbf{D}_2^3-\gamma_2^4\mathbf{D}_2^4\right)&=\mathbf{Q}\times\nabla_{\mathbf{r}_0}\left[\mathbb{E}_2^3(\mathbf{r}_0)-\mathbb{E}_2^4(\mathbf{r}_0)\right]\\&\quad+\mathbf{Q}\cdot\nabla_{\mathbf{r}_0}\left[\mathbf{H}_2^{23}\mathbb{E}_2^3(\mathbf{r}_0)-\mathbf{H}_2^{24}\mathbb{E}_2^4(\mathbf{r}_0)\right]\end{aligned} \tag{6.88}$$

$$5(\gamma_2^4\mathbf{D}_2^4-\gamma_2^5\mathbf{D}_2^5)=\mathbf{Q}\times\nabla_{\mathbf{r}_0}\left[\mathbb{E}_2^4(\mathbf{r}_0)-\mathbb{E}_2^5(\mathbf{r}_0)\right] \\ +\mathbf{Q}\cdot\nabla_{\mathbf{r}_0}\left[\mathbf{H}_2^{24}\mathbb{E}_2^4(\mathbf{r}_0)-\mathbf{H}_2^{25}\mathbb{E}_2^5(\mathbf{r}_0)\right] \tag{6.89}$$

$$5(\gamma_2^5\mathbf{D}_2^5-\gamma_2^3\mathbf{D}_2^3)=\mathbf{Q}\times\nabla_{\mathbf{r}_0}\left[\mathbb{E}_2^5(\mathbf{r}_0)-\mathbb{E}_2^3(\mathbf{r}_0)\right] \\ +\mathbf{Q}\cdot\nabla_{\mathbf{r}_0}\left[\mathbf{H}_2^{25}\mathbb{E}_2^5(\mathbf{r}_0)-\mathbf{H}_2^{23}\mathbb{E}_2^3(\mathbf{r}_0)\right]. \tag{6.90}$$

Inserting the Cartesian forms of $\mathbb{E}_2^i$, $i=3,4,5$ and performing the indicated calculations we obtain

$$\frac{5(\gamma_2^3\mathbf{D}_2^3-\gamma_2^4\mathbf{D}_2^4)}{h_1h_2h_3}=(h_3x_{02}-h_2x_{03})(Q_3\hat{\mathbf{x}}_2-Q_2\hat{\mathbf{x}}_3)+h_3x_{01}(Q_1\hat{\mathbf{x}}_3-Q_3\hat{\mathbf{x}}_1) \\ +h_2x_{01}(Q_1\hat{\mathbf{x}}_2-Q_2\hat{\mathbf{x}}_1)+h_3\mathbf{H}_2^{23}(Q_1x_{02}+Q_2x_{01}) \\ -h_2\mathbf{H}_2^{24}(Q_1x_{03}+Q_3x_{01}). \tag{6.91}$$

$$\frac{5(\gamma_2^4\mathbf{D}_2^4-\gamma_2^5\mathbf{D}_2^5)}{h_1h_2h_3}=(h_2x_{01}-h_1x_{02})(Q_2\hat{\mathbf{x}}_1-Q_1\hat{\mathbf{x}}_2)+h_2x_{03}(Q_3\hat{\mathbf{x}}_2-Q_2\hat{\mathbf{x}}_3) \\ +h_1x_{03}(Q_3\hat{\mathbf{x}}_1-Q_1\hat{\mathbf{x}}_3)+h_2\mathbf{H}_2^{24}(Q_1x_{03}+Q_3x_{01}) \\ -h_1\mathbf{H}_2^{25}(Q_2x_{03}+Q_3x_{02}) \tag{6.92}$$

$$\frac{5(\gamma_2^5\mathbf{D}_2^5-\gamma_2^3\mathbf{D}_2^3)}{h_1h_2h_3}=(h_1x_{03}-h_3x_{01})(Q_1\hat{\mathbf{x}}_3-Q_3\hat{\mathbf{x}}_1)+h_1x_{02}(Q_2\hat{\mathbf{x}}_1-Q_1\hat{\mathbf{x}}_2) \\ +h_3x_{02}(Q_2\hat{\mathbf{x}}_3-Q_3\hat{\mathbf{x}}_2)+h_1\mathbf{H}_2^{25}(Q_2x_{03}+Q_3x_{02}) \\ -h_3\mathbf{H}_2^{23}(Q_1x_{02}+Q_2x_{01}). \tag{6.93}$$

We simply further systems (6.91)–(6.93) by projecting eq. (6.91) on the x_1- axis, eq. (6.92) on the x_3- axis and eq. (6.93) on the x_2- axis, then we obtain the following equations:

$$5\frac{\gamma_2^3\mathbf{D}_{2,1}^3-\gamma_2^4\mathbf{D}_{2,1}^4}{h_1h_2h_3}=-(h_2Q_2+h_3Q_3)x_{01}+h_3H_{2,1}^{23}(Q_1x_{02}+Q_2x_{01}) \\ -h_2H_{2,1}^{24}(Q_1x_{03}+Q_3x_{01}) \tag{6.94}$$

$$5\frac{\gamma_2^4\mathbf{D}_{2,3}^4-\gamma_2^5\mathbf{D}_{2,3}^5}{h_1h_2h_3}=-(h_1Q_1+h_2Q_2)x_{03}+h_2H_{2,3}^{24}(Q_1x_{03}+Q_3x_{01}) \\ -h_1H_{2,3}^{25}(Q_2x_{03}+Q_3x_{02}) \tag{6.95}$$

$$5\frac{\gamma_2^5\mathbf{D}_{2,2}^5-\gamma_2^3\mathbf{D}_{2,2}^3}{h_1h_2h_3}=-(h_1Q_1+h_3Q_3)x_{02}+h_1H_{2,2}^{25}(Q_2x_{03}+Q_3x_{02})$$

$$- h_3 H^{23}_{2,2}(Q_1 x_{02} + Q_2 x_{01}) \tag{6.96}$$

where the subindex after the comma denotes the component of the relative vector.

Similarly, projecting eq. (6.82) on the x_1- axis, eq. (6.83) on the x_2- axis and eq. (6.84) on the x_3- axis, we obtain

$$\frac{3\gamma_1^1 \mathbf{D}^1_{1,1}}{h_1 h_2 h_3} = -\left[\frac{H^{12}_{1,1}}{h_2} Q_2 + \frac{H^{13}_{1,1}}{h_3} Q_3\right] \tag{6.97}$$

$$\frac{3\gamma_1^2 \mathbf{D}^2_{1,2}}{h_1 h_2 h_3} = -\left[\frac{H^{11}_{1,2}}{h_1} Q_1 + \frac{H^{13}_{1,2}}{h_3} Q_3\right] \tag{6.98}$$

$$\frac{3\gamma_1^3 \mathbf{D}^3_{1,3}}{h_1 h_2 h_3} = -\left[\frac{H^{11}_{1,3}}{h_1} Q_1 + \frac{H^{12}_{1,3}}{h_2} Q_2\right]. \tag{6.99}$$

Equations (6.97)–(6.99) form a linear system for the components Q_1, Q_2, Q_3.

Finally, inserting the values of Q_1, Q_2, Q_3 in eqs. (6.94)–(6.96) we arrive at a3×3 linear system for the determination of x_{01}, x_{02}, x_{03}. This completes the determination of $\mathbf{r}_0$ and $\mathbf{Q}$ in terms of the known coefficients of expansion (6.80).

Note that we need the three components $D^1_{1,1}, D^2_{1,2}, D^3_{1,3}$ for the determination of the three components Q_1, Q_2, Q_3, while for the determination of the three components x_{01}, x_{02}, x_{03} we need in addition the six components $D^3_{2,1}, D^3_{2,2}, D^4_{2,1}, D^4_{2,3}, D^5_{2,2}, D^5_{2,3}$.

This analytical procedure can be generalized to any finite number of dipoles, but the details are extremely complicated.

Chapter 7
A minimisation approach to EEG for the spherical and ellipsoidal geometries

As mentioned in the introduction of Chapter 6, we discuss an inversion method for the electroencephalographic (EEG) problem in spherical and ellipsoidal geometries for the case that the current is continuously distributed. It is shown that by imposing the additional condition that the current has a minimum L^2-norm it is possible to determine a unique current.

7.1 Statement of the problem

We consider the three-shell problem with the nested domains $\Omega_c, \Omega_f, \Omega_b, \Omega_s$ and Ω_e defining the cerebrum, the cerebrospinal fluid, the bone, the skin and the exterior space, respectively. The corresponding outer boundaries are denoted by S_c, S_f, S_b, S_s. Similarly we use the symbols $\sigma_c, \sigma_f, \sigma_b, \sigma_s$ for the electric conductivities of the corresponding regions. The neuronal current $\mathbf{J}$ is continuously distributed in the region of the cerebrum Ω_c.

We utilize the Helmholtz decomposition of the vector field $\mathbf{J}$ and obtain

$$\mathbf{J}(\mathbf{r}) = \nabla\Psi(\mathbf{r}) + \nabla \times \mathbf{A}(\mathbf{r}) \tag{7.1}$$

with

$$\nabla \cdot \mathbf{A}(\mathbf{r}) = 0. \tag{7.2}$$

In order to reconstruct a neuronal current uniquely we will impose the condition that $\mathbf{J}^P$ minimizes the L^2-norm, which will be denoted by W. Using the Helmholtz decomposition (7.1), W can be rewritten in the form

$$W = \int_{\Omega_c} |\mathbf{J}|^2 dv = W_1 + W_2 + W_3 \tag{7.3}$$

where

$$W_1 = \int_{\Omega_c} |\nabla\Psi|^2 dv \tag{7.4}$$

$$W_2 = \int_{\Omega_c} |\nabla \times \mathbf{A}|^2 dv \tag{7.5}$$

https://doi.org/10.1515/9783110547535-008

$$W_3 = \int_{\Omega_C} \nabla\Psi \cdot \nabla \times \mathbf{A}\, dv. \tag{7.6}$$

In order to eliminate W_3 we impose the boundary condition (iii) of the introduction, namely,

$$\hat{\mathbf{n}} \times \nabla\mathbf{A}|_{S_C} = 0 \tag{7.7}$$

where $\hat{\mathbf{n}}$ denotes the outer unit normal on S_c. Indeed, using the identities

$$\nabla\Psi \cdot \nabla \times \mathbf{A} = \nabla \cdot (\Psi\nabla \times \mathbf{A}) - \Psi\nabla \cdot (\nabla \times \mathbf{A}) \tag{7.8}$$

and $\nabla \cdot (\nabla \times \mathbf{A}) = 0$ we find

$$W_3 = \int_{\Omega_C} \nabla \cdot (\Psi\nabla \times \mathbf{A}) dv = \oint_{S_C} \hat{\mathbf{n}} \cdot (\Psi\nabla \times \mathbf{A}) ds = 0. \tag{7.9}$$

We impose the following boundary condition on Ψ:

$$\hat{\mathbf{n}} \cdot \nabla\Psi = \frac{\partial\Psi}{\partial n} = 0, \text{ on } S_c. \tag{7.10}$$

This boundary condition (7.10) implies that W_1 can be rewritten in the form

$$W_1 = -\int_{\Omega_C} \Psi(\Delta\Psi) dv. \tag{7.11}$$

Indeed, using the identity

$$|\nabla\Psi|^2 = \nabla \cdot (\Psi\nabla\Psi) - \Psi(\Delta\Psi) \tag{7.12}$$

and employing the Gauss theorem and the condition (7.10), W_1 becomes eq. (7.11).

The fundamental integral representations derived in Chapter 2 state that for EEG we have

$$u_s(\mathbf{r}) = -\frac{1}{4\pi} \int_{\Omega_C} (\Delta\Psi(\mathbf{r}_0))\; \upsilon_s(\mathbf{r}, \mathbf{r}_0) dv(\mathbf{r}_0), \quad \mathbf{r} \in S_C \tag{7.13}$$

where $\upsilon_s(\mathbf{r}, \mathbf{r}_0)$ is the monopolic potential in the shell of the skin.

The inversion process for EEG amounts to evaluating Ψ, which is the part of $\mathbf{J}$ that generates the potential $u_s(\mathbf{r})$ on S_s. We emphasize that the functions υ_s does not depend on the current $\mathbf{J}$, that is, on the representation potentials Ψ. The function υ_s depends only on the conductivities and the geometry of the brain–head model.

Hence, for the inversion algorithm, the function v_s is constant with respect to $\mathbf{J}$ which can be obtained from the solution of the corresponding forward problem. In fact, v_c, v_f, v_b, v_s solve the transmission problem defined by eqs. (2.28)–(2.39).

7.2 Inversion of EEG data for the spherical model

In the case of spherical geometry, the regions $\Omega_c, \Omega_f, \Omega_b, \Omega_s$ and Ω_e are identified by the radial component in the intervals $[0, c), (c,f), (f, b), (b, s)$ and $(s, +\infty)$. Furthermore, the appropriate complete set of orthonormal surface eigenfunctions are the spherical harmonics $Y_n^m(\hat{\mathbf{r}})$. For the spherical case, the inversion process can be implemented as follows. Suppose that we know $u_s(\mathbf{r})$ on the boundary $r = s$, then we expand these data in surface spherical harmonics:

$$u_s(s\hat{\mathbf{r}}) = \sum_{n=1}^{\infty} \sum_{m=-n}^{n} u_n^m Y_n^m(\hat{\mathbf{r}}) \tag{7.14}$$

The actual data is the known sequence $\{u_n^m\}$.

Next we decompose the field u_s in spherical harmonics, namely for the scalar function Ψ we assume the representation

$$\Psi(\mathbf{r}_0) = \sum_{n=1}^{\infty} \sum_{m=-n}^{n} y_n^m(r_0) Y_n^m(\hat{\mathbf{r}}_0). \tag{7.15}$$

The goal is to find the values of $\{y_n^m(r_0)\}$ that minimize the norm $\|\mathbf{J}\|$ with $\{u_n^m\}$ given. It turns out that by using the addition theorem

$$P_n(\hat{\mathbf{r}} \cdot \hat{\mathbf{r}}_0) = \frac{4\pi}{2n+1} \sum_{m=-n}^{n} Y_n^m(\hat{\mathbf{r}}) Y_n^{m*}(\hat{\mathbf{r}}_0) \tag{7.16}$$

in the spectral form (7.15) of $\Psi(\mathbf{r}_0)$, as well as the orthogonality condition

$$\int_0^{2\pi} \int_0^{\pi} Y_n^m(\hat{\mathbf{r}}_0) P_{n'}(\hat{\mathbf{r}} \cdot \hat{\mathbf{r}}_0) \sin\theta_0 d\theta_0 d\varphi_0 = \frac{4\pi}{2n'+1} Y_{n'}^m(\hat{\mathbf{r}}) \delta_{nn'}, \tag{7.17}$$

it is possible to arrive at the expansion

$$u_s(\mathbf{r}) = -\sum_{n=1}^{\infty} \sum_{m=-n}^{n} \frac{k_n}{2n+1} Y_n^m(\hat{\mathbf{r}}) \cdot \left[\int_0^{c_1} \left(\frac{d}{dr_0} r_0^2 \frac{d}{dr_0} y_n^m(r_0) - n(n+1) y_n^m(r_0) \right) r_0^n dr_0 \right] \tag{7.18}$$

where k_n are the constant coefficients of the expansion of the potential u_s, entering representation (7.13) in terms of $r_0^n P_n(\hat{\mathbf{r}}_0 \cdot \hat{\mathbf{r}})$.

Proof of formula (7.18). We have that

$$u_s(\mathbf{r}) = -\frac{1}{4\pi}\sum_{n=1}^{\infty} k_n \int_0^{c_1}\int_0^{2\pi}\int_0^{\pi}\left[\sum_{n'=1}^{\infty}\sum_{m'=-n'}^{n'}\left(\Delta y_{n'}^{m'}(r_0)Y_{n'}^{m'}(\hat{\mathbf{r}}_0)\right)\right]$$
$$\cdot r_0^n P_n(\hat{\mathbf{r}}\cdot\hat{\mathbf{r}}_0) r_0^2 \sin\theta_0 dr_0 d\theta_0 d\varphi_0. \tag{7.19}$$

But for the Beltrami operator

$$\mathrm{B} = \frac{1}{\sin\theta}\partial_\theta \sin\theta\ \partial_\theta + \frac{1}{\sin^2\theta}\partial_{\varphi\varphi} \tag{7.20}$$

we have the spectral relation

$$\mathrm{B}Y_n^m(\hat{\mathbf{r}}) = -n(n+1)Y_n^m(\hat{\mathbf{r}}). \tag{7.21}$$

This implies

$$\Delta y_{n'}^{m'}(r_0)Y_{n'}^{m'}(\hat{\mathbf{r}}_0) = \frac{1}{r_0^2}\left[\frac{d}{dr_0}r_0^2\frac{d}{dr_0}y_{n'}^{m'}(r_0) - n(n+1)y_{n'}^{m'}(r_0)\right]Y_{n'}^{m'}(\hat{\mathbf{r}}_0)$$
$$= \frac{1}{r_0^2}\left[M_n(r_0)y_{n'}^{m'}(r_0)\right]Y_{n'}^{m'}(\hat{\mathbf{r}}_0) \tag{7.22}$$

where

$$M_n(r_0) = \frac{d}{dr_0}r_0^2\frac{d}{dr_0} - n(n+1). \tag{7.23}$$

We rewrite eq. (7.19) in the form

$$u_s(\mathbf{r}) = -\frac{1}{4\pi}\sum_{n=1}^{\infty} k_n \sum_{n'=1}^{\infty}\sum_{m'=-n'}^{n'}\int_0^{c_1}\int_0^{2\pi}\int_0^{\pi}\left(M_{n'}(r_0)y_{n'}^{m'}(r_0)\right)$$
$$\cdot Y_{n'}^{m'}(\hat{\mathbf{r}}_0)P_n(\hat{\mathbf{r}}\cdot\hat{\mathbf{r}}_0)r_0^n \sin\theta_0 dr_0 d\theta_0 d\varphi_0$$
$$= -\sum_{n=1}^{\infty}\frac{k_n}{2n+1}\sum_{n'=1}^{\infty}\sum_{m'=-n'}^{n'}\int_0^{c_1}\left(M_{n'}(r_0)y_{n'}^{m'}(r_0)\right)Y_n^m(\hat{\mathbf{r}})\delta_{nn'}r_0^n dr_0$$
$$= -\sum_{n=1}^{\infty}\sum_{m=-n}^{n}\frac{k_n}{2n+1}\left[\int_0^{c_1}\left(M_n(r_0)y_n^m(r_0)\right)r_0^n dr_0\right]Y_n^m(\hat{\mathbf{r}}) \tag{7.24}$$

which is the right-hand side (RHS) of eq. (7.18).

If we assume the expansion

$$u_s(\mathbf{r}) = \sum_{n=1}^{\infty} \sum_{m=-n}^{n} u_n^m Y_n^m(\hat{\mathbf{r}}) \tag{7.25}$$

for $\mathbf{r} \in S_s$, where $\{u_n^m\}$ are known from the measurements and compare it with expansion (7.19), we obtain, by virtue of orthogonality, the sequence of integral–differential equations

$$\int_0^{c_1} \left(\frac{d}{dr_0} r_0^2 \frac{d}{dr_0} y_n^m(r_0) - n(n+1) y_n^m(r_0) \right) r_0^n dr_0 = -\frac{2n+1}{k_n^m} u_n^m. \tag{7.26}$$

The constants k_n^m are the coefficients of the expansion of u_s in terms of $r_0^n Y_n^m(\hat{\mathbf{r}}_0)$; they are given in eqs. (4.105), (4.104) and (4.103) in which (4.103) is proportional to r_0^n. Note that k_n^m depend only on the radii c, f, b, s and the conductivities $\sigma_c, \sigma_f, \sigma_b, \sigma_s$.

In spherical geometry, eq. (7.10) implies that

$$\frac{\partial}{\partial r_0} \Psi(r_0, \theta_0, \varphi_0) \bigg|_{r_0 = c} = 0 \tag{7.27}$$

and hence eq. (7.15) imposes the boundary conditions

$$\frac{d}{dr_0} y_n^m(r_0) \bigg|_{r_0 = c} = 0. \tag{7.28}$$

Integrating by parts the first term of the integrand of eq. (7.26) we find:

$$\begin{aligned}
&\int_0^c \frac{d}{dr_0} \left(r_0^2 \frac{d}{dr_0} y_n^m(r_0) \right) r_0^n dr_0 = \\
&= r_0^{n+2} \frac{d}{dr_0} y_n^m(r_0) \bigg|_{r_0=0}^{r_0=c} - n \int_0^c \left(\frac{d}{dr_0} y_n^m(r_0) \right) r_0^{n+1} dr_0 \\
&= -n y_n^m(r_0) r_0^{n+1} \bigg|_{r_0=0}^{r_0=c} + n(n+1) \int_0^c y_n^m(r_0) r_0^n dr_0
\end{aligned} \tag{7.29}$$

where we have used boundary conditions (7.28). The second term above cancels with the second term of the integrand of eq. (7.26), and hence eq. (7.26) becomes

$$y_n^m(c) = \frac{(2n+1) u_n^m}{n c^{n+1} k_n^m}. \tag{7.30}$$

Hence, the basic equation (7.13) fixes the values of the unknown coefficients $y_n^m(r_0)$, appearing in eq. (7.15), at $r_0 = c$. This is consistent with eq. (2.51), which expresses the RHS of eq. (7.13) as an integral involving Ψ and $\nabla\Psi$ on the cortical surface S_c. Hence, only $y_n^m(c)$ and $\frac{d}{dr_0} y_n^m(r_0)|_{r_0=0} = c$ affect the EEG data.

The determination of $y_n^m(r_0)$ from boundary conditions (7.28) and (7.30) follows from the minimization of W.

Taking into consideration that W_3 vanishes and W_2 is independent of Ψ, this minimization reduces to minimizing W_1. Considering the expansion

$$\Delta\Psi(\mathbf{r}_0) = \sum_{n=1}^{\infty} \sum_{m=-n}^{n} \left(M_n(r_0) y_n^m(r_0)\right) Y_n^m(\hat{\mathbf{r}}_0) \tag{7.31}$$

where the operator M_n is defined by

$$M_n = \frac{d}{dr_0}\left(r_0^2 \frac{d}{dr_0}\right) - n(n+1) \tag{7.32}$$

as well as the orthogonality conditions

$$\int_{S^2} Y_n^{m*}(\hat{\mathbf{r}}_0) Y_{n'}^{m'}(\hat{\mathbf{r}}_0) dS(\hat{\mathbf{r}}_0) = \delta_{nn'}\delta_{mm'} \tag{7.33}$$

we find

$$W_1 = -\sum_{n=0}^{\infty} \sum_{m=-n}^{n} \int_0^c \left(M_n y_n^m(r_0)\right) y_n^m(r_0) dr_0. \tag{7.34}$$

The integrand of the above integral can be re-written in the form

$$(M_n y_n^m) y_n^m = \frac{d}{dr_0}\left(r_0^2 y_n^m \frac{d}{dr_0} y_n^m\right) - r_0^2\left(\frac{d}{dr_0} y_n^m\right)^2 - n(n+1)\left(y_n^m\right)^2. \tag{7.35}$$

Hence, since $\frac{dy_n^m}{dr_0}$ vanishes at $r_0 = c$, we find

$$W_1 = \sum_{n=0}^{\infty} \sum_{m=-n}^{n} \int_0^c \left[r_0^2 \left(\frac{dy_n^m}{dr_0}\right)^2 + n(n+1)\left(y_n^m\right)^2 \right] dr_0. \tag{7.36}$$

In summary, the problem of determining $y_n^m(r_0)$ reduces to minimizing the functional W_1 defined in eq. (7.36) under the boundary conditions (7.28) and (7.30). This problem is solved in [48], where it is shown that the minimizing functions denoted by $\hat{y}_n^m(r_0)$ are given by

$$\hat{y}_n^m(r_0) = -\frac{2n+1}{c^{2n+1}}\frac{u_n^m}{k_n^m} r_0^n \left[\ln\frac{r_0}{c} - \frac{1}{n}\right]. \tag{7.37}$$

Indeed, in order to minimize W_1 we replace y_n^m by $y_n^m + \eta_n$ and keep terms of order η_n This implies that the minimizing function $\hat{y}_n^m$ satisfies the condition

$$\int_0^c \left[n(n+1)\hat{y}_n^m \eta_n + r_0^2\left(\frac{d}{dr_0}\hat{y}_n^m\right)\left(\frac{d}{dr_0}\eta_n\right)\right] dr_0 = 0. \tag{7.38}$$

Integration by parts, together with the boundary condition $\frac{d}{dr_0}\hat{y}_n^m(c) = 0$ yields

$$\int_0^c (M_n(r_0)\hat{y}_n^m(r_0))\eta_n dr_0 = 0. \tag{7.39}$$

The obvious choice $M_n\hat{y}_n^m = 0$ contradicts the boundary condition (7.30). Thus, in order to satisfy eq. (7.39) we must restrict the η_n to be orthogonal to r_0^n with respect to the associated scalar product in $L^2([0,c])$, namely, η_n satisfies the condition

$$\int_0^c r_0^n \eta_n(r_0) dr_0 = 0. \tag{7.40}$$

The necessity for the above restriction follows from the fact that eq. (7.30) arises from the equation

$$\int_0^c (M_n(r_0)\hat{y}_n^m(r_0)) r_0^n dr_0 = -\frac{2n+1}{k_n^m} u_n^m. \tag{7.41}$$

This equation implies that

$$M_n(r_0)\hat{y}_n^m(r_0) = \beta_n^m r_0^n + f_n(r_0) \tag{7.42}$$

where $f_n(r_0)$ is orthogonal to r_0^n, whereas β_n^m is determined by substituting the above expression in eq. (7.41):

$$\beta_n^m = -\frac{(2n+1)^2}{c^{2n+1}}\frac{u_n^m}{k_n^m}. \tag{7.43}$$

Thus, defining $\hat{y}_n^m$ via the equation

$$M_n(r_0)\hat{y}_n^m(r_0) = \beta_n^m r_0^n \tag{7.44}$$

and restricting the class of allowed variations to satisfy eq. (7.40), eq. (7.39) is satisfied.

The homogeneous version of eq. (7.44) has the solutions r_0^n and $r_0^{-(n+1)}$. A particular solution of the inhomogeneous equation is given by

$$\frac{\beta_n^m}{2n+1} r_0^n \ln r_0.$$

Hence, eliminating the singular solution $r_0^{-(n+1)}$ we obtain

$$\hat{y}_n^m(r_0) = A_n^m r_0^m + \frac{\beta_n^m}{2n+1} r_0^n \ln r_0 \tag{7.45}$$

where the constant A_n^m is determined from the condition $\frac{d}{dr_0} y_n^m(c) = 0$. Thus

$$A_n^m = -\frac{\beta_n^m}{2n+1}\left[\ln c + \frac{1}{n}\right]. \tag{7.46}$$

Inserting, eqs. (7.43) and (7.46) in the solution (7.45) we obtain eq. (7.37).

7.3 Inversion of EEG data for the ellipsoidal model

In this section, we discuss the inverse EEG problem in ellipsoidal geometry. For the notation we refer to Section 3.3. We will use the same notation with the spherical case, now referred to the boundaries $\rho = c_1, f_1, b_1, s_1$ instead of $r = c, f, b, s$.

On the boundary S_s we have from eq. (3.77) the expansion

$$u_s(\mathbf{r})|_{S_s} = -\frac{1}{4\pi}\sum_{n=1}^{\infty}\sum_{m=1}^{2n+1}\frac{C_{sn}^{sm}}{\gamma_n^m} S_n^m(\mu,\nu)\int_{\Omega_c}(\Delta\Psi(\mathbf{r}_0))\mathbb{E}_n^m(\mathbf{r}_0)dv(\mathbf{r}_0) \tag{7.47}$$

where $\mathbf{r} = (\rho,\mu,\nu)$, $\mathbf{r}_0 = (\rho_0,\mu_0,\nu_0)$ and C_{sn}^{sm} are constants depending on $\sigma_c, \sigma_f, \sigma_b, \sigma_s$ and c_1, f_1, b_1, s_1. Their values are given in eq. (3.96).

The expansion of the EEG data on $\rho = s_1$ is

$$u_s(\mathbf{r})|_{\rho=s_1} = \sum_{n=1}^{\infty}\sum_{m=1}^{2n+1}\tilde{u}_n^m S_n^m(\mu,\nu) \tag{7.48}$$

where the coefficient $\tilde{u}_n^m$ is known. The corresponding expansion of Ψ is given by

$$\Psi(\rho_0,\mu_0,\nu_0) = \sum_{n=1}^{\infty}\sum_{m=1}^{2n+1}\tilde{y}_n^m(\rho_0) S_n^m(\mu_0,\nu_0) \tag{7.49}$$

where $\tilde{y}_n^m(\rho_0)$ are unknown functions.

The L^2-norm of the current is given in eqs. (7.3)–(7.6). The boundary condition (7.7) takes the form

$$\hat{\rho}\cdot\nabla\times\mathbf{A}|_{\rho_0=c_1} = 0 \tag{7.50}$$

and following the steps in eqs. (7.8)–(7.9) we obtain $W_3 = 0$.

Equation (7.10) becomes

$$\hat{\rho} \cdot \nabla \Psi|_{\rho_0 = c_1} = \frac{\partial \Psi}{\partial \rho_0}|_{\rho_0 = c_1} = 0. \tag{7.51}$$

Condition (7.51) immediately implies that

$$W_1 = - \int_{\Omega_c} \Psi(\Delta \Psi) dv. \tag{7.52}$$

Applying the Laplace operator in its ellipsoidal form we obtain

$$\Delta \Psi(\rho_0, \mu_0, \nu_0) = \sum_{n=1}^{\infty} \sum_{m=1}^{2n+1} \Delta(\tilde{y}_n^m(\rho_0) S_n^m(\mu_0, \nu_0)) \tag{7.53}$$

and using formula (3.88) we rewrite eq. (7.53) as

$$\Delta \Psi(\rho_0, \mu_0, \nu_0) = \frac{\sqrt{\rho_0^2 - h_3^2}\sqrt{\rho_0^2 - h_2^2}}{(\rho_0^2 - \mu_0^2)(\rho_0^2 - \nu_0^2)} \sum_{n=1}^{\infty} \sum_{m=1}^{2n+1} (L_n^m \tilde{y}_n^m(\rho_0)) S_n^m(\mu_0, \nu_0) \tag{7.54}$$

where

$$L_n^m = \frac{d}{dr_0} \sqrt{\rho_0^2 - h_3^2}\sqrt{\rho_0^2 - h_2^2} \frac{d}{dr_0} + \frac{(h_3^2 + h_2^2) p_n^m - n(n+1)\rho_0^2}{\sqrt{\rho_0^2 - h_3^2}\sqrt{\rho_0^2 - h_2^2}}. \tag{7.55}$$

Multiplying eq. (7.54) by (7.49), integrating over Ω_c, and using the orthogonality (3.47) we obtain

$$W_1 = - \sum_{n=1}^{\infty} \sum_{m=1}^{2n+1} \gamma_n^m \int_{h_2}^{c_1} (L_n^m \tilde{y}_n^m(\rho_0)) \tilde{y}_n^m(\rho_0) d\rho_0. \tag{7.56}$$

Note the reduction of eq. (7.56) to the spherical case, where $h_3 \to 0$, $h_2 \to 0$, $\rho_0 \to r_0$ and $\gamma_n^m \to 1$, recalling that the surface spherical harmonics have been taken as normalized, we obtain

$$L_n^m \to \frac{d}{dr_0} r_0^2 \frac{d}{dr_0} - n(n+1) = M_n. \tag{7.57}$$

Hence, eq. (7.56) is reduced to eq. (7.34).

The integrand in eq. (7.56) can be rewritten in the form

$$(L_n^m\tilde{y}_n^m)\tilde{y}_n^m = \frac{d}{dr_0}\left(\sqrt{\rho_0^2-h_3^2}\sqrt{\rho_0^2-h_2^2}\,\tilde{y}_n^m\frac{d}{dr_0}\tilde{y}_n^m\right)$$
$$-\sqrt{\rho_0^2-h_3^2}\sqrt{\rho_0^2-h_2^2}\left(\frac{d}{dr_0}\tilde{y}_n^m\right)^2+$$
$$+\frac{(h_3^2+h_2^2)P_n^m-n(n+1)\rho_0^2}{\sqrt{\rho_0^2-h_3^2}\sqrt{\rho_0^2-h_2^2}}(\tilde{y}_n^m)^2. \tag{7.58}$$

Introducing the expression (7.58) in (7.56) and using the boundary condition (7.51) we obtain

$$W_1=\sum_{n=1}^{\infty}\sum_{m=1}^{2n+1}\gamma_n^m\int_{h_2}^{c_1}\left[\sqrt{\rho_0^2-h_3^2}\sqrt{\rho_0^2-h_2^2}\left(\frac{d}{dr_0}\tilde{y}_n^m(\rho_0)\right)^2\right.$$
$$\left.-\frac{(h_3^2+h_2^2)p_n^m-n(n+1)\rho_0^2}{\sqrt{\rho_0^2-h_3^2}\sqrt{\rho_0^2-h_2^2}}(\tilde{y}_n^m(\rho_0))^2\right]d\rho_0. \tag{7.59}$$

Utilizing formula (7.54) in expansion (7.47) and using orthogonality we find the expansion

$$u_s|_{S_s}=-\frac{1}{4\pi}\sum_{n=1}^{\infty}\sum_{m=1}^{2n+1}C_{sn}^{sm}S_n^m(\mu_0,\nu_0)\int_{h_2}^{c_1}(L_n^m\tilde{y}_n^m(\rho_0))E_n^m(\rho_0)d\rho_0. \tag{7.60}$$

Comparing eq. (7.60) with (7.48) and using the orthogonality of the surface ellipsoidal harmonics S_n^m we obtain the equations

$$\int_{h_2}^{c_1}(L_n^m\tilde{y}_n^m(\rho_0))E_n^m(\rho_0)d\rho_0=-\frac{4\pi\tilde{u}_n^m}{C_{sn}^{sm}} \tag{7.61}$$

for $n=1,2,\ldots$ and $m=1,2,\ldots,2n+1$. Integration by parts gives

$$\int_{h_2}^{c_1}\left[\frac{d}{d\rho_0}\sqrt{\rho_0^2-h_3^2}\sqrt{\rho_0^2-h_2^2}\left(\frac{d}{d\rho_0}\tilde{y}_n^m(\rho_0)\right)\right]E_n^m(\rho_0)d(\rho_0)$$
$$=\sqrt{\rho_0^2-h_3^2}\sqrt{\rho_0^2-h_2^2}\left(\tfrac{d}{d\rho_0}\tilde{y}_n^m(\rho_0)\right)E_n^m(\rho_0)\Bigg|_{\rho_0=h_2}^{\rho_0=c_1}$$
$$-\int_{h_2}^{c_1}\sqrt{\rho_0^2-h_3^2}\sqrt{\rho_0^2-h_2^2}\left(\frac{d}{d\rho_0}\tilde{y}_n^m(\rho_0)\right)\left(\frac{d}{d\rho_0}E_n^m(\rho_0)\right)d\rho_0$$

$$= -\sqrt{\rho_0^2 - h_3^2}\sqrt{\rho_0^2 - h_2^2}\left(\frac{d}{d\rho_0}E_n^m(\rho_0)\right)\tilde{y}_n^m(\rho_0)\Bigg|_{\rho_0 = h_2}^{\rho_0 = c_1}$$

$$+ \int_{h_2}^{c_1}\left[\frac{d}{d\rho_0}\sqrt{\rho_0^2 - h_3^2}\sqrt{\rho_0^2 - h_2^2}\frac{d}{d\rho_0}E_n^m(\rho_0)\right]\tilde{y}_n^m(\rho_0)d\rho_0$$

$$= -c_2c_3E_n^{m'}(c_1)\tilde{y}_n^m(c_1)$$

$$- \int_{h_2}^{c_1}\frac{(h_3^2 + h_2^2)p_n^m - n(n+1)\rho_0^2}{\sqrt{\rho_0^2 - h_3^2}\sqrt{\rho_0^2 - h_2^2}}E_n^m(\rho_0)\tilde{y}_n^m(\rho_0)d\rho_0. \tag{7.62}$$

Therefore,

$$\int_{h_2}^{c_1}(L_n^m\tilde{y}_n^m(\rho_0))E_n^m(\rho_0)d\rho_0 = -c_2c_3E_n^{m'}(c_1)\tilde{y}_n^m(c_1) = -\frac{4\pi\tilde{u}_n^m}{C_{sn}^{sm}} \tag{7.63}$$

from which we obtain the boundary condition

$$\tilde{y}_n^m(c_1) = \frac{4\pi\tilde{u}_n^m}{c_2c_3E_n^{m'}(c_1)C_{sn}^{sm}}. \tag{7.64}$$

Following a variational argument similar to the one used in the minimization for the sphere we obtain the nonhomogeneous equation

$$\int_{h_2}^{c_1}\left(L_n^m\hat{\tilde{y}}_n^m(\rho_0)\right)E_n^m(\rho_0)d,\rho_0 = -4\pi\frac{\tilde{u}_n^m}{C_{sn}^{sm}} \tag{7.65}$$

for the minimizing function $\hat{\tilde{y}}_n^m$. This equation implies that

$$L_n^m\hat{\tilde{y}}_n^m(\rho_0) = \delta_n^mE_n^m(\rho_0) + d_n^m\rho_0 \tag{7.66}$$

where δ_n^m is a constant and the function $d_n^m\rho_0$ is orthogonal to $E_n^m(\rho_0)$ in the L^2-inner product.

Introducing eq. (7.66) in (7.65) we obtain

$$\delta_n^m = -\frac{4\pi}{||\mathrm{E}_n^m||^2}\frac{u_n^m}{C_{sn}^{sm}}. \tag{7.67}$$

Eq. (7.66) takes the form

$$L_n^m\hat{\tilde{y}}_n^m(\rho_0) = -4\pi\frac{E_n^m(\rho_0)}{||\mathrm{E}_n^m||^2}\frac{u_n^m}{C_{sn}^{sm}}. \tag{7.68}$$

The homogeneous equation

$$L_n^m \hat{y}_n^m(\rho_0) = 0 \tag{7.69}$$

is the Lamé equation, which has the solutions $E_n^m(\rho_0)$ and $F_n^m(\rho_0)$. Using these two linearly independent solutions and the method of Lagrange we obtain the particular solution

$$w_n^m(\rho_0) = \frac{4\pi \tilde{u}_n^m}{(2n+1)||\mathrm{E}_n^m||^2 C_{sn}^{sm}} \left[F_n^m(\rho_0) \int_{h_2}^{\rho_0} E_n^m(\rho')E_n^m(\rho')d\rho' \right.$$
$$\left. + E_n^m(\rho_0) \int_{\rho_0}^{c_1} E_n^m(\rho')F_n^m(\rho')d\rho' \right]. \tag{7.70}$$

The general solution of eq. (7.68) is then given by

$$\hat{y}_n^m(\rho_0) = B_n^m E_n^m(\rho_0) + w_n^m(\rho_0) \tag{7.71}$$

where w_n^m is given by eq. (7.70) and the constant B_n^m is determined by the boundary condition

$$\frac{d}{d\rho_0} \hat{y}_n^m(\rho_0) \bigg|_{\rho_0 = c_1} = 0. \tag{7.72}$$

Hence,

$$\frac{4\pi}{2n+1} \frac{u_n^m}{||\mathrm{E}_n^m||^2 C_{sn}^{sm}} \left[\left(\frac{d}{d\rho_0} E_n^m(c_1) \right) \cdot 0 - (E_n^m(c_1))^2 F_n^m(c_1) \right.$$
$$\left. + \left(\frac{d}{d\rho_0} F_n^m(c_1) \right) \int_{h_2}^{c_1} (E_n^m(\rho'))^2 d\rho' + (E_n^m(c_1))^2 F_n^m(c_1) \right] + B_n^m \frac{d}{d\rho_0} E_n^m(c_1) = 0 \tag{7.73}$$

or

$$B_n^m = -\frac{4\pi}{2n+1} \frac{\tilde{u}_n^m}{||\mathrm{E}_n^m||^2 C_{sn}^{sm}} \frac{(F_n^m(c_1))'}{(E_n^m(c_1))'} \int_{h_2}^{c_1} (E_n^m(\rho'))^2 d\rho' \tag{7.74}$$

where the integral above is equal to $||\mathrm{E}_n^m||^2$.

Finally, the minimizing function takes the form

$$\hat{\tilde{y}}_n^m(\rho_0) = \frac{4\pi}{2n+1} \frac{\tilde{u}_n^m}{||\mathrm{E}_n^m||^2 C_{sn}^{sm}} \left[F_n^m(\rho_0) \int_{h_2}^{\rho_0} E_n^m(\rho') E_n^m(\rho') d\rho' \right.$$

$$\left. + E_n^m(\rho_0) \int_{\rho_0}^{c_1} E_n^m(\rho') F_n^m(\rho') d\rho' \right] - \frac{4\pi}{2n+1} \frac{\tilde{u}_n^m}{C_{sn}^{sm}} \frac{\left(F_n^m(c_1)\right)'}{\left(E_n^m(c_1)\right)'} \cdot E_n^m(\rho_0) \tag{7.75}$$

7.4 Reduction to the sphere

In what follows we will reduce the solution (7.785) to the corresponding solution (7.37) for the sphere. First, we reduce the Lamé functions.

The Lamé equation reads

$$(x^2 - h_3^2)(x^2 - h_2^2) E_n^{m\prime\prime}(x) + x(x^2 - h_3^2 + x^2 - h_2^2) E_n^{m\prime}(x)$$

$$+ \left[(h_3^2 + h_2^2) p_n^m - n(n+1) x^2 \right] E_n^m(x) = 0. \tag{7.76}$$

In the case of the sphere

$$h_2^2 \to 0, h_3^2 \to 0 \tag{7.77}$$

thus the above equation reduces to

$$x^2 R''_n(x) + 2x R'_n(x) - n(n+1) R_n(x) = 0. \tag{7.78}$$

The solutions of eq. (7.76) are the Lamé functions of the first kind $E_n^m(x)$ and those of the second kind

$$F_n^m(x) = (2n+1) E_n^m(x) \int_x^{+\infty} \frac{dt}{\left(E_n^m(t)\right)^2 \sqrt{t^2 - h_3^2} \sqrt{t^2 - h_2^2}}. \tag{7.79}$$

The solutions of eq. (7.78) are the functions

$$S_n^i(x) = x^n \text{ and } S_n^e(x) = \frac{1}{x^{n+1}}. \tag{7.80}$$

Hence,

$$E_n^m(x) \to x^n$$

and

$$F_n^m(x) \to (2n+1)x^n \int_x^{+\infty} \frac{dt}{t^{2n+2}} = (2n+1)x^n \left(- \frac{1}{(2n+1)t^{2n+1}} \Big|_x^{+\infty} \right)$$
$$= x^n \frac{1}{x^{2n+1}} = \frac{1}{x^{n+1}}. \tag{7.81}$$

Using the above reductions from the ellipsoid to the sphere it follows that the form of the coefficients $\hat{\tilde{y}}_n^m(\rho)$ of the expansion of Ψ in surface ellipsoidal harmonics, which are given by eq. (7.75), are reduced as follows:

$$\hat{\tilde{y}}_n^m(\rho) \underset{e \to s}{\longrightarrow} \frac{1}{2n+1} \frac{2n+1}{c_1^{2n+1}} \frac{\tilde{u}_n^m}{C_{sn}^{sm}} \left[\rho^n \int_\rho^{c_1} \rho'^n \frac{1}{\rho'^{n+1}} d\rho' \right.$$
$$\left. + \frac{1}{\rho^{n+1}} \int_0^\rho \rho'^{2n} d\rho' - \rho^n \left(- \frac{n+1}{n} \frac{1}{c_1^{2n+1}} \right) \int_0^{c_1} \rho'^{2n} d\rho' \right]$$
$$= \frac{1}{c_1^{2n+1}} \frac{\tilde{u}_n^m}{C_{sn}^{sm}} \left[\rho^n \ln \frac{c_1}{\rho} + \frac{\rho^n}{2n+1} + \frac{n+1}{n} \frac{\rho^n}{2n+1} \right]$$
$$= - \frac{1}{c_1^{2n+1}} \frac{\tilde{u}_n^m}{C_{sn}^{sm}} \rho^n \left[\ln \frac{\rho}{c_1} - \frac{1}{n} \right]. \tag{7.82}$$

Hence, indeed the reduced form on the RHS of eq. (7.82) coincides with the spherical solution (7.53). The constants are connected to the relation

$$\frac{u_n^m}{k_n^m} = \frac{4\pi}{2n+1} \frac{\tilde{u}_n^m}{C_{sn}^{sm}}. \tag{7.83}$$

Chapter 8
Numerical implementation

In this chapter, we use the previously developed theory to numerically implement the inversion algorithms following refs. [64, 65]. The basic equations to be used in this section are the following:

For electroencephalogram (EEG)

$$u_S(\mathbf{r}) = -\frac{1}{4\pi}\int_{\Omega_c} (\nabla_{\boldsymbol{\tau}} \cdot \mathbf{J}^p(\boldsymbol{\tau}))\upsilon_S(\mathbf{r}, \boldsymbol{\tau})dV(\boldsymbol{\tau}), \quad \mathbf{r} \in S_S, \tag{8.1}$$

or

$$u_S(\mathbf{r}) = \frac{1}{4\pi}\int_{S_c} \hat{\mathbf{n}} \cdot [\Psi(\boldsymbol{\tau})\nabla_{\boldsymbol{\tau}}\upsilon_S(\mathbf{r}, \boldsymbol{\tau}) - \upsilon_S(\mathbf{r}, \boldsymbol{\tau})\nabla_{\boldsymbol{\tau}}\Psi(\boldsymbol{\tau})]dS(\boldsymbol{\tau}), \quad \mathbf{r} \in S_S. \tag{8.2}$$

and for MEG

$$\mathbf{r} \cdot \mathbf{B}(\mathbf{r}) \approx -\frac{\mu}{4\pi}\int_{\Omega_c} (\Delta_\tau(\boldsymbol{\tau} \cdot \mathbf{A}(\boldsymbol{\tau})))\frac{dV(\boldsymbol{\tau})}{|\mathbf{r} - \boldsymbol{\tau}|}, \quad \mathbf{r} \in \Omega_e, \tag{8.3}$$

or

$$\mathbf{r} \cdot \mathbf{B}(\mathbf{r}) \approx \frac{\mu}{4\pi}\int_{S_c} \mathbf{n} \cdot [A^\tau \nabla_{\boldsymbol{\tau}} \frac{1}{|\mathbf{r} - \boldsymbol{\tau}|} - \frac{1}{|\mathbf{r} - \boldsymbol{\tau}|}\nabla_{\boldsymbol{\tau}} A^\tau(\boldsymbol{\tau})]dS(\boldsymbol{\tau}), \tag{8.4}$$

$$\mathbf{r} \in \Omega_e, \quad A^\tau(\boldsymbol{\tau}) = \boldsymbol{\tau} \cdot \mathbf{A}(\boldsymbol{\tau}),$$

where the approximation is justified by the small values of the vector function $\mathbf{H}$.

8.1 Ellipsoidal geometry

The basic notations and needed elements of the theory of ellipsoidal harmonic have been discussed in Section 3.4.

The numerical algorithms we use for the computation of ellipsoidal harmonics are based on the comprehensive papers of Romain and Sean-Pierre [80], and of Bardham and Knepley [9]. For the convenience of the reader we first restate the basic formulas used in [9, 80] in the notation of our previous chapters. We will use the notation

$$\mathbf{r} = (x_1, x_2, x_3) = (\rho, \mu, \nu) \tag{8.5}$$

for the observation point and

$$\boldsymbol{\tau} = (\rho_o, \mu_o, \nu_o) \tag{8.6}$$

https://doi.org/10.1515/9783110547535-009

for a generic point for space where the current is supported. The objective in this section is to present eqs. (8.1) and (8.3) for the particular case of the three-layer ellipsoidal head geometry (ignoring the CSF layer). The ellipsoidal harmonics are computed using the open-source code provided in [9].

Using the results in [9, 47, 54, 62] together with extensive numerical testing, we have obtained the following numerically stable representation for the auxiliary function $\upsilon_s(\mathbf{r},\boldsymbol{\tau})$:

$$\upsilon_s(r,\tau) = 4\pi \sum_{n=1}^{\infty} \sum_{p=1}^{2l+1} \frac{C_n^p}{\gamma_n^p} \mathbb{E}_n^p(\tau)\mathbb{E}_n^p(r), \tag{8.7}$$

where C_n^p are geometry-dependent coefficients. In other words, in order to obtain a numerically stable approach for computing the expansion coefficients of υ_s, it is convenient to express these coefficients in the form $\frac{C_n^p}{\gamma_n^p}$. A numerically robust approach for estimating the coefficients C_l^m using a boundary element method (BEM) solver will be discussed later in this chapter.

Equation (8.7) yields

$$\partial \upsilon_s(r,\tau)\partial \rho_0 = 4\pi \sum_{l=1}^{\infty} \sum_{m=1}^{2l+1} \frac{C_l^m}{\gamma_l^m} \frac{dE_l^m(\rho_0)}{d\rho_0} E_l^m(\mu_0)E_l^m(\nu_0)\mathbb{E}_l^m(r). \tag{8.8}$$

For $\tau \in \Omega_c$, the function $\Psi(\boldsymbol{\tau})$ can be expanded in the form

$$\Psi(\boldsymbol{\tau}) = \sum_{n=1}^{\infty} \sum_{p=1}^{2n+1} \psi_n^p(\rho_0)E_n^p(\mu_0)E_n^p(\nu_0), \tag{8.9}$$

where $\{\psi_n^p(\rho_0)\}$ are unknown functions. Similarly, $\tau A^{\tau}(\boldsymbol{\tau})$ can be expanded in the form

$$\tau A^{\tau}(\boldsymbol{\tau}) = \sum_{n=1}^{\infty} \sum_{m=1}^{2n+1} f_n^m(\rho_0)E_n^m(\mu_0)E_n^m(\nu_0), \quad \boldsymbol{\tau} \in \Omega_c. \tag{8.10}$$

It is shown in [47] that eq. (8.1) yields the following equation for the case of the ellipsoidal geometry:

$$u_s(r) = 4\pi b_3 c_3 \sum_{n=1}^{\infty} \sum_{p=1}^{2n+1} C_n^p \mathbb{E}_n^p(r)[\psi_n^p(a_3)\dot{E}_n^p(a_3) - \dot{\psi}_n^p(a_3)E_n^p(a_3)], \quad r \in S_s. \tag{8.11}$$

In the case of magnetoencephalogram (MEG) and for the particular case of the ellipsoidal geometry, it was shown in [47] that the measurement equation is given by

$$4\pi\mu r \cdot B(r) \approx 4\pi b_3 c_3 \sum_{n=1}^{\infty} \sum_{p=1}^{2n+1} [f_n^p(a_3)\dot{E}_n^p(a_3) - \dot{f}_n^p(a_3)E_n^p(a_3)] \frac{\mathbb{F}_n^p(r)}{2n+1}, \quad r \in \Omega_e. \tag{8.12}$$

8.1.1 Estimating the geometry-dependent coefficients C_l^m

The coefficients C_l^m appearing in eq. (8.7) have been derived in [54]. However, unlike the analogous expression in the case of spherical geometry derived in [88], the expressions derived in [54] are far more complex. In [66], an algorithmic procedure was presented to estimate the geometry-dependent coefficients C_l^m using data generated from a BEM solver and the relation $4\pi u_s(\mathbf{r},\boldsymbol{\tau}) = \mathbf{Q}(\boldsymbol{\tau}) \cdot \nabla_{\boldsymbol{\tau}} v_s(\mathbf{r},\boldsymbol{\tau})$. The algorithmic steps presented in [66] are recalled here. The electric potential $u_s(\mathbf{r},\boldsymbol{\tau},\mathbf{Q}(\boldsymbol{\tau}),C_l^m)$ due to a dipole source $\mathbf{Q}(\boldsymbol{\tau})$ is given by

$$u_s(r^{\cdot},\tau,\mathbf{Q}(\tau),C_l^m) = \sum_{l=1}^{\infty}\sum_{m=1}^{2l+1}\frac{C_l^m}{\gamma_l^m}\mathbf{Q}\cdot\nabla_{\tau}\mathbb{E}_l^m(\tau)\mathbb{E}_l^m(r), \tag{8.13}$$

where $\nabla_{\tau}\mathbb{E}_l^m(\tau)$ can be estimated accurately, using a finite difference approximation (numerical results regarding the accuracy of a finite difference approximation of $\nabla_{\tau}\mathbb{E}_l^m(\tau)$ are depicted in Figure 8.3). For the case of *an arbitrary* head model, there exist numerical solvers to compute the electric potential (volts) on the surface S_s due to a dipole source in Ω_c (Coulomb-meter); see, for example, [56, 57]. The steps for a numerical estimate of C_l^m are as follows:

(i) Choose N source-observation pairs $(\mathbf{r},\boldsymbol{\tau})$ of randomly oriented dipoles $\mathbf{Q}(\boldsymbol{\tau})$.

(ii) Employ a BEM/FEM solver to compute the potentials for each source observation pair $(\mathbf{r},\boldsymbol{\tau})$. Here, we use the freely available openMEEG [56] to obtain these potentials and denote them by $\{\tilde{u}_i : i = 1,\ldots,N\}$. These solutions feature as data in a minimization algorithm.

(iii) It is clear from eq. (8.13) that the relationship between $u_s(\mathbf{r},\boldsymbol{\tau},\mathbf{Q},C_l^m)$ and C_l^m is linear. The least squares estimate of C_l^m is obtained by the minimizing function

$$H(C_l^m) = \min\sum_{i=1}^{N} \| \tilde{u}_i - u_s(\mathbf{r}_i,\boldsymbol{\tau}_i,\mathbf{Q},C_l^m)\|^2, \tag{8.14}$$

where $u_s(\mathbf{r},\boldsymbol{\tau},\mathbf{Q},C_l^m)$ denotes the parametric form of eq. (8.11) and $\{\tilde{u}_i : i = 1,\ldots,N\}$ are the data generated in step (ii). The estimated geometry-dependent coefficients using the proposed numerical procedure are denoted by $\tilde{C}_l^m$ to differentiate them from the exact coefficients denoted by C_l^m.

(iv) The above procedure can be cross-validated as follows: Choose *a new set* of source–observation pairs $(\mathbf{r},\boldsymbol{\tau})$ of randomly oriented dipoles $\mathbf{Q}(\boldsymbol{\tau})$. Employ a BEM/FEM solver to compute the potential for each source observation pair $(\mathbf{r},\boldsymbol{\tau})$ to generate the data set $\{u_i : 1 \le i \le T\}$. Then, employing the estimated $\tilde{C}_l^m$ (step iii) in eq. (8.11) to generate estimated data $u_s(\mathbf{r},\boldsymbol{\tau},\mathbf{Q},\tilde{C}_l^m)$ and compare them with the cross-validation data. Results of such comparison are depicted in Figure 8.4.

8.1.2 Numerical results

Numerical tests relevant to the implementation of ellipsoidal harmonics employed in this study are discussed in [9]. Here we present additional numerical results for several important functions that feature in the inversion formulae in [47] with the relevant root mean square errors (RMSE). Following the recommendations of the authors in [9] regarding numerical stability, we have restricted the order of the ellipsoidal harmonics to $n \leq 12$.

The basic idea, on which the numerical approach in [9, 76] is developed, is that for every x in the interval of interest, the Lame polynomial

$$K_m(x) = \sum_{k=0}^{m} a_k x^{2k} \tag{8.15}$$

is replaced with the polynomial

$$\hat{K}_m(x) = \sum_{k=0}^{m} b_k \left(1 - \frac{x^2}{h_3^2}\right)^k \tag{8.16}$$

in such a way that

$$K_m(x) = \hat{K}_m(x). \tag{8.17}$$

From the numerical point of view, all the coefficients b_k have the same dimensions, and the variable $(1 - x^2/h_3^2)$ is dimensionless. This is not true with the terms appearing in the form (8.15), where only the products $a_k x^{2k}$ have the same dimensions. In the terminology of linear algebra, the vectorial polynomial

$$\mathbf{P}_\tau = (a_0, a_1, \ldots, a_m) \tag{8.18}$$

with components a_k, expressed in the basis $\{x^{2k}\}_{k=0}^{m}$, is also expressed as

$$\hat{\mathbf{P}}_\tau = (b_0, b_1, \ldots, b_m) \tag{8.19}$$

with components b_k, expressed in the basis $\left\{\left(1 - \frac{x^2}{h_3^2}\right)^k\right\}_{k=0}^{m}$. The above basis of representation is more convenient for a numerical treatment.

The parameters τ_a, τ_b, τ_c of the three-layer ellipsoidal head model used in this study are given in Table 8.1. These parameters are obtained by fitting ellipsoidal shells to a realistic head model.

Table 8.1: Ellipsoidal head model and triangular surface mesh parameters used in the openMEEG computations.

Domain	Conductivity (S/m)	*a* (m)	*b* (m)	*c* (m)	Nodes	Triangles
Cerebrum	0.33	0.0702	0.0614	0.0527	3,124	6,244
Skull	0.0042	0.0780	0.0683	0.0585	3,124	6,244
Scalp	0.33	0.0867	0.0758	0.0650	3,126	6,248

8.1.3 Numerical verification

In this section, we verify numerically the known expressions

$$\frac{1}{|\mathbf{r}-\boldsymbol{\tau}|} = \sum_{n=0}^{\infty}\sum_{p=1}^{2n+1} \frac{4\pi}{2n+1}\frac{1}{\gamma_n^p}\mathbb{E}_n^p(\boldsymbol{\tau})\mathbb{F}_n^p(\mathbf{r}), \quad |\mathbf{r}|>|\boldsymbol{\tau}| \tag{8.20}$$

and

$$\nabla_{\boldsymbol{\tau}}\frac{1}{|\mathbf{r}-\boldsymbol{\tau}|}\cdot\hat{\mathbf{n}} = \sum_{n=0}^{\infty}\sum_{p=1}^{2n+1} \frac{4\pi c_2 c_3}{2n+1}\frac{1}{\gamma_n^p}\mathbb{F}_n^p(\mathbf{r})\frac{E_n^p(\nu_0)E_n^p(\mu_0)}{\sqrt{c_1^2-\mu_0^2}\sqrt{c_1^2-\nu_0^2}}\frac{dE_n^p(\rho_0)}{d\rho_0}, \tag{8.21}$$

where the normal differentiation is given by

$$\frac{\partial}{\partial n} = \hat{\mathbf{n}}\cdot\nabla = \frac{c_2 c_3}{\sqrt{c_1^2-\mu_0^2}\sqrt{c_1^2-\nu_0^2}}\frac{\partial}{\partial\rho_0}, \tag{8.22}$$

and $\hat{\mathbf{n}}$ denotes the outward unit normal in the surface $\rho = c_1$.

We compare numerically the left- and right-hand sides of eqs. (8.20) and (8.21). This provides comparisons between analytic expressions and their corresponding ellipsoidal harmonic expansions, where the maximum degree is $n_{\text{max}} = 12$. These comparisons provide important insight into the convergence properties of ellipsoidal harmonics and the accuracy of the implementation described in [9]. We consider two different cases relevant to EEG and MEG.

In the first case, we place the observation point on the surface of the scalp, that is, $\mathbf{r} \in S_s$. This is relevant to EEG, where the sensor array is positioned on the scalp. Results are shown in Figure 8.1. Subplot (a) compares the left- and right-hand sides of eq. (8.20). Subplot (b) compares the left- and right-hand sides of eq. (8.21). The absolute errors between the analytic expressions and the corresponding ellipsoidal harmonic expansions are shown in subplots (c) and (d). As before, subplot (c) depicts the absolute error between the left- and right-hand sides of eq. (8.20). Subplot (d) is the absolute error between the left- and right-hand sides of eq. (8.21).

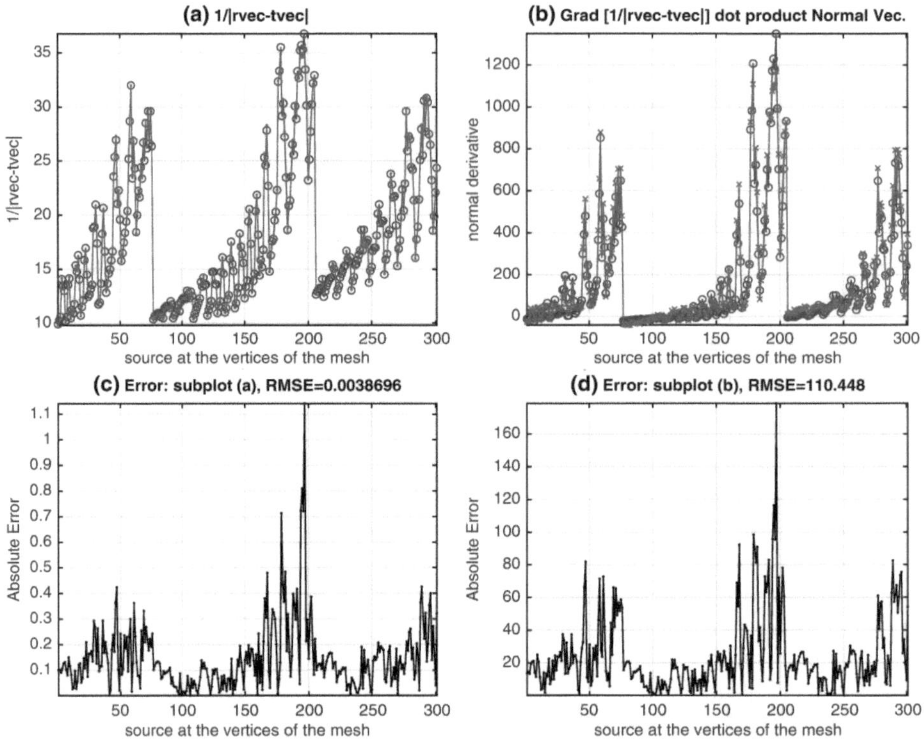

Figure 8.1: Comparison of the left- and right-hand sides of eqs. (8.20) and (8.21) and the corresponding error plots. The observation point is on the surface S_s, that is, $\mathbf{r} \in S_s$. The set of points $\{\tau_j \in S_c\}$ are chosen to be a subset of nodes of the surface mesh S_c. For reasons associated with numerical stability, we choose the limit of the outer sum to be $n_{\max} = 12$. Subplot (a) shows the left-hand side and right-hand side of eq. (8.20), that is, $\frac{1}{|\mathbf{r}-\tau|}$. The blue solid line with circle depicts the left-hand side and the red solid line with dots depicts the right-hand side. Subplot (c) shows the absolute error between the two sides of eq. (8.20), RMSE =0.0039. Subplot (b) shows the left-hand side and right-hand side of eq. (8.21), that is, $\nabla_\tau \frac{1}{|\mathbf{r}-\tau|} \cdot \hat{\mathbf{n}}$. The blue solid line with circle depicts the left-hand side and the red solid line with dots depicts the right-hand side. Subplot (d) shows the absolute error between the two sides of eq. (8.21), RMSE =110.45.

The second test case is identical to the first case except that the observation point is placed outside the head, that is, $\mathbf{r} \in \Omega_e$. This is relevant to MEG, where the sensor array is outside the head. Results are shown in Figure 8.2. Subplot (a) compares the left- and right-hand sides of eq. (8.20). Subplot (b) compares the left- and right-hand sides of eq. (8.21). The absolute errors between the analytic expressions and its corresponding ellipsoidal harmonic expansions are shown in subplot (c) and subplot (d). More precisely, subplot (c) depicts the absolute error between the left- and right-hand sides of eq. (8.20), whereas subplot (d) depicts the absolute error between the left- and right-hand sides of eq. (8.21).

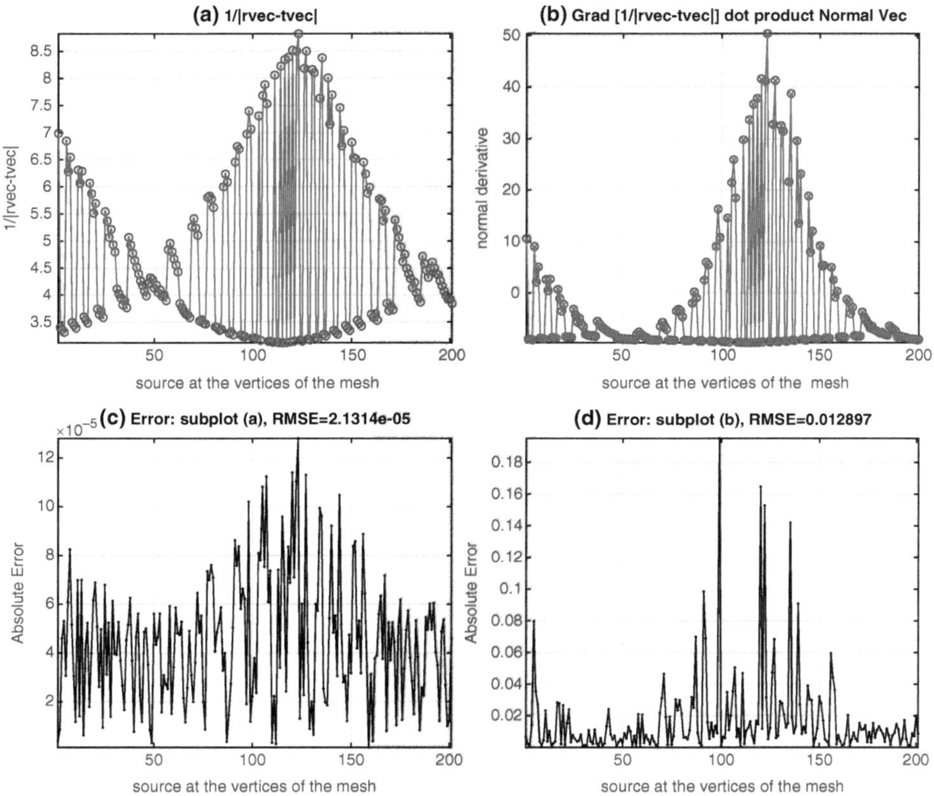

Figure 8.2: Comparison of the left and right sides of eqs. (8.20) and (8.21) and the corresponding error plots. The observation point is $\mathbf{r} \in \Omega_e$. The set of points $\{\tau_j \in S_c\}$ are chosen to be a subset of nodes of the surface mesh S_c. For reasons associated with numerical stability, we choose the limit of the outer sum to be $n_{\max} = 12$. Subplot (a) depicts the left- and right-hand sides of eq. (8.20), that is, $\frac{1}{|\mathbf{r}-\tau|}$. The blue solid line with circle depicts the left-hand side and the red solid line with dots depicts the right-hand side. Subplot (c) depicts the absolute error between the two sides of eq. (8.20), RMSE = 2.1314×10^{-5}. Subplot (b) shows the left- and right-hand sides of eq. (8.21), that is, $\nabla_\tau \frac{1}{|\mathbf{r}-\tau|} \cdot \hat{\mathbf{n}}$. The blue solid line with circle depicts the left-hand side and the red solid line with dots depicts the right-hand side. Subplot (d) depicts the absolute error between the two sides of eq. (8.21), RMSE =0.0129.

A comparison of the absolute errors of Figures 8.1 and 8.2 (subplots (c) and (d)) suggests that the numerical stability in the implementation of ellipsoidal harmonics as outlined in [9] is sensitive to the distance between the source point $\boldsymbol{\tau} \in S_c$ and the observation point $\mathbf{r}$, that is, $d = |\mathbf{r} - \boldsymbol{\tau}|$. More precisely, the errors between the left- and right-hand sides of eqs. (8.20) and (8.21) increase substantially as the distance between the source–observation pair $(\mathbf{r}, \boldsymbol{\tau})$ decreases. Moreover, in tests that increase the number of terms in the ellipsoidal harmonic expansion for the right-hand sides of eqs. (8.20) and (8.21), we have found that the numerics become very unstable for $n_{\max} > 12$, which is consistent with discussions in [9].

8.1.4 The computation of C_l^m

Equation (8.13) requires the computation of the gradient $\nabla_\tau \mathbb{E}_n^p(\tau)$, which we choose to estimate using the simple finite difference technique. A reasonable test in assessing this estimate ($\nabla_\tau \mathbb{E}_n^p(\tau)$) is to consider again a simple test involving eq. (8.20): we consider the scalar product of the gradient of eq. (8.20) with a dipole $\mathbf{Q}(\boldsymbol{\tau})$. We place 462 dipoles $\mathbf{Q}(\boldsymbol{\tau})$ in the cerebrum, $\boldsymbol{\tau} \in \Omega_c$, and choose an arbitrary observation vector on the scalp surface $\mathbf{r} \in S_s$. The test involves comparing $\mathbf{Q} \cdot \nabla_{\boldsymbol{\tau}} \frac{1}{|\mathbf{r}-\boldsymbol{\tau}|}$ (which is straightforward to compute) with its corresponding ellipsoidal harmonics approximation as follows:

$$\frac{\mathbf{Q} \cdot (\mathbf{r} - \boldsymbol{\tau})}{|\mathbf{r} - \boldsymbol{\tau}|^3} \approx \sum_{n=1}^{12} \sum_{p=1}^{2n+1} \frac{4\pi}{2n+1} \frac{1}{\gamma_n^p} \mathbf{Q} \cdot \nabla_{\boldsymbol{\tau}} \mathbb{E}_n^p(\boldsymbol{\tau}) \mathbb{F}_n^p(\mathbf{r}). \tag{8.23}$$

The finite difference technique is used to estimate $\nabla_\tau \mathbb{E}_n^p(\tau)$ occurring in the right-hand side of (8.23). The results are shown in Figure 8.3.

8.1.5 Cross-validation of C_l^m

The estimated coefficients $\tilde{C}_l^m$ are used to generate a *new* data set which is compared with data generated from a BEM solver [55]. This approach is referred to as cross-validation. A total of 9,700 source–observation pairs were used during the estimation step, that is, $N = 9{,}700$ in eq. (8.14). The geometric and mesh parameters used in the BEM computations are given in Table 8.1.

The results of this analysis are shown in Figure 8.4. Subplot (a) depicts the estimated coefficients $\log[\tilde{C}_l^m]$ presented in sequential order. For example, on the x-axis, index 10 refers to the coefficient $\tilde{C}_4^1$. Subplot (b) depicts the comparison between the estimated data using eq. (8.13) and data generated using a BEM solver on a new set of source–observation pairs. Subplot (c) shows the absolute error, that is, $\text{Error}(i) := |u_i - u_s(\mathbf{r}_i, \boldsymbol{\tau}_i, \mathbf{Q}, \tilde{C}_l^m)|$, where $\tilde{C}_l^m$ has been used to generate the estimated data.

8.1.6 EEG inversion matrix

In this subsection, we present reconstructions for EEG using synthetic data. Recall the EEG inversion formula given by eq. (8.13). For reasons related to the sign

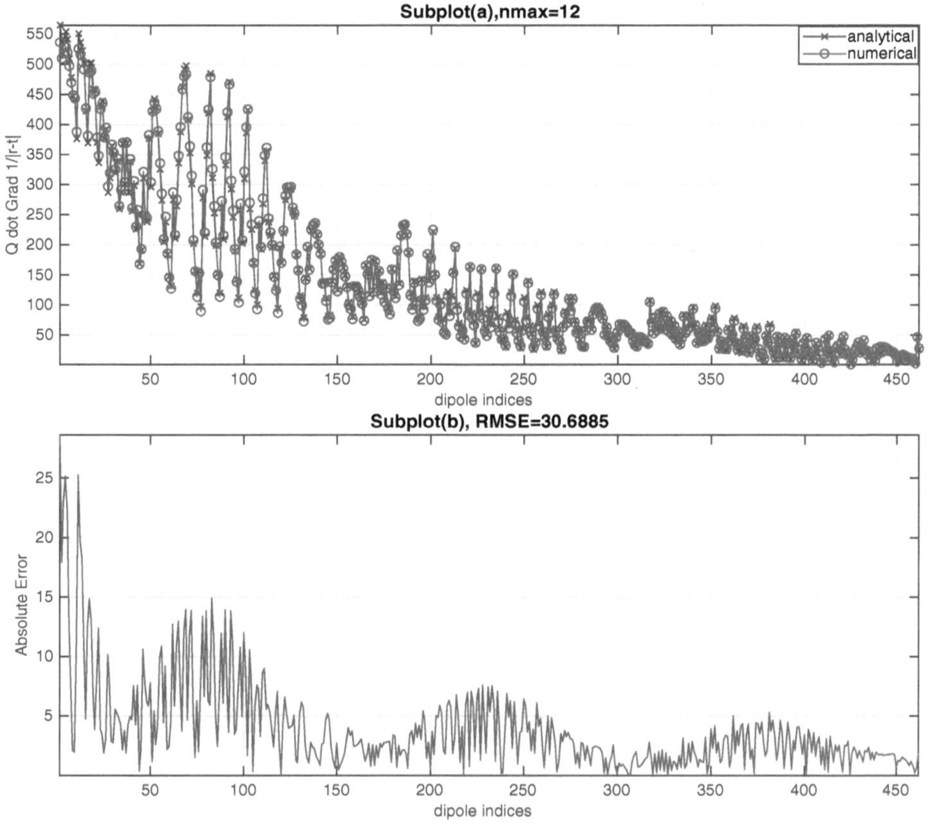

Figure 8.3: In subplot (a) the blue solid line with cross depicts the left-hand side of (8.21) and the red solid line with circle depicts the right-hand side of eq. (8.21). Subplot (b) shows the absolute error with RMSE = 30.7. A total of 462 dipoles inside the cerebrum $\tau \in \Omega_c$ has been selected to perform this test. The observation point $\mathbf{r} \in S_s$ has been chosen on the surface S_s.

ambiguity issue discussed in [9], we rewrite the parametrization of eq. (8.9) as follows:

$$\Psi(\tau) = \sum_{n=1}^{\infty} \sum_{p=1}^{2n+1} \psi_n^p(\lambda) \mathbb{E}_n^p(\tau), \tag{8.24}$$

where the functions $\psi_n^p(\lambda)$ have a λ dependence. For the particular case of the spherical head model, a procedure was outlined in [48] to establish the radial dependence via a minimization procedure. The expansion of data on an ellipsoidal surface fixes the angular dependence; thus, only the radial dependence is not determined. Here, for simplicity we assume that $\psi_n^p(\lambda)$ takes the form $\psi_n^p(\lambda) = \lambda a_n^p$. This dependence can be fixed via the analysis of Chapter 7. Equation (8.24) is employed in the numerical

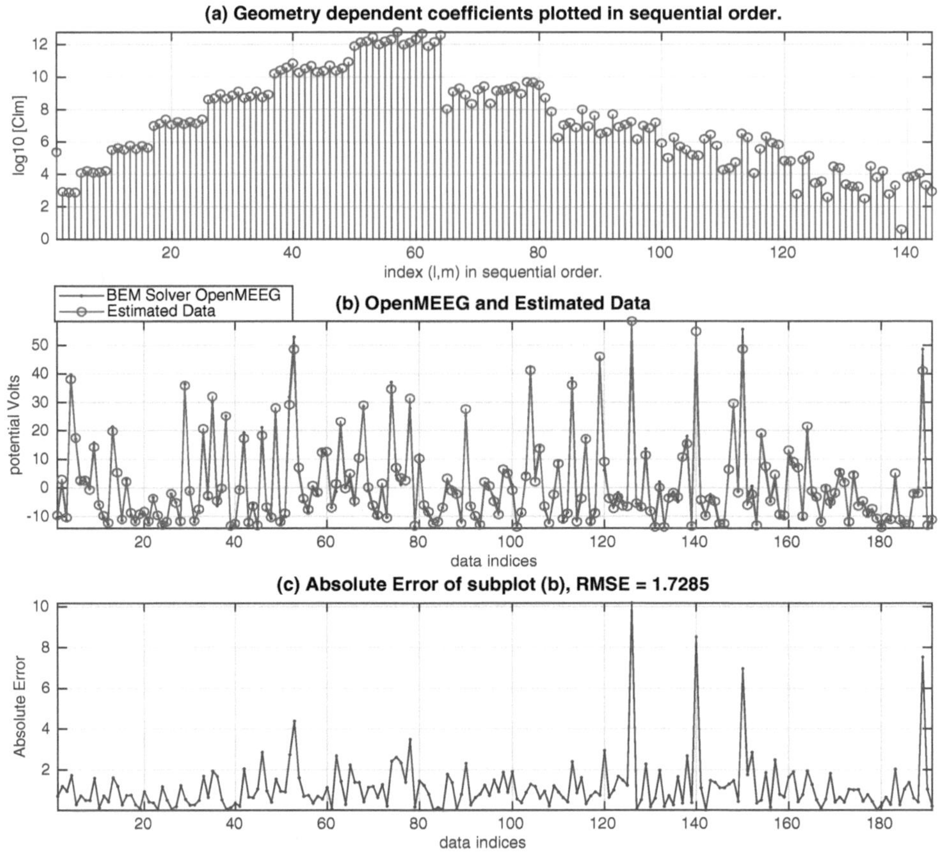

Figure 8.4: Subplot (a) depicts the plot of the logarithm of the geometry-dependent coefficients in sequential order, that is, $\log[C_l^m]$. For example, on the x-axis, index 5 refers to $(l=2, m=0)$ and index 25 on the x-axis refers to $(l=4, m=8)$, that is, C_4^8. In these computations, $n_{\max}=12$. Subplot (b) depicts the comparison between test data and estimated data $u_s(\mathbf{r}, \tau, \mathbf{Q}, \tilde{C}_l^m)$. Subplot(c) depicts the absolute error between test and estimated data depicted in subplot (b) with RMSE = 1.7285.

computations due to problems associated with sign ambiguity in the implementation of ellipsoidal harmonics discussed in [9]. On the surface S_c, $\lambda = a_3$ and $\{a_n^p\}$ are the unknown coefficients linearly related to data. Equation (8.11) gives rise to the system of linear equations

$$\mathbf{u} = \mathrm{E}\alpha, \tag{8.25}$$

where $\mathbf{u}$ is the data (from EEG electrodes), $E \in \mathbb{R}^{N_s \times N_p}$ is the ellipsoidal inversion matrix constructed exactly as stated by eq. (8.11) and $\alpha = \{a_n^p\} \in \mathbb{R}^{N_p}$ is the vector of unknown coefficients to be estimated from the data. We have assumed a sensor array of 312 electrodes distributed evenly around the scalp, that is, $N_s = 312$. The position vectors of the electrodes on the surface S_s are simply a subset of the nodes of the surface S_S. Surface S_S has $N = 3,121$ nodes and one electrode has been placed for every tenth node of this mesh. This is an ideal scenario. In practice, it is physically impossible to place one electrode at every tenth node. Despite this ideal situation, the results will show that this problem is still highly ill-conditioned. Moreover, we have restricted the dimension of the vector of unknown coefficients a_n^p to be $N_p = 100$, that is, $\alpha \in \mathbb{R}^{100}$. So, we are intentionally considering the case that we have more data than unknowns.

Our first numerical test involves a very simple test of generating a vector of random coefficients $\alpha = \{a_n^p\}$ to obtain a data set via eq. (8.25), $\mathbf{u} = \mathrm{E}\alpha$. We then obtain an estimate of α via the same equation, $\hat{\alpha} = \mathrm{E}^{-1}\mathbf{u}$. This test involving no additive noise reveals some interesting properties of the ellipsoidal inversion matrix E. The results are shown in Figure 8.5. Figure 8.5(a) compares the randomly generated vector α and its corresponding estimate $\hat{\alpha}$. It can be observed that we can only recover

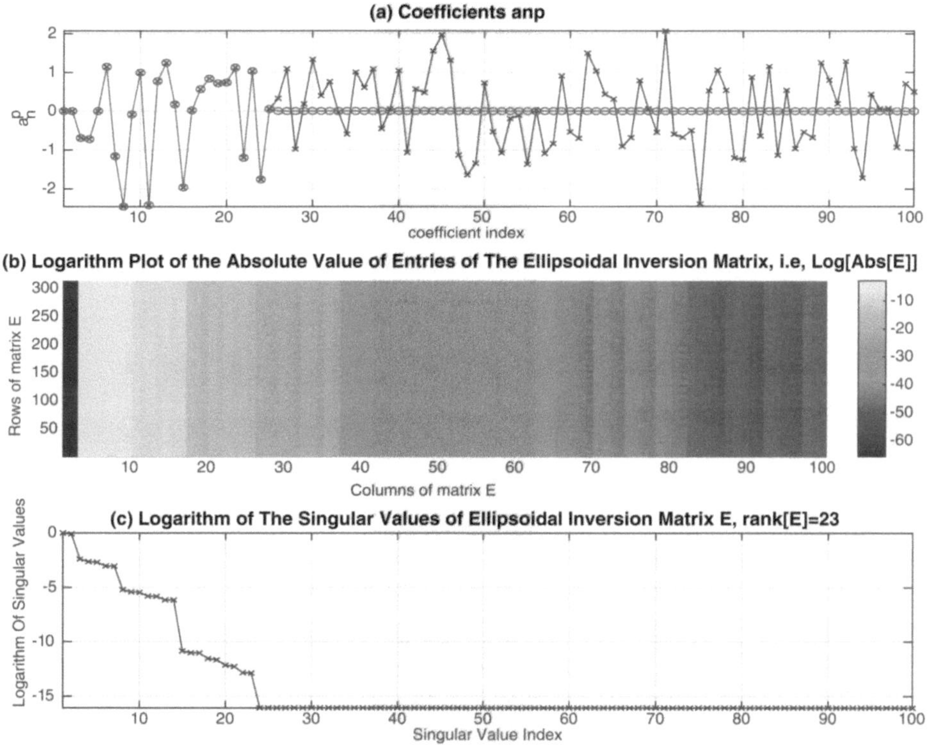

Figure 8.5: Subplot (a) compares the randomly generated vector of $\alpha = \{a_n^p\}$ and its corresponding estimate $\hat{\alpha}$ obtained via eq. (8.25). The blue solid line with cross is $\alpha = \{a_n^p\}$ and the red solid line with circle is $\hat{\alpha}$. Subplot (b) depicts the logarithm of the matrix entries of the ellipsoidal inversion matrix $\{\log [E]_{i,j} : i \leq 312, j \leq 100\}$. The rows correspond to sensor indices and the columns are the corresponding indices to the coefficients a_n^p; see eq. (8.25). Subplot (c) depicts the logarithm of the singular values of the ellipsoidal inversion matrix E. The rank of this ellipsoidal matrix is 23.

around 25 coefficients out of 100. This corresponds to the degree n and order p of the ellipsoidal harmonics $\mathbb{E}_n^p$ being in the set $\{(n,p):0<n\leq 5, 0\leq p\leq 2n\}$. Moreover, $\{(n=0,p=0),(n=1,p=0)\}$ are also excluded from this set. The reason for recovering only a small subset of α is explained in Figure 8.5(b), which depicts the logarithmic plot of all entries of the ellipsoidal inversion matrix, that is, $\{\log[|\mathrm{E}_{i,j}|]:i\leq N_s, j\leq N_p\}$. This figure shows that for all sensors (rows of E), the entries multiplying the relevant coefficients $\{a_n^p\}$ decay fast with increasing degree n and order p. So, the contribution to the data from higher degree ellipsoidal harmonic terms is very small. For example, on average, the matrix entry in $\mathrm{E}_{*,j}$ multiplying the coefficient a_1^1 is 3×10^{12} larger than the entry multiplying a_4^8, that is, $\frac{\mathrm{E}_{k,3}}{\mathrm{E}_{k,25}}\approx 3\times 10^{12}\ \ \forall k\leq N_S$. This fact poses a severe restriction to the class of functions that can be represented by the expansion of eq. (8.9). Figure 8.5(c) depicts the logarithmic plot of the singular values of the ellipsoidal inversion matrix E which is consistent with the conclusion drawn from Figure 5(b). It shows that the singular values exhibit fast decay. The inversion matrix E has a rank of 23 with a condition number matrix $\kappa(\mathrm{E})=4.1\times 10^{62}$. Despite the fact that the above problem is overdetermined, the matrix E is severely ill-conditioned and should not be employed directly for inversion or reconstruction purposes. In the case of the spherical model problem, the inverse problem was numerically solved via the technique of reproducing kernels [50]. However, deriving such kernels for the ellipsoid head model is challenging and is work in progress. It is worth noting that eq. (8.11) may still be employed to generate data for a given restricted class of $\Psi(\boldsymbol{\tau})$ functions.

In summary: first, the expansion in terms of ellipsoidal harmonics is suitable only for a very restricted class of functions, and second, this expansion yields an inversion matrix with a very high condition number.

8.1.7 EEG reconstructions

Due to the above difficulty we will not solve the inverse problem using the ellipsoidal harmonics and eq. (8.25). Instead, we construct an inversion matrix using the surface inversion equation (8.2), and a numerical quadrature technique involving local radial basis functions (RBF) as outlined in [79]. This important paper presents a numerical scheme for the integration of functions on any smooth, closed, triangulated surface. This technique requires only function values of the integrands at the nodes of the triangulated mesh. Moreover, the above paper is complemented with an open-source Matlab implementation, which includes detailed numerical tests. In this setting, we choose to parametrize $\Psi(\boldsymbol{\tau})$ using multiquadratic RBFs, namely,

$$\Psi(\boldsymbol{\tau})=\sum_{n=1}^{N}\beta_j\sqrt{\|\boldsymbol{\tau}-\boldsymbol{\tau}_j\|^2+c^2}, \tag{8.26}$$

where $\{\beta_j\}$ are the radial basis coefficients and c is the shape parameter. The shape parameter should be estimated from the data. However, to avoid high computational costs, we have arbitrarily set this parameter to be $c = 10^{-2}$. In order to avoid the inverse crime [68], we generate the data using the ellipsoidal harmonic inversion equation (8.25). This restricts the class of functions analyzed here, but the inversion method introduced in this section is valid for any smooth function. The construction of the inversion matrix involves employing eq. (8.26) in eq. (8.2). The surface integral is evaluated using the quadrature integration method outlined in [79].

We consider the case that data is contaminated with additive white Gaussian noise (AWGN) with a signal-to-noise ratio of 20 dB (SNR = 20 dB). This leads to the following system of linear equations:

$$\mathbf{u} = \mathrm{L}\beta, \tag{8.27}$$

where $\mathrm{L} \in \mathbb{R}^{N_s \times N}$. Here $N_s = 312$ is the number of electrodes and $N = 3,124$ is the number of nodes in the surface mesh of the cerebrum (see Table 8.1 for details of the surface meshes of the head ellipsoidal head model).

Recall that Figure 8.5(b) depicts $\log[|E_{ij}|]$, which decays quite rapidly with increasing degree n and order p. Moreover, Figure 8.5(a) and (c) suggests that only 25 coefficients can be recovered. So, in this simple test case we consider that data is generated from only the coefficient a_4^8, which corresponds to column entry 25 that multiplies the matrix entries $E_{k,25}$ for all sensor indices k. All other coefficients are set to zero, that is, $\{a_n^p = 0 : n \neq 4, p \neq 8\}$.

The reconstruction steps are outlined as follows:

(i) Set $\{a_n^p\} = 0$, except for $n = 4, p = 8$, which is set to $a_4^8 = 1$, $\lambda = a_3$ on the surface S_c and $\psi_n^p(a_3) = a_3 a_n^p$ in eq. (8.24), that is, $\psi_4^8(a_3) = a_3$. Compute $\{\Psi(\boldsymbol{\tau}_j) : j = 1, \ldots, N\}$ at the nodes of the surface mesh.

(ii) Generate the synthetic data via eq. (8.25), $\mathbf{u} = \mathrm{E}\alpha$ and add noise. The signal-to-noise ratio is set at 20 dB, that is, SNR =20 dB.

(iii) Reconstruct via eq. (8.27), that is, $\mathbf{u} = \mathrm{L}\beta$. Here, we employ the truncated singular value decomposition to invert the matrix L with 40 of the largest singular values.

(iv) Substitute the estimated $\hat{\beta}$ into eq. (8.26) to obtain the estimated $\hat{\Psi}(\boldsymbol{\tau})$.

(v) The reconstructions on an ellipsoidal head model are shown in Figure 8.6. The same reconstructions and the error $= |\Psi(\boldsymbol{\tau}) - \hat{\Psi}(\boldsymbol{\tau})|$ are shown in Figure 8.7.

8.1.8 MEG inversion matrix

In this section, we present reconstructions using synthetic data for MEG. Recall from eq. (8.10) that $\tau A^{\tau}(\boldsymbol{\tau})$ can be parametrized as follows:

$$\tau A^{\tau}(\boldsymbol{\tau}) = \sum_{n=1}^{\infty} \sum_{p=1}^{2n+1} f_n^p(\lambda) \mathbb{E}_n^p(\boldsymbol{\tau}), \quad \boldsymbol{\tau} \in \Omega_c. \tag{8.28}$$

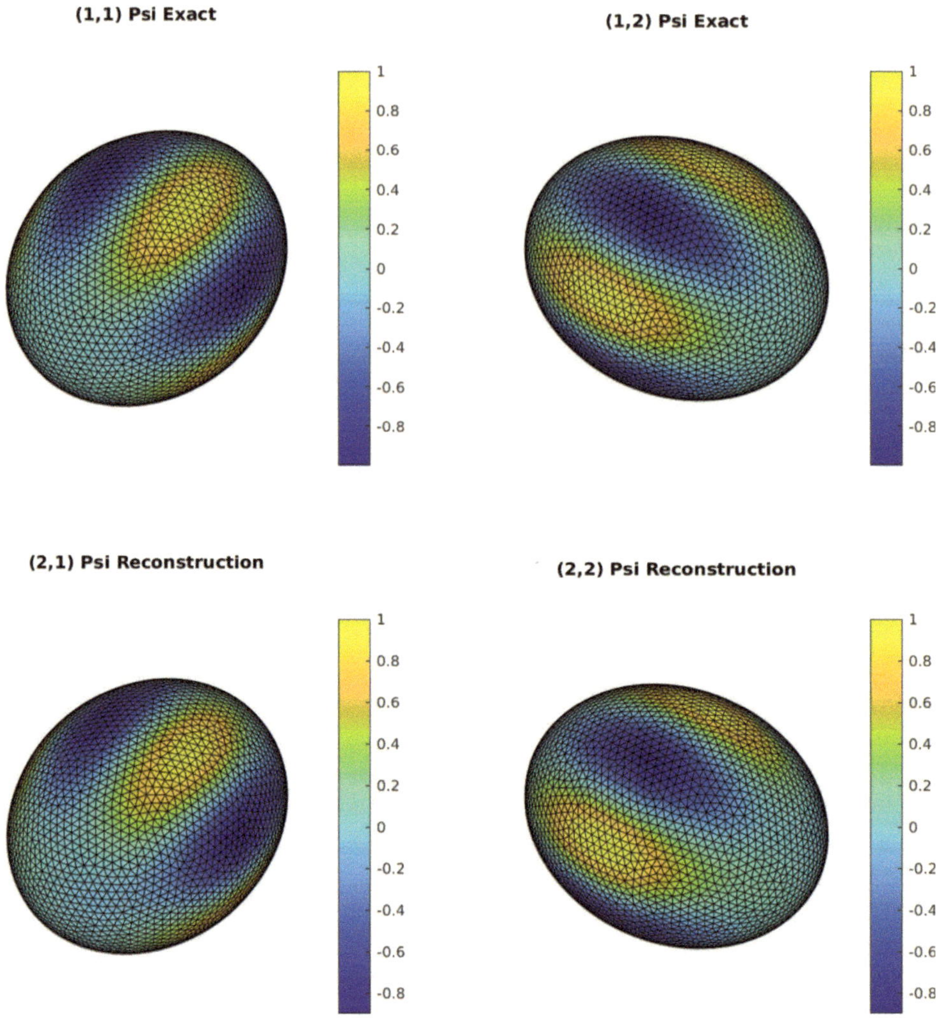

Figure 8.6: A reconstruction on an ellipsoidal head model outlined in Section 8.1 is shown. Subplot (1,1) and Subplot (1,2) depict the normalized $\Psi(\tau)$ corresponding to $\{a_n^p = 0, a_4^8 = 1\}$ from different viewing angles. Subplot (2,1) and subplot (2,2) depict the corresponding reconstructions. The signal-to-noise ratio is SNR =20 dB.

We follow an identical procedure to the one outlined for EEG to study the properties of the MEG inversion matrix. A typical MEG machine has generally 300 magnetometers. Here we assume an array of 518 magnetometers distributed evenly outside the head. On average, we placed one magnetometer $\mathbf{r} \in \Omega_e$ for every six nodes of the surface S_s. Despite employing far more magnetometers than an MEG machine, the results will show that the relevant inversion matrix is still highly ill-conditioned. The inversion formula used for this reconstruction is given by eq. (8.12). The contribution

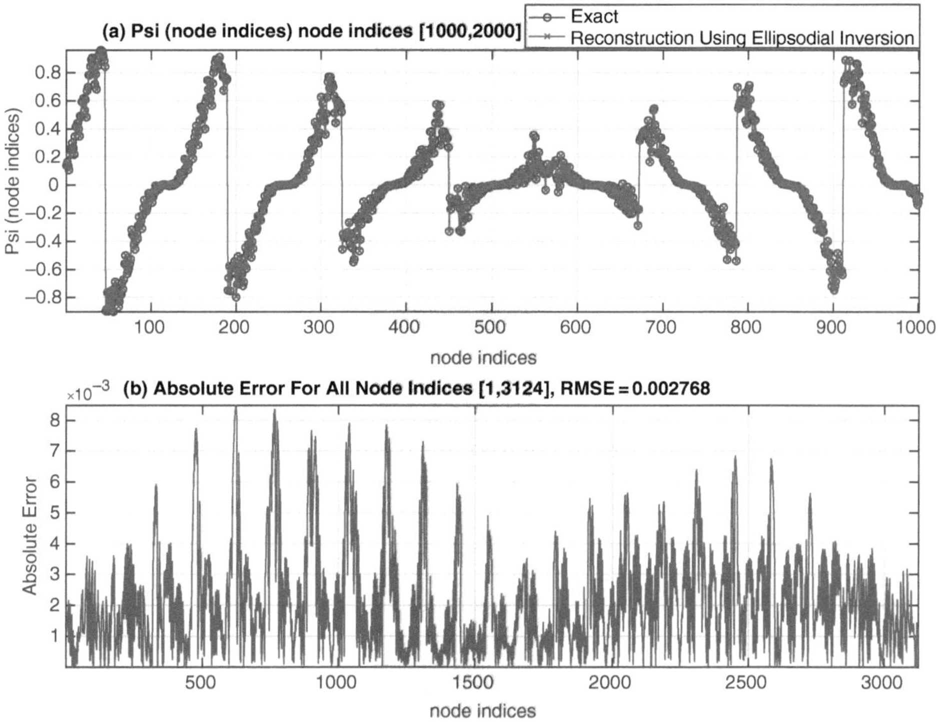

Figure 8.7: A reconstruction on an ellipsoidal head model outlined in Section 8.1 is shown. The signal-to-noise ratio is SNR = 20 dB. Subplot (a) depicts the comparison of $\Psi(\tau)$ and its corresponding reconstruction $\hat{\Psi}(\tau)$. The blue solid line with circle is $\Psi(\tau)$ and the red solid line with cross is $\hat{\Psi}(\tau)$. Subplot (b) depicts the absolute error $E(j) := |\Psi(\tau_j) - \hat{\Psi}(\tau_j)|$ for every node index with RMSE =0.0028.

from the second term in eq. (8.12) involving $\mathbf{r} \cdot \mathbf{H}_n^{m,l}$ is negligible and will be omitted in this analysis; see [50]. For simplicity, we assume that the functions $f_n^p(\lambda)$ take the form $f_n^p(\lambda) = \lambda a_n^p$. On the surface S_c, $\lambda = a_3$ and $\{a_n^p\}$ are the unknown coefficients linearly related to data, that is, to the left-hand side of eq. (8.12). Equation (8.12) gives rise to the following system of linear equations:

$$\mathbf{b} = \mathrm{M}\eta, \tag{8.29}$$

where $\mathbf{b}$ is the data (from MEG electrodes), $\mathrm{M} \in \mathbb{R}^{N_s \times N_p}$ is the MEG inversion matrix constructed from eq. (8.12), neglecting the contribution from the second term. The vector $\boldsymbol{\eta} = \{a_n^p\} \in \mathbb{R}^{N_p}$ is the vector of unknown coefficients to be estimated from data. Most MEG sensor arrays are between 108 and 306 channels. We have restricted the dimension of the vector of unknown coefficients $\boldsymbol{\eta} = \{a_n^p\}$ to be $N_p = 100$, that is, $\boldsymbol{\beta} \in \mathbb{R}^{100}$. So, we are considering the case that we have more data than unknowns.

Similar to the case of EEG, we generate a vector of random coefficients $\eta = \{a_n^p\}$ to obtain a data set via eq. (8.29), $\mathbf{b} = \mathrm{M}\boldsymbol{\eta}$. We proceed to obtain the estimate $\hat{\eta}$ via the same equation, $\hat{\boldsymbol{\eta}} = \mathrm{M}^{-1}\mathbf{b}$. In this test, we have not added noise to the data. The results are shown in Figure 8.8. Figure 8.8(a) shows the comparison of the randomly generated vector β and its corresponding estimate $\hat{\boldsymbol{\eta}}$. It can be observed that

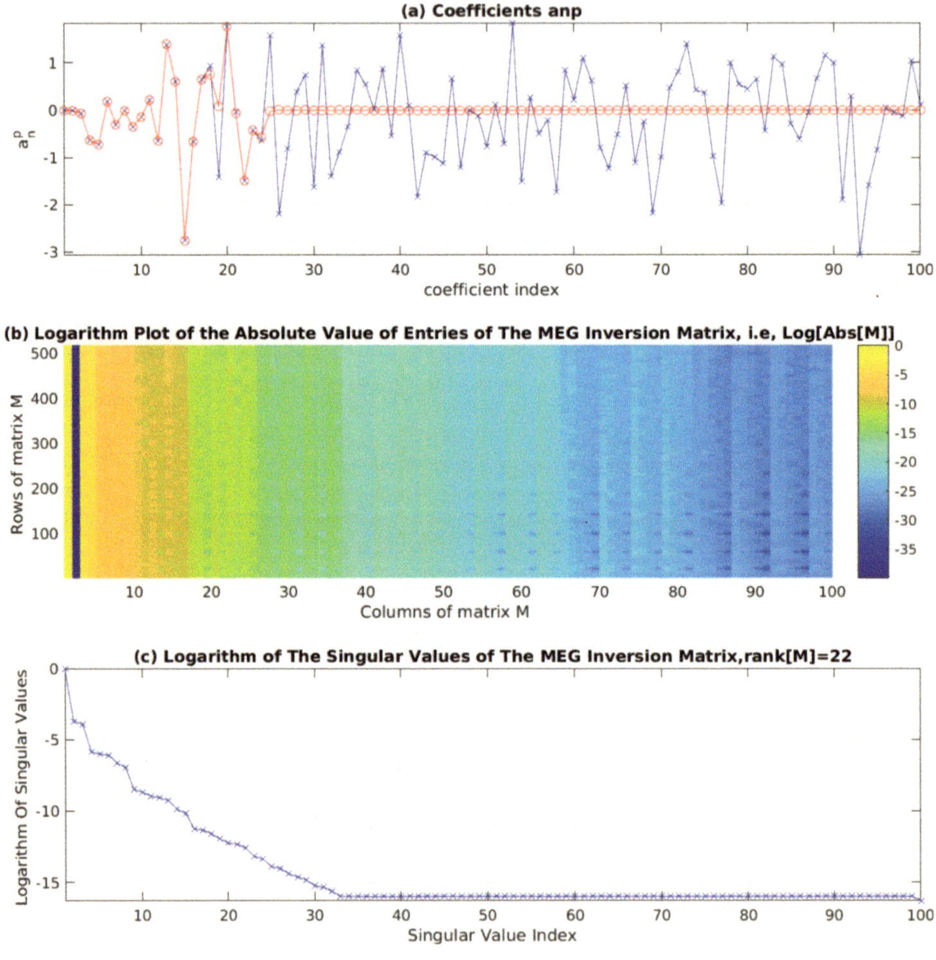

Figure 8.8: Subplot (a) depicts the comparison of the randomly generated vector of $\eta = a_n^p$ and its corresponding estimate $\hat{\eta}$ via eq. (8.29). The blue solid line with cross is $\eta = \{a_n^p\}$ and the red solid line with circle is $\hat{\eta}$. Subplot (b) depicts the logarithm of the absolute values of the matrix entries of the ellipsoidal inversion matrix $\{\log[|M|_{i,j|}: i \leq 518, j \leq 100\}$. The rows correspond to sensor indices and the columns are the corresponding indices to the coefficients a_n^p; see eq. (8.29). Moreover, $n_{\max} = 12$, in eq. (8.13). Subplot (c) depicts the logarithm of the singular values of the ellipsoidal inversion matrix M. The rank of this ellipsoidal matrix is 22 with a condition number $\kappa(M) = 2.38 \times 10^{29}$.

we can only recover at most around 22 coefficients out of 100. This corresponds to the degree n and order p of the ellipsoidal harmonics $\mathbb{E}_n^p$ being in the set $\{(n,p):0<n\leq 4, 0\leq p\leq 2n\}$. Moreover, $\{(n=0,p=0),(n=1,p=0)\}$ are also excluded from this set. This result can be explained by Figure 8.8(b), which shows the logarithmic plot of all entries of the matrix, that is, $\{\log[|\mathrm{M}_{i,j}|]:i\leq N_s, j\leq N\}$. This figure shows that for all sensors (rows of M), the entries multiplying the relevant coefficients $\{a_n^p\}$ exhibit fast decay. So, the contribution of higher degree ellipsoidal harmonic terms on the data are very small. For example, on average, the matrix entry in $\mathrm{E}_{*,j}$ multiplying the coefficient a_1^1 is 2.8×10^8 larger than the entry multiplying a_4^5, that is, $|\frac{\mathrm{M}_{k,3}}{\mathrm{M}_{k,22}}|\approx 2.8\times 10^8 \quad \forall k\leq N_s$. This limits the class of functions that can be represented by the expansion of eq. (8.10). Figure 8.8(c) shows the logarithmic plot of the singular values of the matrix M that is consistent with the conclusion drawn from Figure 8.8 (b): It shows that the singular values decay quickly and the rank of this matrix is 22 with a condition number of $\kappa(M)=2.38\times 10^{29}$. As in the case of EEG, this matrix is highly ill-conditioned and difficult to invert.

The reconstruction steps are the same steps taken for the case of EEG. However, here we set $a_4^3=1$ and the remaining $a_n^p=0$ (Figure 8.10). The data is generated via eq. (8.29). For the inversion step, an inversion matrix is constructed using the surface inversion equation (8.4) and a numerical quadrature technique involving local RBFs as outlined in [79].

8.1.9 Conclusion

The modeling of dipole sources for an ellipsoidal head geometry has been covered rigorously in the important studies of [54, 60, 61]. Here, we analyze the case that $\mathbf{J}^p(\boldsymbol{\tau})$ is continuously distributed in Ω_c [13]. In the particular case of the ellipsoidal head model, the assumption that ellipsoidal harmonics represent a suitable basis for the expansion of the unknown functions $\Psi(\boldsymbol{\tau})$, $\mathbf{A}^\tau(\boldsymbol{\tau})$ may be interpreted as implicit regularization, providing an attempt to mitigate the ill-posedness of the associated inverse problems. However, we have shown that the inversion matrices constructed from eqs. (8.11) and (8.12) are highly ill-conditioned and unsuitable for reconstructions. This is the case even for a highly overdetermined system of equations (more data than unknowns). Robust inversion techniques such as the reproducing kernel method were successful for the EEG and MEG inversion problem in the case of the spherical head model [48]. However, deriving these kernels for the ellipsoidal geometry is more challenging and is currently under study. The above difficulties can be overcome by using an alternative approach to the inversion equation. This involves reformulating the volume integral equations (8.1) and (8.3) as the surface integrals of (8.2) and (8.4). In this setting, $\Psi(\boldsymbol{\tau})$ and $A^\tau(\boldsymbol{\tau})$ were parametrized in terms of multiquadratic RBFs. The surface integrals were computed using the powerful numerical quadrature technique outlined in [79]. This gives rise to inversion matrices suitable for reconstructions. In order to avoid the "inverse crime," data were generated using the ellipsoidal harmonic expansions and eqs. (8.11) and (8.12). Reconstructions shown in Figures 8.6 and 8.9 suggest that the above approach based on the inversion equations (8.2) and (8.4) is a new, robust, reconstruction approach to the distributed inverse source problem in EEG and MEG.

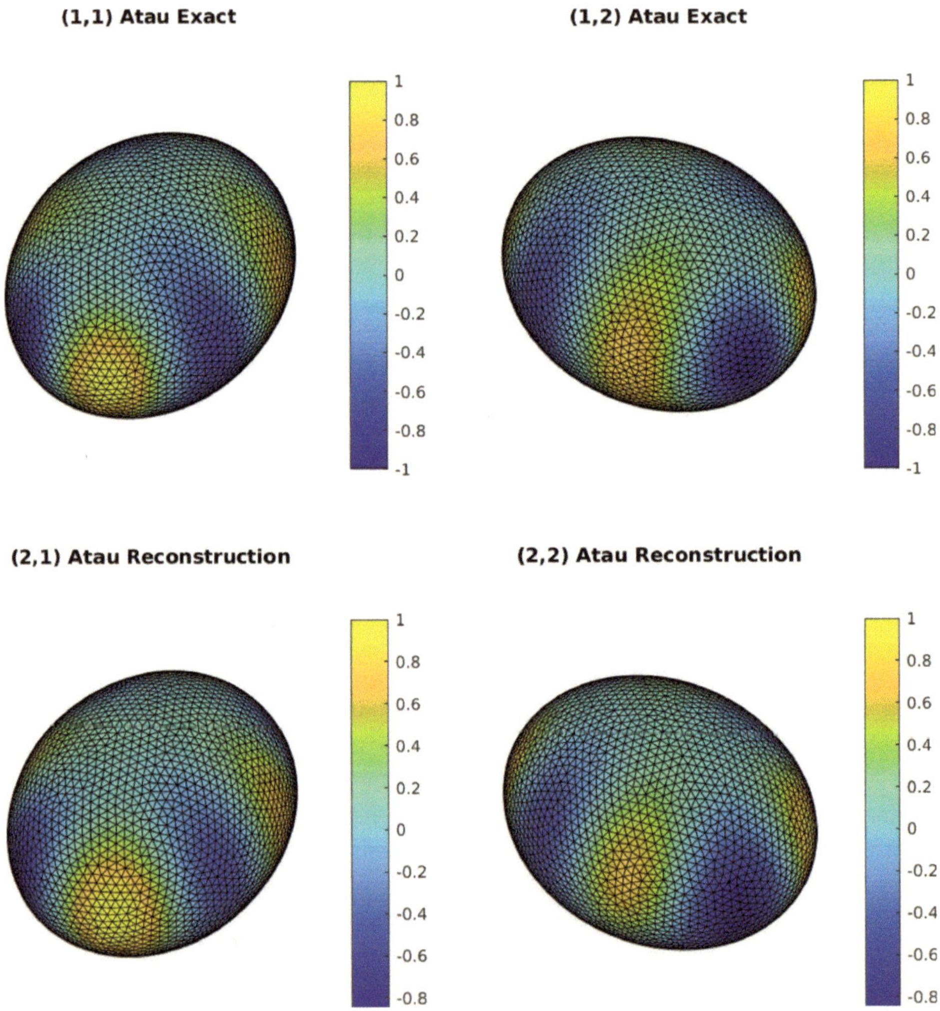

Figure 8.9: A reconstruction test case on an ellipsoidal head model for MEG is depicted. Subplot (1,1) and subplot(1,2) depict the normalized $A^{\tau}(\tau)$ corresponding to $\{a_n^p = 0, a_4^3 = 1\}$ from different viewing angles. Subplot (2,1) and subplot (2,2) depict the corresponding reconstructions, respectively. The signal-to-noise ratio is SNR =20 dB.

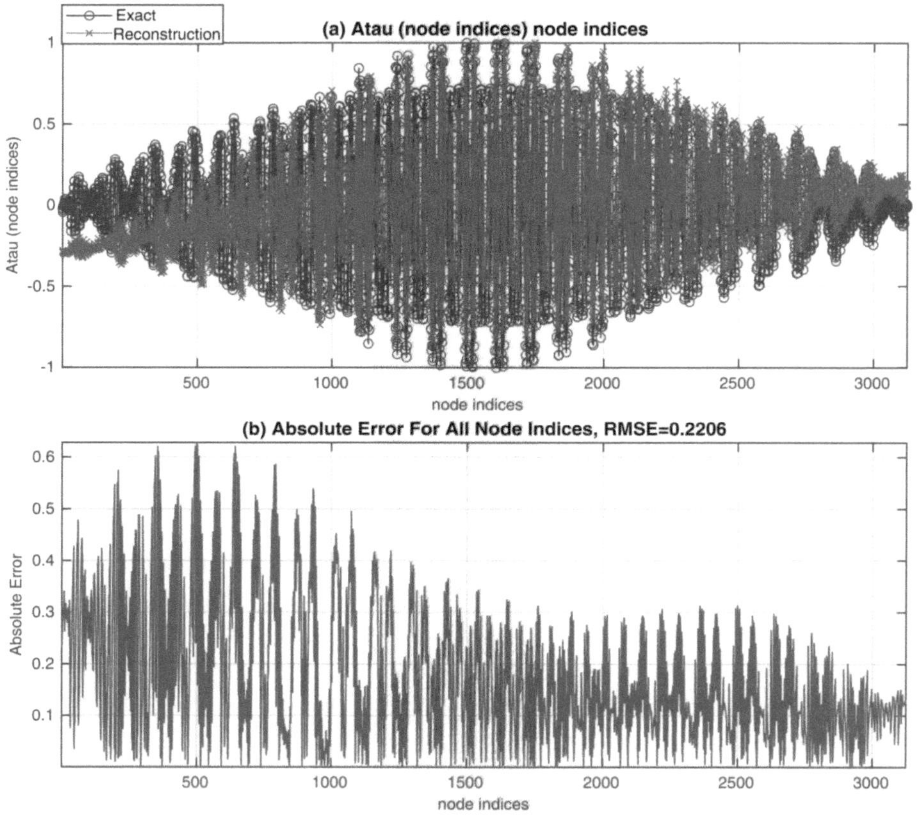

Figure 8.10: A reconstruction test case on an ellipsoidal head model for MEG is depicted. The signal-to-noise ratio is SNR = 20 dB. Subplot (a) depicts the comparison of $A^\tau(\tau)$ and its corresponding reconstruction $\hat{A}^\tau(\tau)$. The blue solid line with circle is $A^\tau(\tau)$ and the red solid line with cross is $\hat{A}^\tau(\tau)$. Subplot (b) depicts the absolute error $E(j) := |A^\tau(\tau_j) - \hat{A}^\tau(\tau_j)|$ for every node index with RMSE =0.2206.

8.2 Arbitrary geometry

It was shown in [29, 42, 47] that for the particular case of the spherical head model, $\upsilon_s(\mathbf{r}, \boldsymbol{\tau})$ is given by

$$\upsilon_s(\mathbf{r}, \boldsymbol{\tau}) = \sum_{n=1}^{\infty} H_n \frac{2n+1}{4\pi} \frac{\tau^n}{r^{n+1}} P_n(\hat{\boldsymbol{\tau}} \cdot \hat{\mathbf{r}}), \tag{8.30}$$

where the coefficients H_n depend only on the conductivities and the geometry and do not depend on the neuronal current. Equation (8.30) shows that the auxiliary function $\upsilon_s(\mathbf{r}, \boldsymbol{\tau})$ depends on the three variables $(r, \tau, \hat{\mathbf{r}} \cdot \hat{\boldsymbol{\tau}})$. It is worth noting that in the particular case of the spherical head model, $\upsilon_s(\mathbf{r}, \mathbf{0}) = 0$.

In the case of a realistic head model, OpenMEEG [57] accurately solves for the electric potential (volts) due to a dipole source [57]. A hybrid analytical-numerical algorithm was outlined in [68] to compute the auxiliary function $\upsilon_s(\mathbf{r},\boldsymbol{\tau})$. It involves employing the fundamental theorem of line integrals to approximate $\upsilon_s(\mathbf{r},\boldsymbol{\tau})$ from solutions of $4\pi u_s(\mathbf{r},\boldsymbol{\tau}) = \mathbf{Q}\cdot\nabla_{\boldsymbol{\tau}}\upsilon_s(\mathbf{r},\boldsymbol{\tau})$ computed by OpenMEEG: suppose that C is a smooth curve given by $\underline{\mathbf{r}}(t)$, $a \le t \le b$ and let $f(\underline{\mathbf{r}}(t))$ be a function whose gradient $\nabla f(\underline{\mathbf{r}}(t))$ is continuous on C. Then,

$$f(\underline{\mathbf{r}}(b)) - f(\underline{\mathbf{r}}(a)) = \int_C \nabla f(\underline{\mathbf{r}}(t))\cdot d\underline{\mathbf{r}}(t), \tag{8.31}$$

where $\underline{\mathbf{r}}(a)$ and $\underline{\mathbf{r}}(b)$ represent the initial and final points on C, respectively (the above theorem holds in any number of dimensions).

If we let $f = \upsilon_s(\mathbf{r},\boldsymbol{\tau})$ and $\underline{\mathbf{r}}(t) = \hat{\boldsymbol{\tau}}t$, $0 \le t \le \tau$ in eq. (8.29), then we find

$$\upsilon_s(\mathbf{r},\boldsymbol{\tau}) - \upsilon_s(\mathbf{r},\mathbf{0}) = 4\pi\int_0^{\tau} u_s(\mathbf{r}, t\hat{\boldsymbol{\tau}}, \hat{\boldsymbol{\tau}})dt, \tag{8.32}$$

where $\upsilon_s(\mathbf{r},\mathbf{0})$ is a constant and $4\pi u_s(\mathbf{r},\boldsymbol{\tau},\hat{\boldsymbol{\tau}}) = \hat{\boldsymbol{\tau}}\cdot\nabla_{\boldsymbol{\tau}}\upsilon_s(\mathbf{r},\boldsymbol{\tau})$. It is straightforward to compute numerically the one-dimensional integral of eq. (8.32). In the case of the spherical head model, $\upsilon_s(\mathbf{r},\mathbf{0}) = 0$ (see eq. (8.30)). If the centroid of the brain mesh of a realistic head model is denoted by $\mathbf{c}$, then $\upsilon_s(\mathbf{r},\mathbf{c})$ is an additive constant and has no effect on the inversion formulas of eqs. (8.1) and (8.2). If we estimate $\mathbf{c}$ from the triangulated surface mesh of the brain by fitting a sphere to its nodes, then a reasonable assumption is $\upsilon_s(\mathbf{r},\mathbf{c}) \approx 0$.

The accuracy of reconstructing $\Psi(\boldsymbol{\tau})$ is dependent on $\upsilon_s(\mathbf{r},\boldsymbol{\tau})$. One approach is to employ a large number of sources $\mathbf{Q}(\boldsymbol{\tau})$, to generate the required data set $u_s(\mathbf{r},\boldsymbol{\tau},\hat{\boldsymbol{\tau}})$ and to approximate the line integral of eq. (8.32), using a simplified integration method. However, this approach is computationally expensive and requires certain level of memory management. A computationally tractable alternative that can mitigate these errors involves a carefully constructed regression model of $u_s(\mathbf{r},\boldsymbol{\tau},\hat{\boldsymbol{\tau}})$ using data generated via OpenMEEG. The details of this construction are discussed below.

8.2.1 Regression model of $u_s(\mathbf{r},\tau,\hat{\tau})$ for a realistic head model for EEG

Regression models constructed using data generated by PDE solvers are referred to as *surrogate models* [52, 53, 58, 76, 82]. They are essentially machine learning models that *approximate* a mapping between a set of inputs and outputs of a simulation. Consider a finite sample pairs of data, $(\mathbf{x}_n, \mathbf{y}_n)_{n=1}^N$ (the training data), where $\mathbf{x} \in \mathbb{R}^{d_{\text{in}}}$ and $\mathbf{y} \in \mathbb{R}^{d_{\text{out}}}$. These pairs of data represent the inputs and outputs of a computationally expensive simulation model $\mathbf{y} = f(\mathbf{x})$. The goal is to construct a function $\hat{f}:\mathbb{R}^{d_{\text{in}}} \to \mathbb{R}^{d_{\text{out}}}$, which is as close as possible to the true function f. In this study, the function $f = u_s(\mathbf{r}, t\hat{\boldsymbol{\tau}}, \hat{\boldsymbol{\tau}})$ on the right-hand side of eq. (8.32). A vital step in constructing an accurate surrogate model is the correct identification of the inputs to the model. In this connection, we use our insight from the analysis of the spherical head model: eq. (8.30) shows that $v_s(\mathbf{r}, \boldsymbol{\tau})$ is a function of r which is the radial distance from the center of the coordinate system to the sensor position, of τ which is the radial distance from the center of the coordinate system to the source position vector and of $(\hat{\mathbf{r}} \cdot \hat{\boldsymbol{\tau}})$ which is the cosine of the angle between the source and the observation unit vectors. Thus, the necessary inputs to the surrogate model are $(r, \tau, \hat{\mathbf{r}} \cdot \hat{\boldsymbol{\tau}})$. An accurate *learning* of the function $u_s(\mathbf{r}, t\hat{\boldsymbol{\tau}}, \hat{\boldsymbol{\tau}})$ depends on a dense sampling of the inputs $(r, \tau, \hat{\mathbf{r}} \cdot \hat{\boldsymbol{\tau}}) \in [r_{\min}, r_{\max}] \times [\tau_{\min}, \tau_{\max}] \times [-1, 1]$. The position vectors of the observation points and source points from the centroid $\mathbf{c}$ of the brain are denoted by $\mathbf{r}$ and $\boldsymbol{\tau}$, respectively. The position vectors of the nodes of the brain with respect to $(0, 0, 0)$ are denoted by $\{\mathbf{x}_j \in \mathbb{R}^3 : 1 \le j \le V_b\}$; similarly the position vectors of the nodes of the scalp are denoted by $\left\{\mathbf{y}_j \in \mathbb{R}^3 : 1 \le j \le V_s\right\}$. For any source–observation pair $(\mathbf{r}, \boldsymbol{\tau})$, OpenMEEG employs a current dipole source with moment $\mathbf{Q}(\boldsymbol{\tau})$ Coulomb-meter. This solution, which is represented in volts, will be denoted by $u_s(\mathbf{r}, \boldsymbol{\tau}, \mathbf{Q}(\boldsymbol{\tau}))$. The steps needed for generating the data points required for constructing the surrogate model are given below:

Algorithm: Steps for generating the data set $u_s(\mathbf{r}, \boldsymbol{\tau}, \mathbf{Q}(\boldsymbol{\tau}))$

(1) Fit a sphere or an ellipsoid to the realistic head model to estimate the centroid (center). The position vector of this centroid is denoted by $\mathbf{c}$.
(2) Translate the coordinate system from $(0, 0, 0)$, that is, $\forall j$, let $\boldsymbol{\tau}_j = \mathbf{x}_j - \mathbf{c}$, $\mathbf{r}_j = \mathbf{y}_j - \mathbf{c}$.
(3) For all the nodes of the brain mesh, compute $\hat{\boldsymbol{\tau}}_j := \frac{\boldsymbol{\tau}_j}{|\boldsymbol{\tau}_j|}$. This is needed to define radially oriented sources.
(4) Find the minimum and maximum distances of the scalp nodes from the centroid $\mathbf{c}$, that is, $r_{\min} = \min\{|\mathbf{r}_j| : 1 \le j \le V_s\}$, $r_{\max} = \max\{|\mathbf{r}_j| : 1 \le j \le V_s\}$.
(5) Select M nodes on the scalp (observation vectors) $\mathbf{r}_k$ in the interval $r \in [r_{min}, r_{max}]$.
(6) It is known that sources placed too close to the surface of the cerebrum S_c can result in inaccuracies in the numerical computation [55, 57]. Choose V_b position vectors 3 mm from each node of the brain mesh as source position vectors $\{\boldsymbol{\tau}_j^s := \boldsymbol{\tau}_j - 0.003\hat{\boldsymbol{\tau}}_j : j \le V_b\}$ and set the corresponding dipole moment for every node as $\mathbf{Q}(\boldsymbol{\tau}_j) = \hat{\boldsymbol{\tau}}_j$ to compute the corresponding electric potential, $\{u(\mathbf{r}_i, \boldsymbol{\tau}_j^s, \hat{\boldsymbol{\tau}}_j) : i \le M, j \le V_b\}$.
(7) Add V_b sources at the centroid $\mathbf{c}$, with the dipole moment $\mathbf{Q}(\mathbf{c}) = \hat{\boldsymbol{\tau}}_j$, to compute $\{u(\mathbf{r}_i, \mathbf{c}, \hat{\boldsymbol{\tau}}_j) : i \le M, j \le V_b\}$.
(8) Add V_b sources at the position vectors $\frac{\boldsymbol{\tau}_j}{2}\hat{\boldsymbol{\tau}}_j$ with dipole moments $\mathbf{Q}\left(\frac{\boldsymbol{\tau}_j}{2}\right) = \hat{\boldsymbol{\tau}}_j$, to compute $\left\{u\left(\mathbf{r}_i, \frac{\boldsymbol{\tau}_j}{2}, \hat{\boldsymbol{\tau}}_j\right) : i \le M, j \le V_b\right\}$.

The above steps will result in a total of $M \times 3V_b$ data points. For the purpose of constructing the surrogate model it is convenient to organize the data as a matrix

$\mathbf{d} \in \mathbb{R}^{3V_b M \times 4}$. The first three columns $(r, \tau, \hat{\boldsymbol{\tau}} \cdot \hat{\mathbf{r}})$ feature as inputs and the last column (electric potential) is the output.

We employ the commercial software package pSeven by datadvance [43] to construct the regression model (machine learning model). A discussion of the features of pSeven and its comparison to open-source software is presented in [11].

It is noted that pSeven has a *smart selection* option for scanning through a set of algorithms to select the best model among the set. It performs for each training set a numerical optimization of the technique as well as its parameters [12, 83] by minimizing the cross-validation error; see [11]. Among the algorithms scanned by pSeven are the following: ridge regression [43], stepwise regression [71], elastic net [90], Gaussian processes [78], sparse Gaussian processes [18, 77], *high-dimensional approximation (HDA)* [11, 45] and *high-dimensional approximation combined with Gaussian processes* (this technique is related to artificial neural networks and, more specifically, to the two-layer perceptron with a nonlinear activation function [50]). Two desirable features of pSeven are: (i) all data manipulation is done via graphical user interface and (ii) it can export the constructed surrogate model as a stand-alone function in a number of scientific computing languages, including Matlab, C source for MEX, C source for stand alone program, C header for library, C source for library, functional mock-up interface for Co-simulation 1.0 and executable.

8.2.2 Minimization and a numerical solution of the inverse problem

It was shown in [42, 47, 66, 67] that the only component of the neuronal current $\mathbf{J}^p(\boldsymbol{\tau})$ that affects EEG data is $\Psi(\boldsymbol{\tau})$. However, as stated earlier, even this scalar function cannot be computed uniquely, unless one imposes an appropriate constraint. It was shown in [48, 49] that for a spherical conductor the L_2 norm of $\mathbf{J}^p(\boldsymbol{\tau})$ (minimal energy) yields a unique solution. The generalization of these results to ellipsoidal geometry is presented in Chapter 7. Here, these results are generalized further to the case of a realistic head model.

Let us define the functional E (energy) by

$$E = \int_{\Omega_C} |\mathbf{J}^p|^2 dV(\boldsymbol{\tau}). \tag{8.33}$$

By letting $\mathbf{J}^p = \nabla\Psi + \nabla \times \mathbf{A}$, we find that

$$\begin{aligned} |\mathbf{J}^p|^2 &= (\nabla\Psi + \nabla \times \mathbf{A}) \cdot (\nabla\Psi + \nabla \times \mathbf{A}). \\ &= |\nabla\Psi|^2 + |\nabla \times \mathbf{A}|^2 + 2\nabla\Psi \cdot \nabla \times \mathbf{A}. \end{aligned} \tag{8.34}$$

Hence,

$$E = \int_{\Omega_C} (|\nabla\Psi|^2 + |\nabla \times \mathbf{A}|^2 + 2\nabla\Psi \cdot \nabla \times \mathbf{A}) dV(\boldsymbol{\tau}). \tag{8.35}$$

However, it is shown in Proposition 9.1 and lemma 1 of [19] that

$$\int_{\Omega_C} \nabla\Psi(\boldsymbol{\tau}) \cdot \nabla \times \mathbf{A}(\boldsymbol{\tau}) dV(\boldsymbol{\tau}) = 0. \tag{8.36}$$

Thus,

$$E = \int_{\Omega_C} (|\nabla\Psi|^2 + |\nabla \times \mathbf{A}|^2) dV(\boldsymbol{\tau}). \tag{8.37}$$

Taking into consideration that u_S does not depend on $\mathbf{A}$ and that the second term of the right-hand side of the above equation is positive, it follows that the minimization of eq. (8.37) is equivalent to minimizing

$$\int_{\Omega_C} |\nabla\Psi|^2 dV(\boldsymbol{\tau}).$$

Thus, we have the following constrained minimization problem:

$$\min_{\Psi(\boldsymbol{\tau})} \int_{\Omega_C} |\nabla\Psi(\boldsymbol{\tau})|^2 dV(\boldsymbol{\tau}) \tag{8.38}$$

$$\text{s.t.} \qquad u_S(\mathbf{r}) = -\frac{1}{4\pi} \int_{\Omega_C} \nabla_{\boldsymbol{\tau}}^2 \Psi(\boldsymbol{\tau}) v_s(\mathbf{r}, \boldsymbol{\tau}) dV(\boldsymbol{\tau}), \quad \mathbf{r} \in S_S,$$

The requirement of the minimum norm of $\nabla\Psi(\boldsymbol{\tau})$ is a well-documented choice of regularization, known as Tikhonov regularization [69, 84]. It forces the estimate $\Psi\mathbb{H} \in {}^1(\Omega_C)$. The continuum form of the constrained optimization problem given by eq. (8.38) does not take into account the impact of measurement noise on data $u_S(\mathbf{r})$. However, additive measurement noise is modeled.

8.2.3 Expansion of $\Psi(\tau)$ using radial basis functions

For the numerical minimization of the constrained optimization problem given in eq. (8.38), RBF [46] is employed. We consider an EEG electrode cap with M electrodes and discretize the domain Ω_C using N cubic voxels. The function $\Psi(\boldsymbol{\tau})$ is expanded using inverse multiquadric RBFs [46], namely

$$\Psi(\boldsymbol{\tau}) = \sum_{j=1}^{N} \lambda_j \phi(\| \boldsymbol{\tau} - \boldsymbol{\tau}_j \|), \tag{8.39}$$

where $\{\boldsymbol{\tau}_j \in \Omega_c, 1 \leq j \leq N\}$ are the position vectors of the center of each voxel, and $\{\lambda_j : 1 \leq j \leq N\}$ denote the RBF coefficients. The inverse multiquadric basis function $\phi(\| \boldsymbol{\tau} - \boldsymbol{\tau}_j \|)$ is defined by

$$\phi(\| \boldsymbol{\tau} - \boldsymbol{\tau}_j \|) := \frac{1}{\sqrt{\alpha^2 + \| \boldsymbol{\tau} - \boldsymbol{\tau}_j \|^2}}. \tag{8.40}$$

The coefficient α in eq. (8.40) is referred to as the *shape parameter* and needs to be estimated from appropriate data. In this study, we propose a robust nonlinear least squares strategy for estimating α from a data set $\{u(\mathbf{r}_j) : 1 \leq j \leq M\}$ generated by a known $\Psi(\boldsymbol{\tau})$ which has support in Ω_c is

$$\Psi = \mathrm{A}\lambda, \tag{8.41}$$

where $\lambda \in \mathbb{R}^N$ are the RBF coefficients. The entries of the matrix $\mathrm{A} \in \mathbb{R}^{N \times N}$ are given by

$$\mathrm{A}[i,j] = \phi(\| \boldsymbol{\tau}_i - \boldsymbol{\tau}_j \|). \tag{8.42}$$

If we assume that measurements are contaminated by additive noise, then the discretized form of the measurement equation in the constrained optimization problem given by (8.38) is given by

$$\mathbf{u} = \mathrm{V_s}\Psi_\mathrm{L} + \mathbf{w}, \tag{8.43}$$

where $\mathbf{u}[i] := u(\mathbf{r}_i)$, $\mathrm{V_s}[i,j] := -\frac{v_s(\mathbf{r}_i, \boldsymbol{\tau}_j)}{4\pi N}$, $\Psi_\mathrm{L}[j] := \nabla^2\Psi(\boldsymbol{\tau}_j)$, $j = 1, \ldots, N$ and $\mathbf{w} \in \mathbb{R}^M$ denotes the measurement noise vector. The only unknowns in eq. (8.43) are the discrete Laplacian values $\Psi_\mathrm{L} \in \mathbb{R}^N$. The action of the Laplacian operator L on the inverse multiquadric basis function of eq. (8.40) yields

$$\mathrm{L}[i,j] = -\frac{3\alpha^2}{(\alpha^2 + \| \boldsymbol{\tau}_i - \boldsymbol{\tau}_j \|^2)^{\frac{5}{2}}}, \quad 1 \leq i,j \leq N. \tag{8.44}$$

The optimization problem defined in eq. (8.36) requires the computation of $\nabla\Psi(\boldsymbol{\tau})$. Partial derivative operators enable $\nabla\Psi(\boldsymbol{\tau})$ to be computed at discrete points $\{\boldsymbol{\tau}_j\}_{j=1}^N$. The operators for computing $\frac{\partial\Psi}{\partial\tau_x}, \frac{\partial\Psi}{\partial\tau_y}, \frac{\partial\Psi}{\partial\tau_z}$ are, respectively, given by

$$\mathrm{D}_x[i,j] = -\frac{1}{\sqrt{N}} \frac{(\tau_{x,i} - \tau_{x,j})}{(\alpha^2 + \| \boldsymbol{\tau}_i - \boldsymbol{\tau}_j \|^2)^{\frac{3}{2}}}, \quad i,j \leq N, \tag{8.45}$$

$$\mathrm{D}_y[i,j] = -\frac{1}{\sqrt{N}} \frac{(\tau_{y,i} - \tau_{y,j})}{(\alpha^2 + \| \boldsymbol{\tau}_i - \boldsymbol{\tau}_j \|^2)^{\frac{3}{2}}}, \quad i,j \leq N, \tag{8.46}$$

$$\mathrm{D}_z[i,j] = -\frac{1}{\sqrt{N}} \frac{(\tau_{z,i} - \tau_{z,j})}{(\alpha^2 + \| \boldsymbol{\tau}_i - \boldsymbol{\tau}_j \|^2)^{\frac{3}{2}}}, \quad i,j \leq N. \tag{8.47}$$

8.2.4 Reconstructing $\Psi(\tau)$

The implementation of the constrained optimization problem given by eq. (8.38) is well documented; see [69]. The shape parameter α that features in eqs. (8.40)–(8.47) for computing the operators A, L, D_x, Dy and Dz plays a central role in the accuracy and stability of reconstructions. In fact, the shape parameter α can be interpreted as the regularization parameter [69]. Extensive numerical tests suggest that the optimal α is not sensitive to the choice of $\Psi(\boldsymbol{\tau})$ but it depends on the geometry of the head model, position vectors $\{\boldsymbol{\tau}_j \in \mathbb{R}^3 : j = 1, \ldots, N\}$. Recall the Laplacian operator L is given by eq. (8.44). By letting $\Psi_L = \mathrm{L}\lambda$ in eq. (8.43), we find

$$\mathbf{u} = \mathrm{V_s L}\lambda + \mathbf{w}, \tag{8.48}$$

The model given by eq. (8.48) is linear in $\lambda \in \mathbb{R}^N$ and nonlinear in the shape parameter α; importantly, it is a separable model. The least squares error may be minimized with respect to λ; thus, it is reduced to a function of α only. The discretized form of the optimization problem of eq. (8.40) is given by

$$\begin{aligned} &\min_{\alpha,\lambda} \| \mathrm{G}(\alpha)\lambda \|^2 \\ \text{s.t.} \quad &\mathbf{u} = \mathrm{V_s L}\lambda + \mathbf{w}, \end{aligned} \tag{8.49}$$

where $\| \,.\|^2$ is the Euclidean norm and $\mathrm{G}(\alpha) \in \mathbb{R}^{3N\times N}$ is defined as

$$\mathrm{G}(\alpha) := \begin{bmatrix} \mathrm{D}_x \\ \mathrm{D}_y \\ \mathrm{D}_z \end{bmatrix}.$$

We employ the well-documented weighted least squares (WLS) method to model the impact of the measurement noise $\mathbf{w}$ on the parameters (α, λ). The WLS method involves employing the covariance matrix of the measurement noise $\mathrm{C}_\mathbf{w} = \mathbb{E}[\mathbf{w}\mathbf{w}^T]$ in the cost function to be minimized. For convenience, we introduce the following notation:

$$\mathrm{H}(\alpha) = \begin{bmatrix} \mathrm{C}_\mathbf{w}^{\frac{1}{2}} \mathrm{V_s L} \\ \mathrm{D}_x \\ \mathrm{D}_y \\ \mathrm{D}_z \end{bmatrix} \mathrm{b} = \begin{bmatrix} \mathrm{C}_\mathbf{w}^{\frac{1}{2}} \mathbf{u} \\ \mathbf{0} \\ \mathbf{0} \\ \mathbf{0} \end{bmatrix}, \tag{8.50}$$

where $\mathrm{H}(\alpha) \in \mathbb{R}^{(M+3N)\times N}$, $\mathbf{0} \in \mathbb{R}^N$ and $\mathbf{b} \in \mathbb{R}^{M+3N}$. The optimization problem of eq. (8.49) is reduced to the following system of linear equations:

$$\mathbf{b} = \mathrm{H}(\alpha)\lambda. \tag{8.51}$$

The dependency of $\mathrm{H}(\alpha)$ in eq. (8.51) on the parameter α is via the RBF matrices L, D_x, D_y and D_z, respectively, given by eqs. (8.44)–(8.47). The WLS error of eq. (8.51) reads as

$$J(\alpha,\lambda) = (\mathbf{b} - \mathrm{H}(\alpha)\lambda)^T(\mathbf{b} - \mathrm{H}(\alpha)\lambda). \tag{8.52}$$

The $\lambda \in \mathbb{R}^N$ that minimizes eq. (8.52) for a given α is given by

$$\hat{\lambda} = (\mathrm{H}^T(\alpha)\mathrm{H}(\alpha))^{-1}\mathrm{H}^T(\alpha)\mathbf{b}. \tag{8.53}$$

The resulting WLS error is given by

$$J(\alpha,\hat{\lambda}) = \mathbf{b}^T[\mathbf{I} - \mathrm{H}(\alpha)(\mathrm{H}^T(\alpha)\mathrm{H}(\alpha))^{-1}\mathrm{H}^T(\alpha)]\mathbf{b}, \tag{8.54}$$

where $\mathbf{I} \in \mathbb{R}^{(M+3N)\times(M+3N)}$ is the identity matrix. Hence, the problem of finding the optimal value of the shape parameter α is reduced to the one-dimensional maximization problem

$$\alpha_{\text{opt}} = \arg\max_{\alpha}[\mathbf{b}^T\mathrm{H}(\alpha)(\mathrm{H}^T(\alpha)\mathrm{H}(\alpha))^{-1}\mathrm{H}^T(\alpha)\mathbf{b}]. \tag{8.55}$$

The λ that minimizes eq. (8.52) for the optimal shape parameter α_{opt} is given by

$$\hat{\lambda} = (\mathrm{H}^T(\alpha_{\text{opt}})\mathrm{H}(\alpha_{\text{opt}}))^{-1}\mathrm{H}^T(\alpha_{\text{opt}})\mathbf{b}. \tag{8.56}$$

The optimal shape parameter α_{opt} is employed in eq. (8.40) to compute the RBF matrix A. Finally, $\hat{\Psi}$ is given by

$$\hat{\Psi} = \mathrm{A}\hat{\lambda}. \tag{8.57}$$

8.2.5 Numerical results

In the interest of reproducibility and to encourage further research in this direction, the code (surrogate model and reconstruction routines), head model (triangulated meshes) and the electrode positions are made available via Github https://github.com/parham1976. The realistic head model considered in this study is taken from the sample data set in the OpenMEEG package. The number of nodes and triangles of the surface meshes is shown in Table 8.2. The centroid of the brain mesh is $\mathbf{c} = [-0.0043, 0.0169, 0.0672]$ (this provides the center of the coordinate system for constructing the surrogate model).

Table 8.2: The number of nodes and triangles in the realistic head model.

Surface meshes	Number of nodes	Number of triangles
Cerebrum S_c	2,562	5,120
Skull S_b	2,562	5,120
Scalp S_c	2,562	5,120

8.2.6 Surrogate model

Using the steps outlined earlier for generating the data set required for constructing the surrogate model, we find $3V_bM$ data points. In this study, we have 122 electrodes, that is, $M = 122$ and $V_b = 2,562$ as given in Table 8.2. So, a total of $3V_bM = 937,692$ data points are generated via OpenMEEG. The data set is split into a training and a test data set. A total of $890,807$ data points were randomly selected for training and the remaining $46,885$ data points are used for testing. For the problem under study and the $890,807$ samples chosen as training data, the *smart selection* setting in pSeven package [12] chose a two-layer neural network. We employ the error metrics relative mean distance measure (RDM)$\in [0,2]$, where minimum $\text{RDM} = 0$, as well as the natural logarithm of the magnification factor (ln(MAG)) [59, 86]. RDM is defined as

$$\text{RDM} = \sqrt{\sum_{i=1}^{n}\left(\frac{u_i^s}{\sum_{j=1}^{n}(u_j^s)^2} - \frac{u_i^o}{\sum_{j=1}^{n}(u_j^o)^2}\right)^2}, \tag{8.58}$$

where u_j^s, u_i^o, respectively, denote the electric potential (volts) estimated by the surrogate model and the OpenMEEG package (reference). The natural logarithm of the magnification factor is given by

$$\ln[\text{MAG}] = \ln\left[\sqrt{\frac{\sum_{i=1}^{n}(u_i^s)^2}{\sum_{i=1}(u_i^o)^2}}\right]. \tag{8.59}$$

Figure 8.11 depicts the comparison of the solution obtained via OpenMEEG with the surrogate model. For clarity of presentation only $1,000$ points from the test data set of $46,885$ have been plotted. However, the values of $\text{RDM} = 0.2102$ and $\ln[\text{MAG}] = -0.0208$ are computed using all $46,885$ samples of the test data. In Figure 8.11, the blue solid line with cross is the solution obtained via OpenMEEG and the red solid line with circle is the output of the surrogate model.

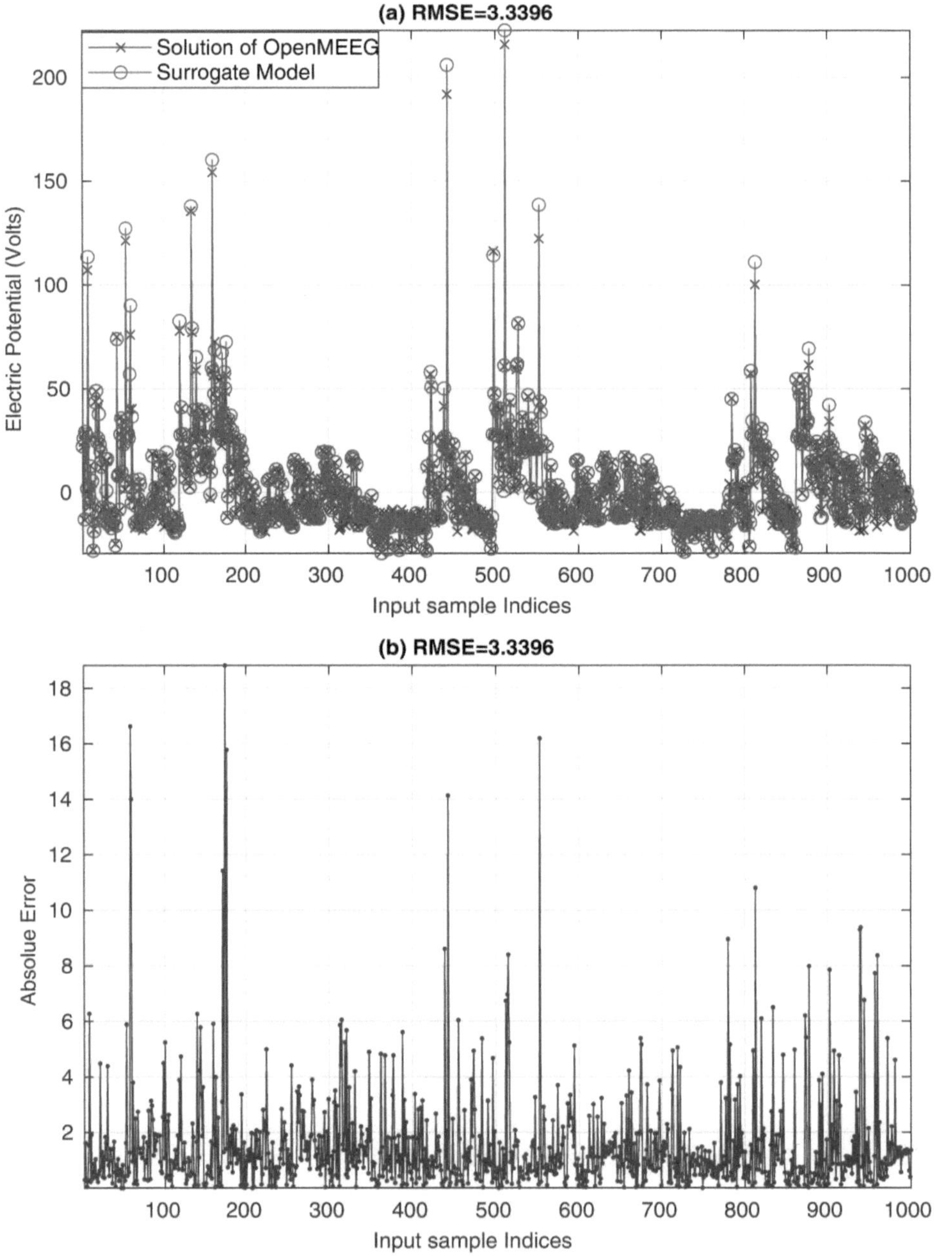

Figure 8.11: The comparison of the solution obtained via OpenMEEG(BEM) and via the surrogate model on the test data set for 1,000 sample points is depicted. It shows the comparison of the solution of $\hat{\tau}_i \cdot \nabla_\tau u_s(\mathbf{r}_i, \tau_i)$ obtained via OpenMEEG and via the surrogate model. The solid blue line with cross is the OpenMEEG solution and the solid red line with circles is the surrogate model solution. The relative mean distance measure (RDM) is 0.2102 and the natural logarithm of the magnification factor (ln(MAG)) is -0.0208.

8.2.7 Reconstruction of $\Psi(\tau)$

We present numerical results using synthetic data. In order to avoid the inverse crime, we generate the data using a different model to that employed for reconstruction. We assume that $\Psi(\boldsymbol{\tau})$ takes the form

$$\Psi(\boldsymbol{\tau}) = \sum_{i=1}^{4} \alpha_i e^{-\beta_i \|\boldsymbol{\tau}-\boldsymbol{\tau}_i\|^2}, \tag{8.60}$$

where the values of $\{\alpha_i, \beta_i, \boldsymbol{\tau}_i : 1 \le i \le 4\}$ are given in Table 8.3.

Table 8.3: The values of the parameters $\{\alpha_i, \beta_i, \tau_i : 1 \le i \le 4\}$ used for generating data.

α_i	β_i	$\tau_{x,i}$	$\tau_{y,i}$	$\tau_{z,i}$
2	1e3	− 0.0638	− 0.0185	0.0546
− 2	1e3	−0.0424	0.0630	0.0324
2	1e3	0.0076	−0.0185	0.1139
− 2	1e3	0.0433	−0.0366	0.0398

The Laplacian of eq. (8.60) is given by

$$\nabla^2\Psi(\boldsymbol{\tau}) = \sum_{i=1}^{4} 2\alpha_i\beta_i e^{-\beta_i\|\boldsymbol{\tau}-\boldsymbol{\tau}_i\|^2}\left(-3 + 2\beta_i \,\|\, \boldsymbol{\tau}-\boldsymbol{\tau}_i\|^2\right). \tag{8.61}$$

The steps for generating the numerical results presented in this section are as follows:

(i) The head model will have a sensor array of $M = 122$ sensors. Set up a uniform Cartesian grid in the domain Ω_c. Discretize Ω_c using $N = 3,293$ voxels. The length of the edge for each voxel is $h = 0.0074$ m.

(ii) Compute and store $\mathrm{V_s}[i,j] := -\frac{V_s(\mathbf{r}_i, \boldsymbol{\tau}_j)}{4\pi N}$. The auxiliary function matrix $\mathrm{V_s} \in \mathbb{R}^{M\times N}$.

(iii) Use in eq. (8.61) the values of $\{\alpha_i, \beta_i, \boldsymbol{\tau}_i : 1 \le i \le 4\}$ given in Table 8.3 to compute $\{\nabla^2\Psi(\boldsymbol{\tau}_j) : j = 1, \ldots, N\}$. Use in eq. (8.43) $\mathrm{V_s}$ and $\Psi_L[j] := \{\nabla^2\Psi(\boldsymbol{\tau}_j)\}$ to generate synthetic data, that is, $\mathbf{u} = \mathrm{V_s}\Psi_L + \mathbf{w}$. Here, we have employed AWGN, with a signal-to-noise ratio of 20 dB.

(iv) Solve the optimization problem of eq. (8.55) to find the optimal shape parameter, α_{opt} and estimate $\hat{\Psi}_L$. The optimal shape parameter is found to be $\alpha_{\text{opt}} = 0.0169$.

(v) Use α_{opt} to compute the linear operator A given by eq. (8.42). The reconstruction is given by eq. (8.57), that is, $\hat{\Psi} = \mathrm{A}\hat{\lambda}$.

The matrix of auxiliary function values $V_s \in \mathbb{R}^{122\times 3293}$. It has a rank($V_s$)=122 and condition number $\kappa(V_s) = 7.9804\times 10^3$. The matrix $H(\alpha_{opt}) \in \mathbb{R}^{10,001\times 3,293}$, which features in eq. (8.49) has a rank($H(\alpha_{opt})$) = 3293 and a condition number $\kappa(H(\alpha_{opt})) = 8.4017\times 10^7$.

Figure 8.12(a)–(d) depicts the comparison between Ψ and $\hat{\Psi}$ for position vectors inside the domain $\{\boldsymbol{\tau}_j \in \Omega_c : 1\le j\le N\}$ (internal nodes). To facilitate visualization of the function values Ψ and $\hat{\Psi}$ inside the domain Ω_c, the position vectors $\boldsymbol{\tau}_j \in \Omega_c$ are divided into four nonintersecting sets. The blue solid line with cross shows the function values Ψ given by eq. (8.60) and the red solid line with dots shows the

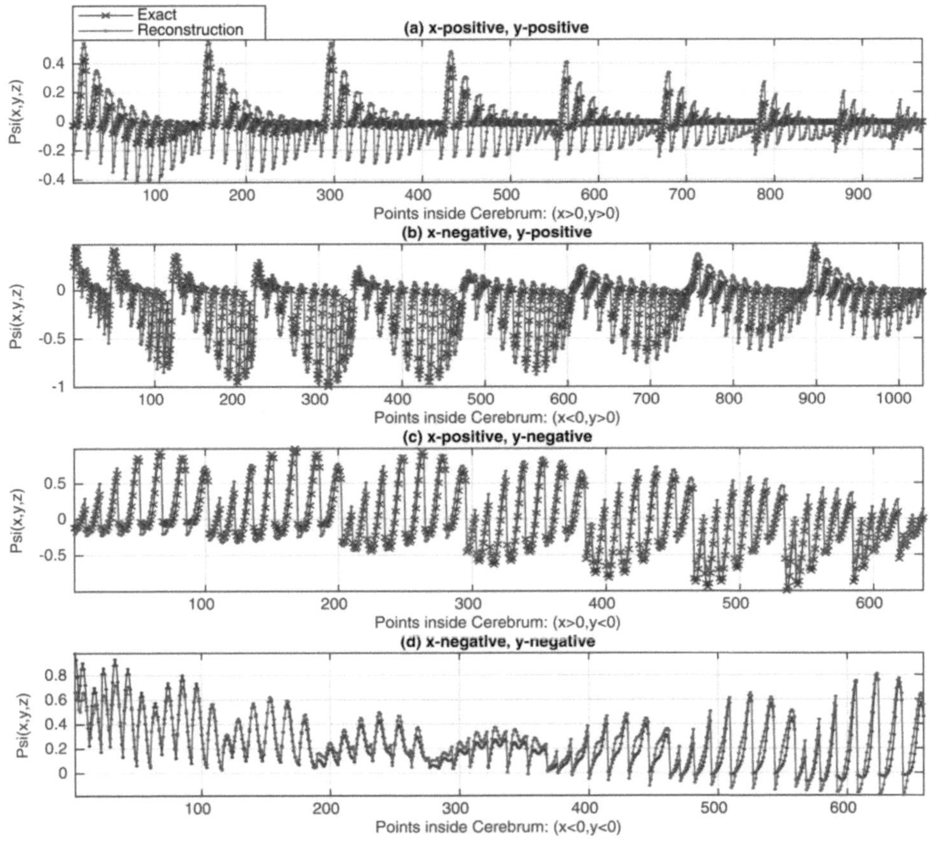

Figure 8.12: This figure depicts the comparison of eqs. (8.60) and (8.57). The blue solid line with cross is $\{\Psi(\tau_j):\tau_j \in \Omega_c\}$ as given by eq. (8.60) and the red solid line with dots is $\{\hat{\Psi}(\tau_j):\tau_j \in \Omega_c\}$ as given by eq. (8.57). Subplot (a) depicts the comparison of the function values of Ψ and $\hat{\Psi}$ for the set of position vectors $\{\tau_j \in \Omega_c:\tau_x > 0, \tau_y > 0\}$. Subplot (b) depicts the comparison of the function values of Ψ and $\hat{\Psi}$ for the set of position vectors $\{\tau_j \in \Omega_c:\tau_x < 0, \tau_y > 0\}$. Subplot (c) depicts the comparison of the function values of Ψ and $\hat{\Psi}$ for the set of position vectors $\{\tau_j \in \Omega_c:\tau_x > 0, \tau_y < 0\}$. Subplot (d) depicts the comparison of the function values of Ψ and $\hat{\Psi}$ for the set of position vectors $\{\tau_j \in \Omega_c:\tau_x < 0, \tau_y < 0\}$. RMSE =0.1122.

estimate $\hat{\Psi}$ given by eq. (8.57). The domain Ω_c has been discretized using a uniform Cartesian grid, that is, $N = 3,293$ cubic voxels. Subplot (a) depicts the comparison of the function values of Ψ and $\hat{\Psi}$ for the set of position vectors $\{\boldsymbol{\tau}_j \in \Omega_c{:}\boldsymbol{\tau}_x > 0, \boldsymbol{\tau}_y > 0\}$. Subplot (b) depicts the comparison of the function values of Ψ and $\hat{\Psi}$ for the set of position vectors $\{\boldsymbol{\tau}_j \in \Omega_c{:}\tau_x < 0, \tau_y > 0\}$. Subplot(c) depicts the comparison of the function values of Ψ and $\hat{\Psi}$ for the set of position vectors $\{\boldsymbol{\tau}_j \in \Omega_c{:}\tau_x > 0, \tau_y < 0\}$. Finally, subplot (d) depicts the comparison of the function values of Ψ and $\hat{\Psi}$ for the set of position vectors $\{\boldsymbol{\tau}_j \in \Omega_c{:}\tau_x < 0, \tau_y < 0\}$. RMSE of Ψ and $\hat{\Psi}$ is given by

$$\text{RMSE} = \left[\frac{1}{N}\sum_{i=1}^{N}\left(\Psi(\boldsymbol{\tau}_j) - \hat{\Psi}(\boldsymbol{\tau}_j)\right)^2\right]^{\frac{1}{2}}. \tag{8.62}$$

Using the analytic expression of $\Psi(\boldsymbol{\tau})$ given by eq. (8.58) and $\hat{\Psi}$ given by eq. (8.57) in eq. (8.62), we find that RMSE = 0.1122.

Figure 8.13 depicts the comparison of Ψ as given by eq. (8.58) on the surface S_c with $\hat{\Psi}$ given by eq. (8.55) on the surface S_c.

The reconstructions in Figure 8.13 are plotted using the open-source fieldtrip package [74].

8.2.8 Conclusion

OpenMEEG package solves the electric potential (volts) $4\pi u_s(\mathbf{r}, \boldsymbol{\tau}) = \mathbf{Q}(\boldsymbol{\tau}) \cdot \nabla_{\boldsymbol{\tau}} v_s(\mathbf{r}, \boldsymbol{\tau})$ for a given dipole source but not for $v_s(\mathbf{r}, \boldsymbol{\tau})$. To this end we propose a strategy involving one-dimensional line integrals to estimate $v_s(\mathbf{r}, \boldsymbol{\tau})$ by constructing a surrogate model from data generated via OpenMEEG. The inputs to the surrogate model were selected based on insight provided by the analytic equation (8.30), that is, $v_s(\mathbf{r}, \boldsymbol{\tau})$ for the spherical head model. The surrogate model was constructed using the pSeven Datadvance surrogate modeling toolbox [43]. A total of 890, 807 data points were randomly selected for training and the remaining 46, 885 data points are used for testing. For this construction, the smart selection routine of Datadvance chose the HDA [11], which is essentially a two-layer neural network. RDM is 0.2102 and the natural logarithm of the magnification factor (ln(MAG)) is − 0.0208.

The irrotational component of the neuronal current denoted by the scalar function $\Psi(\boldsymbol{\tau})$ has been parametrized using inverse multiquadric RBFs on a uniform Cartesian grid inside the cerebrum of a realistic head model. By employing synthetic data, the shape parameter denoted by α is estimated using a computationally efficient approach; it was estimated to be $\alpha_{\text{opt}} = 0.0169$. Furthermore, it is found that the shape parameter α is not sensitive to data, but depends on the configuration of the position vectors $\{\boldsymbol{\tau}_j \in \Omega_c{:}1 \le j \le N\}$ of the centers of the RBF expansion. The choice of RBF and α_{opt} yields the condition number $\kappa(\mathrm{H}(\alpha_{\text{opt}})) = 8.4017 \times 10^7$ for the inversion

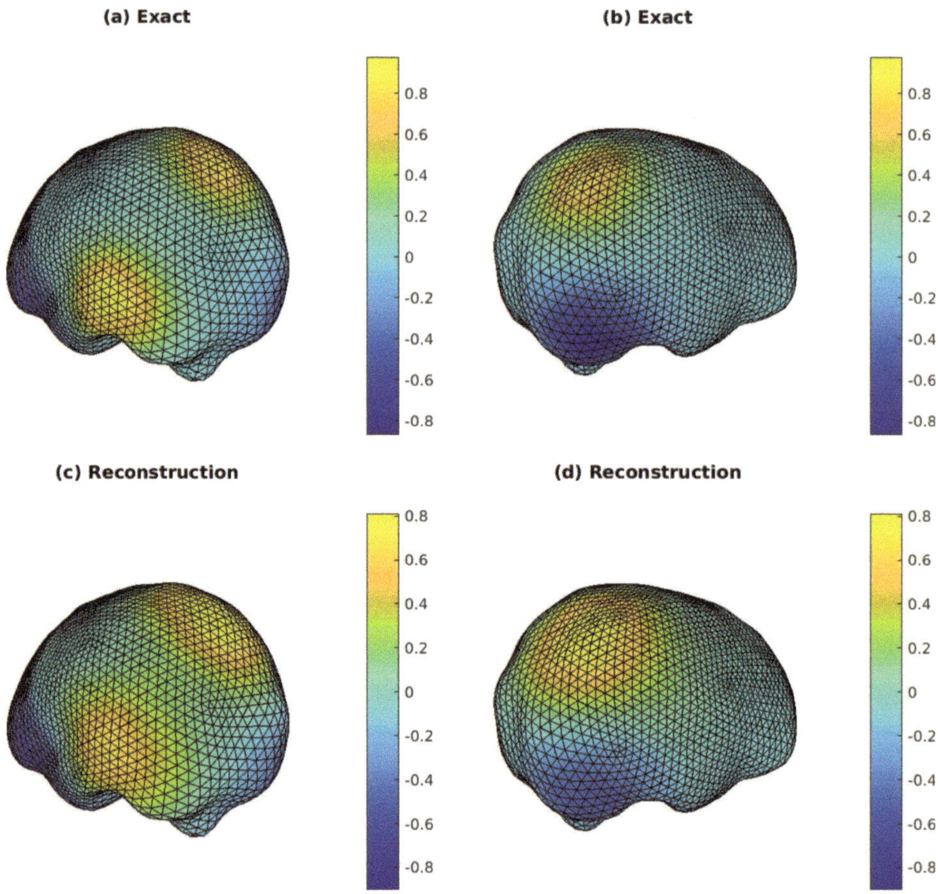

Figure 8.13: The comparison of eqs. (8.60) and (8.57) on the surface of the cerebrum S_c is depicted. Subplots (a) and (b) depict Ψ on the surface S_ι as given by eq. (8.60) and subplots (c) and (d) depict $\hat{\Psi}$ on S_c as given by (8.57). The root mean square error (RMSE) = 0.1122.

matrix $\mathrm{H}(\alpha_{\mathrm{opt}})$. Moreover, the regularization strategy involving minimizing the energy is in fact equivalent to Tikhonov regularization. Reconstructions are shown using synthetic data with RMSE = 0.1122. The complete set of MATLAB files (surrogate model, inversion code and head model) as well as data set for reproducing the results in this chapter are available at https://github.com/parham1976.

Chapter 9
Current confined on the cortical surface

As stated in the introduction, there is a physiologically important case that the inverse problems of both EEG and MEG can be solved uniquely; namely, when the current is confined on the cortical surface and moreover its direction is normal to the surface. The physiological significance of this particular case is discussed in [63], where it is noted that the predominant columnar organization of the pyramidal neurons results in coherent summation of the dendritic currents normal to the surface. These currents produce external magnetic fields measured by MEG and scalp potentials measured by EEG. In contrast, other cells do not have this organization so they do not generate this type of coherent summation.

We first recall formulas (2.48) and (2.97) in the form of the proposition below.

Proposition 9.1 Consider a four-compartments model specified by the domains Ω_c, Ω_f, Ω_b, Ω_s, modelling the spaces occupied by the cortex, the cerebrospinal fluid, the skull (bone) and the scalp, which have conductivities, $\sigma_c, \sigma_f, \sigma_b, \sigma_s$, respectively. Let μ be the permeability of all four compartments, which is the same as the permeability of Ω_e which denotes the space outside the head. Let the primary current $\mathbf{J}^p$ be continuously distributed inside Ω_c. Let $\mathbf{J}^p$ be expressed in terms of the scalar and vector functions Ψ and $\mathbf{A}$ via the Helmholtz decomposition formula.

The electric potential $u_s(\mathbf{r})$ on the external surface of the scalp which is denoted by S_s, and the radial component of the magnetic flux $\mathbf{r}\cdot\mathbf{B}(\mathbf{r})$ in Ω_e satisfy the following equations:

$$u_s(\mathbf{r}) = -\frac{1}{4\pi}\int_{\Omega_c}(\Delta_\tau\Psi(\tau))\upsilon_s(\mathbf{r},\tau)dV(\tau) + \frac{1}{4\pi}\int_{S_c}(\,\hat{\mathbf{n}}(\tau)\cdot\mathbf{J}^p(\tau))\upsilon_s(\mathbf{r},\tau)dS(\tau), \quad \mathbf{r}\in S_s, \tag{9.1}$$

and

$$\begin{aligned}\frac{4\pi}{\mu}\mathbf{r}\cdot\mathbf{B}(\mathbf{r}) = &-\int_{\Omega_c}\Delta_\tau(\tau\cdot A(\tau))\frac{dV(\tau)}{|\mathbf{r}-\tau|} + \frac{1}{4\pi}\int_{\Omega_c}(\Delta_\tau\Psi(\tau))(\mathbf{r}\cdot H(\mathbf{r},\tau))dV(\tau)\\ &+\int_{S_c}\nabla_\tau\cdot[|\mathbf{r}-\tau|(\,\hat{\mathbf{n}}(\tau)\times\mathbf{J}^p(\tau))]dS(\tau) - \frac{1}{4\pi}\int_{S_c}(\,\hat{\mathbf{n}}(\tau)\cdot\mathbf{J}^p(\tau))(\mathbf{r}\cdot\mathbf{H}(\mathbf{r},\tau))dS(\tau)\\ &-\int_{S_c}\tau\cdot(\,\hat{\mathbf{n}})\times\mathbf{J}^{\mathbf{P}}(\tau))\frac{dS(\tau)}{|\mathbf{r}-\tau|}, \quad \mathbf{r}\in\Omega_e,\end{aligned} \tag{9.2}$$

where S_c denotes the boundary of Ω_c, Δ_τ denotes the Laplacian in the variable, τ, υ_s denotes the auxiliary function associated with the scalp potential υ_s and $\mathbf{H}$ is defined in Chapter 2.

https://doi.org/10.1515/9783110547535-010

Proposition 9.2 Suppose that the cortical surface S_c is defined by the equation $\lambda = a$, where $\tau = (\lambda, \mu, \nu)$ and λ, μ, ν are orthogonal coordinates. Assume that $\mathbf{J}^p$ is supported on S_c. Then, $\mathbf{J}^p$ can be expanded in the form

$$\mathbf{J}^p(\tau) = J_1(\mu,\nu)\tau_1(\mu,\nu) + J_2(\mu,\nu)\tau_2(\mu,\nu) + J_3(\mu,\nu)\,\hat{\mathbf{n}}(\mu,\nu), \tag{9.3}$$

where

$$\tau_1(\mu,\nu) = \frac{\partial \tau(a,\mu,\nu)}{\partial \mu}, \quad \tau_2(\mu,\nu) = \frac{\partial \tau(a,\mu,\nu)}{\partial \nu}, \quad \hat{\mathbf{n}} = \frac{\tau_1 \times \tau_2}{|\tau_1 \times \tau_2|}. \tag{9.4}$$

The electric potential on the scalp is given by

$$u_s(\mathbf{r}) = \frac{1}{4\pi}\int_{S_c} \left\{ -\left[\frac{\partial(|\tau_1||\tau_2|J_1(\tau))}{\partial \mu} + \frac{\partial(|\tau_1||\tau_2|J_2(\tau))}{\partial \nu}\right]\frac{h_\lambda(\tau)}{|\tau_1||\tau_2|} + J_3\frac{\partial \upsilon_s(\mathbf{r},\tau)}{\partial \lambda}\right\} dS(\tau),$$
$$\mathbf{r} \in S_s, \tag{9.5}$$

where h_λ is defined by

$$h_\lambda(\mu,\nu) = \left|\frac{\partial \tau(a,\mu,\nu)}{\partial \lambda}\right|. \tag{9.6}$$

Proof. Suppose that $\tilde{\mathbf{J}}^P$ is independent of λ and has support in a small shell Ω of thickness δ in the interior of Ω_c. Then,

$$u_s(\mathbf{r}) = \frac{1}{4\pi}\int_\Omega \tilde{\mathbf{J}}^p(\tau)\cdot(\nabla_\tau \upsilon_s(\mathbf{r},\tau))dV(\tau), \quad \mathbf{r} \in S_s, \tag{9.7}$$

or

$$u_s(\mathbf{r}) = -\frac{1}{4\pi}\int_\Omega (\nabla_\tau \cdot \tilde{\mathbf{J}}^P(\tau))\upsilon_s(\mathbf{r},\tau)dV(\tau)$$
$$+\frac{1}{4\pi}\int_{\partial\Omega}(\hat{\mathbf{n}}(\tau)\cdot\mathbf{J}^p(\tau))\upsilon_s(\mathbf{r},\tau)dS(\tau), \quad \mathbf{r} \in S_s. \tag{9.8}$$

We will compute the limit of (9.8) as $\delta \to 0$. In this connection, we recall that $dV(\tau) = h_\lambda h_\mu h_\nu d\lambda d\mu d\nu$, where

$$h_\lambda(\lambda,\mu,\nu) = \left|\frac{\partial \tau(\lambda,\mu,\nu)}{\partial \lambda}\right|, h_\mu(\lambda,\mu,\nu) = \left|\frac{\partial \tau(\lambda,\mu,\nu)}{\partial \mu}\right|, \quad h_\nu(\lambda,\mu,\nu) = \left|\frac{\partial \tau(\lambda,\mu,\nu)}{\partial \nu}\right|. \tag{9.9}$$

Using the mean value theorem, the volume integral in the right hand side of (9.8) yields

$$-\frac{\delta}{4\pi}\int_{S_c}\left[\left(\nabla_\tau \cdot \tilde{\mathbf{J}}^p\right)\upsilon_s h_\lambda h_\mu h_\nu\right]_{\lambda=a} d\mu d\nu = -\frac{1}{4\pi}\int_{S_c}(\nabla_\tau \cdot \mathbf{J}^p)\upsilon_s h_\lambda(a,\mu,\nu)|\tau_1||\tau_2|d\mu d\nu,$$
$$\tag{9.10}$$

where $\mathbf{J}^p = \delta\tilde{\mathbf{J}}^p$. The lateral sides of $\partial\Omega$ yield a zero contribution as $\delta \to 0$. Furthermore, $\hat{\mathbf{n}}$ is the same on the top surface where $\lambda = a$ and on the bottom surface where $\lambda = a - \delta$. Thus, recalling that $dS = h_\mu h_\nu d\mu d\nu$, the second term of the right hand side of (9.8) yields

$$\frac{1}{4\pi}\int_{S_C} (\hat{\mathbf{n}} \cdot \tilde{\mathbf{J}}^p)\{[\upsilon_s h_\mu h_\nu]_{\lambda=a} - [\upsilon_s h_\mu h_\nu]_{\lambda=a-\delta}\} d\mu d\nu$$

$$= \frac{1}{4\pi}\int_{S_C} (\hat{\mathbf{n}} \cdot \mathbf{J}^p)\left[\frac{\partial(h_\mu h_\nu)}{\partial\lambda}\right]_{\lambda=a} d\mu d\nu. \tag{9.11}$$

Using in (9.8) eqs. (9.10) and (9.11), we find

$$u(\mathbf{r}) = -\frac{1}{4\pi}\int_{S_C} (\nabla \cdot \mathbf{J}^p)\upsilon_s h_\lambda(\alpha,\mu,\nu)|\tau_1||\tau_2| d\mu d\nu$$

$$+ \frac{1}{4\pi}\int_{S_C} (\hat{\mathbf{n}} \cdot \mathbf{J}^p)\left\{|\tau_1||\tau_2|\left[\frac{\partial \upsilon_s}{\partial\lambda}\right]_{\lambda=a} + \upsilon_s\left[\frac{\partial(h_\mu h_\nu)}{\partial\lambda}\right]_{\lambda=a}\right\} d\mu d\nu, \quad \mathbf{r} \in S_C. \tag{9.12}$$

Employing in the aforementioned equation the identities [87]

$$\hat{\mathbf{n}} \cdot \mathbf{J}^p = J_3, \quad \nabla_T \cdot \mathbf{J}^p = \frac{1}{H}\frac{\partial(HJ_1)}{\partial\mu} + \frac{1}{H}\frac{\partial(HJ_2)}{\partial\nu} - \zeta J_3, \quad H = |\tau_1||\tau_2|, \tag{9.13}$$

where ζ denotes the mean curvature of S_C, eq. (9.12) becomes

$$u(\mathbf{r}) = \frac{1}{4\pi}\int_{S_C}\left\{\left[-\frac{1}{H}\frac{\partial(HJ_1)}{\partial\mu} - \frac{1}{H}\frac{\partial(HJ_2)}{\partial\nu} + .\zeta J_3\right]\upsilon_s h_\lambda(\alpha,\mu,\nu)|\tau_1||\tau_2|\right.$$

$$\left. + J_3\left[|\tau_1||\tau_2|\left[\frac{\partial \upsilon_s}{\partial\lambda}\right]_{\lambda=a} + \upsilon_s\left[\frac{\partial(h_\mu h_\nu)}{\partial\lambda}\right]_{\lambda=a}\right]\right\} d\mu d\nu, \quad \mathbf{r} \in S_S. \tag{9.14}$$

Using in the aforementioned equation the identity,

$$\zeta = -\left[\frac{1}{h_\lambda h_\mu h_\nu}\frac{\partial(h_\mu h_\nu)}{\partial\lambda}\right]_{\lambda=a}, \tag{9.15}$$

and recalling that $dS = |\tau_1||\tau_2| d\mu d\nu$, eq. (9.14) simplifies to (9.5). Hence, the proof of Proposition 9.2 is completed.

Corollary 9.1 In the particular case that the current is normal to S_C, i.e, $J_1 = J_2 = 0$, eq. (9.7) becomes

$$u_s(\mathbf{r}) = \frac{1}{4\pi}\int_{S_C} J_3(\tau)\frac{\partial \upsilon_s(\mathbf{r},\tau)}{\partial\lambda} dS(\tau), \quad \mathbf{r} \in S_S. \tag{9.16}$$

Remark 9.1 If $\tilde{\mathbf{J}}^p = \tilde{J}_3\, \hat{\mathbf{n}}$, recalling that the normal component of ∇ is given by $\frac{1}{h_\lambda}\frac{\partial}{\partial\lambda}$, eq. (9.7) becomes

$$u_s(\mathbf{r}) = \frac{1}{4\pi}\int_\Omega \tilde{J}_3 \frac{\partial v_s}{\partial\lambda} h_\mu h_\nu d\lambda d\mu d\nu. \tag{9.17}$$

Letting $\delta \to 0$, and employing the mean value theorem, the aforementioned equation yields

$$u_s(\mathbf{r}) = \frac{1}{4\pi}\int_{S_C} J_3 \frac{\partial v_s}{\partial\lambda} |\tau_1||\tau_2| d\mu d\nu, \tag{9.18}$$

which is eq. (9.16).

The analogous result for MEG is slightly more complicated that the one for EEG, thus for simplicity we only consider the case that $\mathbf{J}^p$ is normal to S_c.

Proposition 9.3 Assume that the current is supported on the cortical surface S_c and that it is normal to this surface, that is,

$$\mathbf{J}^p(\tau) = J_3(\tau)\, \hat{\mathbf{n}}(\tau). \tag{9.19}$$

Then, the radial component of the magnet flux satisfies the following equation:

$$\begin{aligned}
&\frac{4\pi}{\mu}\mathbf{r}\cdot\mathbf{B}(\mathbf{r}) \\
&= \int_{S_C} \tau\cdot\left[\frac{\partial J_3(\tau)}{\partial\nu}\tau_1 + \frac{\partial J_3(\tau)}{\partial\mu}\tau_2\right]\frac{h_\lambda(\tau)\; dS(\tau)}{|\tau_1||\tau_2|\,|\mathbf{r}-\tau|} - \frac{1}{4\pi}\int_{S_C} J_3(\tau)\left(\mathbf{r}\cdot\frac{\partial \mathbf{H}(\mathbf{r},\tau)}{\partial\lambda}\right) dS(\tau), \quad \mathbf{r}\in\Omega_e,
\end{aligned} \tag{9.20}$$

where $\tau_1, \tau_2,\ \hat{\mathbf{n}}, h_\lambda$ are defined in Proposition 9.2.

Proof. Using in the basic eq. (9.2) the substitutions,

$$\mathbf{J}^p = J_3\, \hat{\mathbf{n}}, \quad \Delta_\tau\Psi = \nabla_\tau\cdot\mathbf{J}^p, \quad \Delta_\tau(\tau\cdot\mathbf{A}) = \tau\cdot\Delta_\tau\mathbf{A} = -\tau\cdot(\nabla_\tau\times\mathbf{J}^p), \tag{9.21}$$

we find,

$$\begin{aligned}
&\frac{4\pi}{\mu}\mathbf{r}\cdot\mathbf{B}(\mathbf{r}) \\
&= \int_{\Omega_C} \tau\cdot(\nabla_\tau\times\mathbf{J}^p(\tau))\frac{dV(\tau)}{|\mathbf{r}-\tau|} + \frac{1}{4\pi}\int_{\Omega_C}(\nabla_\tau\cdot\mathbf{J}^P(\tau))(\mathbf{r}\cdot\mathbf{H}(\mathbf{r},\tau))dV(\tau) \\
&\quad - \frac{1}{4\pi}\int_{S_C} J_3(\tau)(\mathbf{r}\cdot\mathbf{H}(\mathbf{r},\tau))dS(\tau), \quad \mathbf{r}\in\Omega_e.
\end{aligned} \tag{9.22}$$

Employing in the aforementioned equation the same limiting procedure as the one used in Proposition 9.2, eq. (9.22) implies the following equation:

$$\frac{4\pi}{\mu}\mathbf{r}\cdot\mathbf{B}(\mathbf{r}) = \int_{S_C} \tau\cdot(\nabla_\tau\times\mathbf{J}^p(\tau))h_\lambda\frac{dS(\tau)}{|\mathbf{r}-\tau|} + \frac{1}{4\pi}\int_{S_C}(\nabla_\tau\cdot\mathbf{J}^p(\tau))(\mathbf{r}\cdot\mathbf{H}(\mathbf{r},\tau))h_\lambda dS(\tau)$$
$$-\frac{1}{4\pi}\int_{S_C} J_3\tau\frac{\partial}{\partial\lambda}\left[(\mathbf{r}\cdot\mathbf{H}(\mathbf{r},\tau))h_\mu h_\nu\right]d\mu d\nu, \quad \mathbf{r}\in\Omega_e. \tag{9.23}$$

However, if $\mathbf{J}^p$ is given by the first of eq. (9.21), then the following equations are valid [87]:

$$\nabla_\tau\cdot\mathbf{J}^p = -\zeta J_3, \quad \nabla\times\mathbf{J}^p = \frac{1}{|\tau_1||\tau_2|}\left(\frac{\partial J_3}{\partial\nu}\tau_1 - \frac{\partial J_3}{\partial\mu}\tau_2\right). \tag{9.24}$$

Employing the aforementioned equations in (9.23), we find after simplifications eq. (9.20).

In summary, under the physiologically reasonable assumption that the current is supported only on the cortical surface and it is normal to this surface, the current can be determined uniquely from either EEG or MEG via eqs. (9.16) or (9.20), respectively. The numerical implementation of these equations in the case of an ellipsoidal head model is discussed in [66].

References

[1] Albanese R., and Monk, P. B. The inverse source problem for Maxwell's equations. Inverse Problems, vol. 22, 2006, pp. 1023–35.

[2] Ammari H. An Introduction to Mathematics of Emerging Biomedical Imaging, Berlin, Springer-Verlag, 2008.

[3] Ammari H., and Kang, H. Reconstruction of Small inhomogeneities from Boundary Measurements. Lecture Notes in Mathematics, Vol. 1846, Springer-Verlag, Berlin, 2004.

[4] Ammari H., and Kang, H. Polarization and Moment Tensors: with Applications to Inverse Problems and Effective Medium Theory. Applied Mathematical Sciences Series, Vol. 162, Springer-Verlag, New York, 2007.

[5] Badia A.E., and Duong, T.H. Some remarks on the problem of source identification from boundary measurements. Inverse Problems, vol. 14, 1998, pp. 883–891.

[6] Baillet S., Mosher, J. C., and Leahy, R. M. Electromagnetic brain mapping. IEEE Signal Processing Magazine, vol. 18, 2001, pp. 14–30.

[7] Balanis C.A. Advanced Engineering Electromagnetics, John Wiley & Son, 1989.

[8] Baratchart L., Leblond, J, and Marmorat, J. P. Inverse source problems in a 3D ball from best meromorphic approximation on 2D slices. Electronic Transactions in Numerical Analysis, vol. 25, 2005, pp. 41–53.

[9] Bardhan J.P., and Knepley, M. G. Computational science and re-discovery: Open source implementation of ellipsoidal harmonics for problems in potential theory. Computational Science & Discovery, vol. 5, 2012, pp. 014006.

[10] Bear M.F., Connors, B.W., and Paradiso, M.A. Neuroscience. Exploring the Brain, Williams and Wilkins, Baltimore, 1996.

[11] Belyaev M., Burnaev, E, Kapushev, E, Panov, M, Prikhodko, P, Vetrov, D, and Yarotsk Y, D. GTApprox: Surrogate modeling for industrial design. Advances in Engineering Software, vol. 102, 2016, pp. 29–39.

[12] Bergstra J., Bardenet, R., Bengio, Y., and Kegl, B. Algorithms for hyper-parameter optimization. NIPS'11 Proceedings of the 24th International Conference on Neural Information Processing Systems, pp. 2546–2554, 2011.

[13] Berg P., and Scherg, M. A fast method for forward computation of multiple-shell spherical head models. Electroencephalography and Clinical Neurophysiology, vol. 90, 1994, pp. 58–64.

[14] Bondeson A., Rylander, T, and Ingelsraom, P. Computational Electromagnetics, Springer, 2005.

[15] Brand L. Vector and Tensor Analysis, John Wiley and Sons, London, 1947.

[16] Bronzan J.B. The magnetic scalar potential. Journal of Applied Physics, vol. 39, 1971, pp. 1357–1359.

[17] Buzsaki G., Anastassiou, C.A, and Koch, C. The origin of extracellular fields and currents EEG, ECOG, LFP and spikes. Nature Reviews Neuroscience, vol. 13, 2012, pp. 417.

[18] Burnaev E.V., and Zaytsev, A.A. Surrogate modeling of multifidelity data for large samples". Journal of Communication Technology and Electronics,vol., vol. 60, 2015, pp. 1348–1355.

[19] Cantarella J., DeTurck, D, and Gluck, H. Vector calculus and the topology of domains in 3-Space. The American Mathematical Monthly, vol. 109, 2002, pp. 409–442.

[20] Chauveau N., Franceries, X, Doyon, B, Rigaud, B, Morucci, J. P, and Celsis, P. Effects of skull thickness, anisotropy, and inhomogeneity on forward EEG/ERP computations using a spherical three-dimensional resistor mesh model. Human Brain Mapping, vol. 21, 2004, pp. 86–97.

[21] Cuffin B. N. A method for localizing EEG sources in realistic head models. IEEE Transaction in Biomedical Engineering, vol. 42, 1995, pp. 68–70.

https://doi.org/10.1515/9783110547535-011

[22] Dassios G. What is recoverable in the inverse magnetoencephalography. AMS. Contemporary Mathematics, vol. 408, 2006, pp. 181–200.
[23] Dassios G. On the image systems for the electric and the magnetic potential of the brain. Bulletin of the Greek Mathematical Society, vol. 54, 2007, pp. 47–58.
[24] Dassios G. The scalar magnetic potential in magnetoencephalography. Journal of Physics: Conference Series, vol. 124, 2008, pp. 1–7.
[25] Dassios G. Neuronal currents and EEG-MEG fields. Mathematical Medicine and Biology, vol. 25, 2008, pp. 133–139.
[26] Dassios G. Ellipsoidal Harmonics. Theory and Applications, Cambridge University Press, Cambridge, 2012.
[27] Dassios G., and Kariotou, F. Magnetoencephalography in ellipsoidal geometry. Journal of Mathematical Physics, vol. 44, 2003, pp. 220–241.
[28] Dassios G., and Kariotou, F. The direct MEG problem in the presence of an ellipsoidal shell inhomogeneity. Quarterly of Applied Mathematics, vol. 63, 2005, pp. 601–618.
[29] Dassios G., and Fokas, A. S. Electro-magneto-encephalography for a three-shell model: dipoles and beyond for the spherical geometry. Inverse Problems, vol. 25, 2009, pp. 1–20.
[30] Dassios G., and Fokas, A. S. Electro-magneto-encephalography for the three-shell model: A single dipole in ellipsoidal geometry. Mathematical Methods in the Applied Sciences, vol. 35, 2012, pp. 1415–1422.
[31] Dassios G., and Satrazemi, K. EEG identification of a localized 1-D neuronal current proceeding of the 13th IEEE international conference on bioinformatics and bioengineering (BIBE-2013). (978-1-4799-7/13-paper 47, 2013 IEEE). Chania. 2013.
[32] Dassios G., and Satrazemi, K. Inversion of electroencephalography data for a 2-D current distribution. Mathematical Methods in the Applied Sciences, vol. 38, 2015, pp. 1098–1105.
[33] Dassios G., and Satrazemi, K. On the inverse MEG problem with a 1-D current distribution. Applied Mathematics, vol. 6, 2015, pp. 95–105.
[34] Dassios G., and Satrazemi, K. Inversion of MEG data for a 2-D current distribution. Journal of Applied Mathematics and Physics, vol. 2, 2014, pp. 771–782.
[35] Dassios G., Hadjiloizi, D, and Kariotou, F. The octapolic ellipsoidal term in magnetoencephalography. Journal of Mathematical Physics, vol. 50, 2009, pp. 1–18.
[36] Dassios G., Doschoris, M., and Fragoyannis, G. The influence of surface deformation on the EEG recordings. Proceeding of the 13th IEEE International Conference on Bioinformatics and Bioengineering (BIBE-2013), (978-1-4799-7/13-paper 58, 2013 IEEE), Chania 2013.
[37] Dassios G., Doschoris, M., and Satrazemi, K. Localizing brain activity from multiple distinct sources via EEG. Journal of Applied Mathematics, vol. 2014, 2014, Article ID 232747.
[38] Dassios G., Fragoyiannis, G., and Satrazemi, K. On the inverse EEG problem for 1-D current distribution. Journal of Applied Mathematics, vol. 2014, 2014, article ID 715785.
[39] Dassios G., Fragoyiannis, G., and Satrazemi, K. Discriminating simple from double sources via EEG and MEG measurements. Mathematical Methods in the Applied Sciences, Special Issue, vol. 2017, 2017, pp. 6187–6191.
[40] Dassios G., Doschoris, M, and Fragoyannis, G. Sensitivity analysis of the forward EEG problem depending on head shape variations. Mathematical Problems in Engineering, vol. 2015, 2015, ID 612568.
[41] Dassios G., Doschoris, M, and Satrazemi, K. On the resolution of synchronous dipolar excitations via MEG measurements. Quarterly of Applied Mathematics, vol. 76, 2017, pp. 39–45.
[42] Dassios G, Fokas, A.S., Hashemzadeh, P., and Leahy, R.M. EEG for current with two dimensional support. IEEE Transactions on Biomedical Engineering, vol. 65, 2018, pp. 2101–2108.

[43] Datadvance pSeven, https://www.datadvance.net/product/pseven/

[44] Dwivedi R., Ramanujam, B., Chandra, P., Sapra, S., Gulati, S., Kalaivani, M., Garg, A., Bal, C., Tripathi, M., Dwivedi, S., and Sagar, R. Surgery for drug-resistant epilepsy in children. New England Journal of Medicine, vol. 377, 2017, pp. 1639–1647.

[45] Efroymson M.A. Multiple Regression Analysis. Mathematical Methods for Digital Computers, John Wiley, New York, 1960.

[46] Flyer N., Wright, G.B., and Fornberg, B. Radial basis function-generated finite differences: A mesh-free method for computational geosciences. (eds) Freeden W., Nashed M.Z., and Sonar T. Handbook of Geomathematics, Springer, Amsterdam, 2015.

[47] Fokas A.S. Electro-magneto-encephalography for a three-shell model: distributed current in arbitrary, spherical and ellipsoidal geometries. Journal of the Royal Society Interface, vol. 6, 2008, pp. 479–488.

[48] Fokas A.S., and Kurylev, Y. Electro-magneto-encephalography for the three-shell model: minimal -norm in spherical geometry. Inverse Problems, vol. 28, 2012. DOI: org/10.1088/0266-5611/28/3/035010.

[49] Fokas A. S., Gelfand, I. M., and Kurylev, Y. Inversion method for magnetoencephalography. Inverse Problems, vol. 12, 1996, pp. L9–L11.

[50] Fokas A. S., Hauk, O., and Mishel, V. Electromagnetoencephalography for the three-shell model: numerical implementation via splines for distributed current in spherical geometry. Inverse Problems, vol. 28, 2012, pp. 035009.

[51] Fokas A., Hashemzadeh, P, and Leahy, R. M. Which part of the neuronal current Can be determined by EEG?. Oxford Handbook of Clinical Magnetoencephalography, Oxford University Press, 2019.

[52] Forrester A. I. J., and Keane, A.J. Recent advances in surrogate-based optimization. Progress in Aerospace Sciences, vol. 45, 2009, pp. 50–79.

[53] Forrester A., Sobester, A, and Keane, A. Engineering Design via Surrogate Modelling: A Practical Guide, John Wiley & Sons, 2008.

[54] Giapaki S.N., and Kariotou, F. The complete ellipsoidal shell-model in EEG imaging. Abstract and Applied Analysis, vol. 2006, 2006, pp. 1–18.

[55] Gramfort A., Papadopoulo, T., Olivi, E., and Clerc, M. Forward field computation with OpenMEE. Computational Intelligence and Neuroscience, 2011, Article ID923703.

[56] Gramfort A., Luessi, M, Larson, E, Engemann, D, Strohmeier, D, Brodbeck, C, Parkkonen, L, and Hamalainen, M. MNE software for processing MEG and EEG data. NeuroImage, vol. 86, 2014, Feb, pp. 446460.

[57] Gramfort A., Papadopoulo, T, Olivi, E, and Clerc, M. OpenMEEG opensource software for quasistatic bioelectromagnetics. BioMedical Engineering OnLine, vol. 9, No. 45, 2010.

[58] Grihon S., Burnaev, E, Belyaev, M, and Prikhodko, P. Surrogate Modeling of Stability Constraints for Optimization of Composite Structures. (eds) Koziel S., and Leifsson L. Surrogate-Based Modeling and Optimization, Springer, New York, 2013, pp. 350–391.

[59] Gullmara D., Jiirgen, J, and Reichenbach, R. Influence of anisotropic electrical conductivity in white matter tissue on the EEG/MEG forward and inverse solution. A high-resolution whole head simulation study. NeuroImage, vol. 51, 2010, pp. 145–163.

[60] Gutierrez D., and Nehorai, A. Array response kernels for EEG/MEG in single-shell ellipsoidal geometry. 1st IEEE Int. Workshop Computational Advances in Multi-Sensor Adaptive Processing, 2005, pp. 225–228.

[61] Gutierrez D., Nehorai, A., and Preissl, H. Ellipsoidal head model for fetal magnetoencephalography: forward and inverse solutions. Physics in Medicine and Biology, vol. 50, 2005, pp. 2141–2157.

[62] Gutierrez D., and Nehorai, A. Array response kernels for EEG and MEG in multilayer ellipsoidal geometry. IEEE Transactions on Biomedical Engineering, vol. 55, 2008, pp. 1103–1111.
[63] Hallez H., Vanrumste, B, Grech, R, Muscat, J, Clercq, W, Vergult, A, D'Asseler, Y, Camilleri, K, Fabri, S, Huffel, S, and Lemahieu, I. Review on solving the forward problem in EEG source analysis. Journal of NeuroEngineering and Rehabilitation, vol. 4, 2007, article 46.
[64] Hansen P.C. Analysis of discrete ill-posed problems by means of the L-curve. SIAM Review, vol. 34, No. 4, 1992.
[65] Hari R., and Ilmoniemi, R. J. Cerebral magnetic fields. Critical Reviews in Biomedical, vol. 14, 1986, pp. 93–126.
[66] Hashemzadeh P., and Fokas, A. S. Helmholtz decomposition of the neuronal current for the ellipsoidal head model". Inverse Problems. vol. 35, 2019.
[67] Hashemzadeh P., Fokas, A.S., and Schonlieb, C.B. A hybrid analytical-numerical algorithm for determining the neuronal current via EEG, https://arxiv.org/pdf/1906.01403.pdf,2019.
[68] Jatoi M. A., Kamel, N., Malik, A. S., Faye, I., and Begum, T. A survey of methods used for source localization using EEG signals. Biomedical Signal Processing and Control, vol. 11, 2014, pp. 42–52.
[69] Kaipio J., and Somersalo, E. Statistical and Computational Inverse Problems, Springer, 2004.
[70] Kwan P., Arzimanoglou, A., Berg, A., Brodie, M., Allen Hauser, W., Mathern, G., Moshé, S., Perucca, E., Wiebe, S., and French, J. Definition of drug resistant epilepsy: consensus proposal by the ad hoc task force of the ILAE commission on therapeutic strategies. Epilepsia, vol. 51, 2010, pp. 1069–1077.
[71] Langheim F.J., Leuthold, A.C., and Georgopolous, A. P. Synchronous dynamic brain networks revealed by Magnetoencephalography. Proceedings of the National Academy of Sciences of the United States of America, vol. 103, 2006, pp. 455–459.
[72] Morse P.M., and Feshbach, H. Methods of Theoretical Physics I, II, McGraw-Hill, New York, 1953.
[73] Mosher J.C., Lewis, P.S., and Leahy, R.M. Multiple dipole modelling and localization from spatio-temporal MEG data. IEEE transactions on Biomedical Engineering, vol. 39, 1992, pp. 541–557.
[74] Oostenveld R., Fries, P., Maris, E., and Schoffelen, J.M. FieldTrip: open source software for advanced analysis of MEG, EEG, and invasive electrophysiological data. Computational Intelligence and Neuroscience, vol. 2011, 2011, 2011, pp. 156869.
[75] Pataraia E., Simos, P., Castillo, E., Billingsley, R., Sarkari, S., Wheless, J., Maggio, V., Maggio, W., Baumgartner, J., Swank, P., and Breier, J. Does magnetoencephalography add to scalp video-EEG as a diagnostic tool in epilepsy surgery?. Neurology, vol. 62, 2004, pp. 943–948.
[76] Pruett W. A., and Hester, R. L. The creation of surrogate models for fast estimation of complex model outcomes. PLOS One, vol. 11, 2016, pp. e0156574.
[77] Quifionero-Candela J., and Rasmussen, C.E. A unifying view of sparse approximate Gaussian process regression. Journal of Machine Learning Research, vol.6, 2005, pp. 1939–1959.
[78] Rasmussen C.E., and Williams, C. K. I. Gaussian Processes for Machine Learning., MIT Press, 2005.
[79] Reeger A, Fornberg,, and Watts, M. L. Numerical quadrature over smooth, closed surfaces. Proceedings of the Royal Society A, vol. 472, 2016, pp. 20160401.
[80] Romain G., and Jean-Pierre,. Ellipsoidal harmonic expansions of the gravitational Potential: theory and applications. Celestial Mechanics and Dynamical Astronomy, vol. 79, 2001, pp. 235–275.
[81] Sarvas J. Basic mathematical and electromagnetic concepts of the biomagnetic inverse problem. Physics in Medicine and Biology, vol. 32, 1987, pp. 11–22.

[82] Simpson T., Toropov, V., Balabanov, V., and Viana, F. Design and analysis of computer experiments in multidisciplinary design optimization: a review of how far we have come – or not. 12th AIAA/ISSMO Multidisciplinary Analysis and Optimization Conference Victoria, British Columbia, Canada, 2008, pp. 1–12.
[83] Snoek J., Larochelle, H., and Adams, R.P. Practical Bayesian optimization of machine learning algorithms. NIPS'12 Proceedings of the 25th International Conference on Neural Information Processing Systems, vol. 2, 2012, pp. 2951–2959.
[84] Tikhonov A.N. On the stability of inverse problems. Doklady Akademii Nauk SSSR, vol. 39, 1943, pp. 195–198.
[85] Vesanen P., Nieminen, J., Zevenhoven, K., Dabek, J., Parkkonen, L., Zhdanov, A., Luomahaara, J., Hassel, J., Penttilä, J., Simola, J., and Ahonen, A. Hybrid ultra-low-field MRI and magnetoencephalography system based on a commercial whole-head neuromagnetometer. Resonance in Medicine, vol. 69, 2013, pp. 1795–1804.
[86] Vorwerk J., Clerc, M., Burger, M., and Wolters, C. H. Comparison of boundary element and finite element approaches to the EEG forward problem. Biomedizinische Technik/Biomedical Engineering, 2012.
[87] Weatherburn C. E. Differential Geometry of Three Dimensions, Vol. II, Cambridge University Press, 1930.
[88] Zhang Z. A fast method to compute surface potentials generated by dipoles within multilayer anisotropic geometry. Physics in Medicine and Biology, vol. 40, No. 3, 1995, pp. 335–349.
[89] Zimmerman J. E., Thiene, P., and Harding, J. T. Design and operation of stable RF-biased superconducting point-contact quantum devices and a note on the properties of perfectly clean metal contacts. Journal of Applied Physics, vol. 41, 1970, pp. 1572.
[90] Zou H., and Hastie, T. Regularization and variable selection via the elastic net. Journal of the Royal Statistical Society. Series B, vol. 67, 2005, pp. 301–320.

For further study Electroencefalography and Magnetoencephalography we refer to...

Acharya U.R., Sree, S.V., Swapna, G., Martis, R. J., and Suri, J. S. Automated EEG analysis of epilepsy: a review. Knowledgebase Systems, vol. 45, 2013, pp. 147–165.

Ammari H., Bao, G., and Fleming, J.L. An inverse source problem for Maxwell's equations in magnetoencephalography. SIAM Journal of Applied Mathematics, vol. 62, 2002, pp. 1369–1382.

Angenent S., Pichon, E, and Tannenbaum, A. Mathematical methods in medical image processing. Bulletin of the American Mathematical Society, vol. 43, 2006, pp. 365–396.

Chew W.C., Michielssen E, Jin J.M., and Song J. Eds. Fast and efficient algorithms in computational electromagnetics. Artech House, 2001.

Churchland P.S., and Sejnowski, T.J. The Computational Brain, MIT Press, Cambridge, 1996.

Clarke J. Squids. Scientific American, vol. 271, 1994, pp. 46–53.

Cohen D. Magnetoencephalography: evidence of magnetic fields produced by alpha-rhythm currents. Science, vol. 161, 1968, pp. 784–786.

Cohen D. Magnetoencephalography: detection of brain's electrical activity with a super-conducting magnetometer. Science, vol. 175, 192, pp. 664–666.

Crease R. Images of conflict: MEG vs. EEG. Science, vol. 253, 1991, pp. 374–375.

Cuffin B. N. EEG localization accuracy improvements using realistically shaped head models. IEEE Transaction in Biomedical Engineering, vol. 43, 1996, pp. 299–303.

Cuffin B. N., and Cohen, D. Magnetic fields of a dipole in special volume conductor shapes. IEEE Transaction in Biomedical Engineering, vol. BME-24, 1977, pp. 372–381.

Cuffin N. B., and Cohen, D. Comparison of the magnetoencephalogram and electroencephalogram. Electroencephalography and Clinical Neurophysiology, vol. 47, 1979, pp. 132–146.

Damasio A.R. The Brain. Collection of Articles from the Scientific American, The Lyons Press, New York, 1999.

Darvas F., Ermer, J.J, Mosher, J.C, and Leahy, R.M. Generic head models for atlas-based EEG source analysis. Hum Brain Mapp, vol. 27, 2006, pp. 129–143.

Dassios G. On the Hidden Electromagnetic Activity of the Brain. Fotiadis D., and Massalas C. V. (Editors). Mathematical Methods in Scattering Theory and Biomedical Engineering, Ioannina, 2005, pp. 297–303.

Dassios G. The magnetic potential for the ellipsoidal MEG problem. Journal of Computational Mathematics, vol. 25, 2007, pp. 145–156.

Dassios G. Electric and Magnetic Activity of the Brain in Spherical and Ellipsoidal Geometry. Ammari Habib (Editor). Chapter 4 in Mathematical Modeling in Biomedical Imaging, Springer, 2009, pp. 133–202.

Dassios G., and Kariotou, K. On the Geselowitz formula in biomagnetics. Quarterly of Applied Mathematics, vol. 61, 2003, pp. 387–400.

Dassios G., and Kariotou, F. The effect of an ellipsoidal shell on the direct EEG problem. Advances in Scattering and Biomedical Engineering, 2003, pp. 495–503. Ioannina.

Dassios G., and Kariotou, F. On the exterior magnetic field and silent sources in magnetoencephalography. Abstract and Applied Analysis, vol. 2004, 2004, pp. 307–314.

Dassios G., and Hadjiloizi, D. On the non-uniqueness of the inverse problem associated with electroencephalography. Inverse Problems, vol. 25, 2009, pp. 1–18.

Dassios G., and Tsampas, M. Vector ellipsoidal harmonics and neuronal current decomposition in the brain. Inverse Problems and Imaging, vol. 3, 2009, pp. 243–257.

Dassios G., and Fokas, A.S. Electro-magneto-encephalography and fundamental solutions. Quarterly of Applied Mathematics, vol. 67, 2009, pp. 771–780.

Dassios G., and Hadjiloizi, D. On the character of neuronal currents as sources for EEG and MEG fields. Bulletin of the Greek Mathematical Society, vol. 57, 2010, pp. 149–159.
Dassios G., and Fokas, A. S. The definite nonuniqueness results for deterministic EEG and MEG data. Inverse Problems, vol. 29, 2013, pp. 1–10.
Dassios G., and Chatjigeorgiou, I. Computation of higher order ellipsoidal harmonics with an application in electroencephalography. Quarterly Journal of Mechanics and Applied Mathematics, vol. 68, 2015, pp. 145–172.
Dassios G., Fokas, A. S, and Kariotou, F. On the non-uniqueness of the inverse MEG problem. Inverse Problems, vol. 21, 2005, pp. L1–L5.
Dassios G., Fokas, A. S., and Hadjiloizi, D. On the complementarity of electroencephalography and magnetoencephalography. Inverse Problems, vol. 23, 2007, pp. 2541–2549.
Dassios G., Kariotou, F, Kotzialis, K, and Kostopoulos, V. Magnetoencephalography for a realistic geometrical model of the brain. 3rd European Symposium on Biomedical Engineering and Medical Physics, pp.22, Patras. 2002.
Dassios G., Giapalaki, S, Kandili, A. N, and Kariotou, F. Analysis of EEG images. Mathematical Methods in Scattering Theory and Biomedical Engineering. Nymphaio, 2005, pp. 304–311.
Dassios G., Giapalaki, S, Kandili, A. N, and Kariotou, F. The exterior magnetic field for the multilayer ellipsoidal model of the brain. Quarterly of Mechanics and Applied Mathematics, vol. 60, 2007, pp. 1–25.
Dassios G., Doschoris, M., Kariotou, F., and Panagiotopoulou, B. Analysis of errors and bounds in electroencephalography. Proceeding of the 13th IEEE International Conference on Bioinformatics and Bioengineering (BIBE-2013), (978-1-4799-7/13-paper 50, 2013 IEEE). Chania 2013.
Dassios G., Doschoris, M, Vafeas, P, and Kariotou, F. Developing an algorithmic frame work tackling boundary value problems in ellipsoidal geometry: The case of EEG, International Conference on Recent Advances in Pure and Applied Mathematics. Instabul. 2015.
Dassios G., Doschoris, M, Vafeas, P, Kariotou, F, and Chatjigeorgiou, I. K. Revisiting a numerical implementation of the EEG problem in ellipsoidal geometry. Pioneer Journal of Advances Applied Mathematics, vol. 14, 2015, pp. 35–51.
El Badia A., and Ha Dung, T. Some remarks on the problem of source identification from boundary measurements. Inverse Problems, vol. 114, 1998, pp. 883–981.
El Badia A., and Ha Dung, T. On some inverse EEG problems. Inverse Problems Engineering Mechanics II, Japan, 2000, pp. 537–44.
El Badia A., and Ha Dung, T. An inverse source problem in potential analysis. Inverse Problems, vol. 16, 2000, pp. 651–63.
Ellenrieder von N, Muravchik, C. H, and Nehorai, A. Effects of geometric head model perturbations on the EEG forward and inverse problems. IEEE Transactions on Biomedical Engineering, vol. 53, No. 3, 2006, pp. 421–429.
Ellenrieder von N, Muravchik, C. H, Wagner, M, and Nehorai, A. Effect of head shape variations among individuals on the EEG/MEG forward and inverse problems. IEEE Transactions on Biomedical Engineering, vol. 56, No. 3, 2009, pp. 587–597.
Epstein C. L. Introduction to the Mathematics of Medical Imaging. Society of Industrial and Applied Mathematics, Philadelphia, 2008.
Ferdjallah M., Bostick, F.X, and Barr, R.E. Potential and current density distributions of cranial electrotherapy stimulation (CES) in a four-concentric-spheres model. IEEE Transactions on Biomedical Engineering, vol. 43, 1996, pp. 939–943.
Ferguson A.S., and Durand, D. Magnetic fields of current monopoles in special volume conductors. IEEE Transactions on Magnetics, vol. 27, 1991, pp. 758–767.
Ferguson A. S., and Durand, D. A theory of the magnetic field from current monopoles. Journal of Applied Physics, vol. 71, 1992, pp. 3107–3113.

Fischl B. FreeSurfer. Neuroimage, vol. 62, 2012, pp. 774–781.
Fletcher D. J. Improved method for computation of potentials in a realistic head shape model. IEEE Transactions on Biomedical Engineering, vol. 42, 1995, pp. 1094–103.
Fokas A. S., Kurylev, Y., and Marinakis, V. The unique determination of neuronal currents in the brain via magnetoencephalography. Inverse Problems, vol. 20, 2004, pp. 1067–1082.
Fokas A. S. An effective approach to electromagnetoencephalography. Plenary talk in the 8th IEEE International Conference on Bioinformatics and Bioengineering. Athens. 2008.
Gabriel S, Lau, R. W., and Gabriel, C. The dielectric properties of biological tissues: II. Measurements in the frequency range 10 Hz to 20 GHz. Physics in Medicine and Biology, vol. 41, 1996, Nov, pp. 2251–69.
Geselowitz D. B. On bioelectric potentials in an inhomogeneous volume conductor. Biophysical Journal, vol. 7, 1967, pp. 1–11.
Geselowitz D. B. On the magnetic field generated outside an inhomogeneous volume conductor by internal current sources. IEEE Transactions in Magnetism, vol. 6, 1970, pp. 346–347.
Grave de Peralta Menendez R, Gonzalez Andino, S. L, Morand, S, Michel, C.M, and Landis, T. Imaging the electrical activity of the brain: ELECTRA. Human Brain Mapp, vol. 9, 2000, pp. 1–12.
Grech R., Cassar, T., Muscat, J., Camilleri, K. P., Fabri, S. G., Zervakis, M., Xanthopoulos, P., Sakkalis, V., and Vanrumste, B. Review on solving the inverse problem in EEG source analysis. Journal of NeuroEngineering and Rehabilitation, vol. 5, No. 25, 2008. DOI: 10.1186/1743-0003-5-25.
Grynszpan F., and Geselowitz, D.B. Model studies of the magnetocardiogram. Biophysical Journal, vol. 13, 1973, pp. 911–925.
Gulrajani R. M. Bioelectricity and Biomagnetism, John Wiley, New York, 1998.
Gutierrez D., and Nehorai, A. Array response kernel for EEG in four-shell ellipsoidal geometry. Pro. 40th Asilomar Conference. Signals, Systems and Computer. 2006, pp. 151–155.
Hadjiloizi D., and Kariotou, F. The fluid-core model in electromagnetic brain activity. Bulletin of the Greek Mathematical Society, vol. 54, 2007, pp. 257–267.
Hadjiloizi D., and Kariotou, F. The fluid core model for MEG revisited: a comparison with the homogenous case". Proceedings of the 8th International Conference on Mathematical Methods in Scattering Theory and Biomedical Engineering. pp.141–149. 2007.
Hämäläinen M.S, and Sarvas, J. Realistic conductivity geometry model of the human head for interpretation of neuromagnetic data. IEEE Transactions on Biomedical Engineering, vol. 36, 1989, pp. 165–171.
Hämäläinen M. S., and Ilmoniemi, R. J. Interpreting magnetic fields in the brain: minimum norm estimates. Medical and Biological Engineering and Computing, vol. 32, 1994, pp. 35–42.
Hämäläinen M., Hari, R., Ilmoniemi, R.J., Knuutila, J., and Lounasmaa, O. V. Magnetoencephalography-theory, instrumentation, and applications to noninvasive studies of the working human brain. Reviews of Modern Physics, vol. 65, 1993, pp. 413–497.
Harrop R., Weinberg, H., Brickett, P., Dykstra, C., Robertson, A., Cheyne, D., Baff, M., and Crisp, D. The biomagnetic inverse problem: some theoretical and pratical considerations. Physics in Medicine and Biology, vol. 32, 1987, pp. 1545–1557.
Hauk O. Keep it simple: A case for using classical minimum norm estimation in the analysis of EEG and MEG data. Neuroimage, vol. 21, 2004, pp. 1612–21.
Haykin S.O. Neural Networks and Learning Machines, Pearson International Edition, 2008.
He B., and Cohen, R. J. Body surface laplacian EEG mapping. IEEE Transactions on Biomedical Engineering, vol. 39, 1992, pp. 1179–1191.
He S., and Romanov, V.G. Identification of dipole sources in a bounded domain for Maxwell's equations. Wave Motion, vol. 28, 1998, pp. 25–40.

He S., and Norgren, M. Magnetostatic image current and its application to an analytic identification of a current dipole inside a conducting sphere. IEEE Transactions on Biomedical Engineering, vol. 47, 2000, pp. 183–191.
He S., and Romanov, V. G. Identification of dipole sources in a bounded domain for Maxwell's equations. Wave Motion, vol. 28, 1998, pp. 25–40.
He B., Yao, D, Lian, J, and Wu, D. An equivalent current source model and Laplacian weighted minimum norm current estimates of brain electrical activity. IEEE Transactions on Biomedical Engineering, vol. 49, 2002, pp. 277–88.
Heller L., Ranken, D., and Best, E. The magnetic field inside special conducting geometries due to internal current. IEEE Transactions on Biomedical Engineering, vol. 51, 2004, pp. 1310–1318.
Helmholtz H. Ueber Einige Gesetze der Vertheiiung Elektrischer Strome in Korperlichen Leitern mit Anwendung auf Diethierisch-Elektrischen Versuche. Annual Review of Physical Chemistry, vol. 89, 1853, pp. 211–233.
Hoekema R., Wieneke, G.H, Leijten, F.S, van Veelen, C.W., van Rijen, P.C., Huiskamp, G.J., Ansems, J., and van Huffelen, A.C. Measurement of the conductivity of skull, temporarily removed during epilepsy surgery. Brain Topography, vol. 16, 2003, pp. 29–38.
Ilmoniemi R. J., Hämäläinen, M.S., and Knuutila, J. The forward and inverse problems in the spherical model. Edited by Weiberg H., Stroink G., and Katila T. Biomagnetism: Applications and Theory, Pergamon, New York, 1985, pp. 278–282.
Irimia A. Electric field and potential calculation for a bioelectric current dipole in an ellipsoid. Journal of Applied Physics, vol. 38, 2005, pp. 8123–8138.
Irimia A., van Horn, J. D., and Halgren, E. Source cancellation profiles of electroencephalography and magnetoencephalography. Neuroimage, vol. 59, 2012, pp. 2464–2474.
Jackson J.D. Classical Electrodynamics, John Wiley, New York, 1962.
Jerbi K., Mosher, J.C, Baillet, S., and Leahy, R.M. On MEG forward modeling using multipolar expansions. Physics in Medicine and Biology, vol. 47, 2002, pp. 523–555.
Jin J. The Finite Element Method in Electromagnetics, Second, John Wiley & Sons,INC.
Kamvyssas G., and Kariotou, F. On the electroencephalography (EEG) problem for the ellipsoidal brain model. Proceedings of the 6th International Conference of the HSTAM, pp.222–226, 2001.
Kamvyssas G., and Kariotou, F. Confocal ellipsoidal boundaries in EEG modeling. Bulletin of the Greek Mathematical Society, vol. 50, 2005, pp. 119–133.
Kariotou F. On the mathematics of EEG and MEG in spheroidal geometry. Bulletin of the Greek Mathematical Society, vol. 47, 2003, pp. 117–135.
Kariotou F. Electroencephalography in ellipsoidal geometry. Journal of Mathematical Analysis and Applications, vol. 290, 2004, pp. 324–342.
Katila T.E. On the current multipole presentation of the primary current distributions. IV international workshop on biomagnetism. Il Nuovo Cimento, vol. 2D, 1983, pp. 660–664.
Kybic J., Clerc, M, Abboud, T, Faugeras, O, Keriven, R, and Papadopoulo, T. A common formalism for the integral formulations of the forward EEG problem. IEEE Transactions on Biomedical Engineering, vol. 24, 2005, pp. 12–28.
Landau L.D., and Lifshifz, E.M. Electrodynamics of Continuous Media, Pergamon. Press, Oxford, 1960.
Lanfer B., Scherg, M., Dannhauer, M., Knösche, T. R., Burger, M., and Wolters, C. H. Influences of skull segmentation inaccuracies on EEG source analysis. Neuroimage, vol. 62, 2012, pp. 418–431.
Law S. K., Nunez, P. L., and Wijesinghe, R. S. High-resolution EEG using spline generated surface Laplacians on spherical and ellipsoidal surfaces. IEEE Transactions on Biomedical Engineering, vol. 40, 1993, pp. 145–53.
Leblond J., Paduret, C., Rigat, S., and Zghal, M. Source localization in ellipsoids by the best meromorphic approximation in planar sections. Inverse Problems, vol. 24, 2008, pp. 1–20.

Leweke S., Michel., V, and Fokas, A.S. Electro-Magnetoencephalography for the sphericalshell model: Novel integral operators with singular value decomposition. Inverse Problems, in press.

Malmivuo J., and Plonsey, P. Bioelectromagnetism, Oxford University Press, Oxford, 1995.

Manola L., Roelofsen, B.H., Holsheimer, J., Marani, E., and Geelen, J. Modelling motor cortex stimulationfor chronic pain control: electrical potential field, activating functions and responses of simple nerve fibre models. Med. Biol. Eng. Comput, vol. 43, 2005, May, pp. 335–43.

Maxwell J.C. A Treatise on Electricity and Magnetism I, II, Oxford University Press, Oxford, 1998.

Menendez G. P. R., and Gonzalez, A. S. L. A critical analysis of linear inverse solutions. IEEE Transactions on Biomedical Engineering, vol. 45, 1998, pp. 440–8.

Michel C.M., Murray, M., Lantz, G., Gonzalez, S., Spinelli, L., and Peralta, R.G. EEG source imaging. Clinical Neurophysiology, vol. 115, 2004, pp. 2195–2222.

Mosher J.C., Leahy, R.M., and Lewis, P.S. EEG and MEG: forward solutions for inverse methods. IEEE Transactions on Biomedical Engineering, vol. 46, 1999, pp. 245–259.

de Munck J. C. A linear discretization of the volume conductor boundary integral equation using analytically integrated elements. IEEE Transactions on Biomedical Engineering, vol. 39, 1992, pp. 986–90.

de Munck J.C. The potential distribution in a layered anisotropic spheroidal volume conductor. Journal of Applied Physics, vol. 64, 1988, pp. 464–470.

de Munck J.C., and Peters, M.J. A fast method to compute the potential in the multi-sphere model. IEEE Transactions on Biomedical Engineering, vol. 40, No. 11, 1993, pp. 1166–1174.

de Munck J. C., and Peters, M. J. A fast method to compute the potential in the multi sphere model. IEEE Transactions on Biomedical Engineering, vol. 40, 1993, pp. 1166–74.

Nara T., and Ando, S. A projective method for an inverse source problem of the Poisson equation. Inverse Problems, vol. 19, 2003, pp. 355–69.

Nara T., Oohama, J., Hashimoto, M., Takeda, T., and Ando, S. Direct reconstruction algorithm of current dipoles for vector magnetoencephalography and electroencephalography. Physics in Medicine and Biology, vol. 52, 2007, pp. 3859–3879.

Nolte G. Perturbative analytical solutions of the magnetic forward problem for realistic volume conductors. Journal of Applied Physics, vol. 89, 2001, pp. 2360–2369.

Nolte G. The magnetic lead field theorem in the quasi-static approximation and its use for MEG forward calculation in realistic volume conductors. Physics in Medicine and Biology, vol. 48, 2003, pp. 3637–52.

Nolte G., Fieseler, T., and Curio, G. Perturbative analytical solutions of the magnetic forward problem for realistic volume conductors. Journal of Applied Physics, vol. 89, 2001, pp. 2360–2369.

Nolte G., Bai, O, Wheaton, L., Mari, S, Vorbach, Z, and Hallett, M. Identifying true brain interaction from EEG data using the imaginary part of coherency. Clinical Neurophysiology, vol. 115, 2004, pp. 2292–307.

Nolte G. Brain Current Multipole Expansion in Realistic Volume Conductors: Calculation, Interpretation and Results. "Recent Research Development in Biomedical Engineering". Vol. 1, 2002, pp. 171–211.

Nolte G., and Curio, G. On the calculation of magnetic fields based on multipole modeling of focal biological current sources. Biophysical Journal, vol. 73, 1997, pp. 1253–1262.

Nolte G., and Curio, G. Perturbative analytical solutions of the electric forward problem for realistic volume conductors. Journal of Applied Physics, vol. 86, 1999, pp. 2800–2811.

Nolte G., and Dassios, G. Semi-analytic calculation for EEG in multi-shell realistic volume conductors based on the lead field theorem. Poster Presentation in BIOMAG. Boston, 2004.

Nolte G., and Dassios, G. Analytic expansion of the EEG lead field for realistic volume conductors. Physics in Medicine and Biology, vol. 50, 2005, pp. 3807–3823.

Nunez P.L. Neocortical Dynamics and Human EEG Rhythms, Oxford University Press, Oxford, 1995.

Nunez P. L., and Srinivasan,. Electric Fields of the Brain. The Neurophysics of EEG, Oxford University Press, Oxford, 2006.
Nunez P. L., and Srinivasan, R. A theoretical basis for standing and traveling brain waves measured with human EEG with implications for an integrated consciousness. Clinical Neurophysiology, vol. 117, 2006, pp. 2424–2435.
Olejniczak P. Neurophysiologic basis of EEG. Journal of Clinical Neurophysiology, vol. 23, No. 3, 2006, pp. 186–189.
Papanicolaou A. C. The Amensias: A Clinical Textbook of Memory Disorders, Oxford University Press, Oxford UK, 2006.
Parsopoulos K. E., Kariotou, F., Dassios, G., and Vrahatis, M. N. Tackling magnetoencephalography with particle swam optimization. International Journal of Bio-Inspired Computation, vol. 1, 2009, pp. 32–49.
Philips C., Mattout, J., Rugg, M. D., Maquet, P., and Friston, K. J. An empirical Bayesian solution to the source reconstruction problem in EEG. Neuroimage, vol. 24,, 2005, pp. 997–1011.
Plonsey R. Current dipole images and reference potentials. IEEE Transactions on Biomedical Electronics, vol. BME 10, 1963, pp. 3–8.
Plonsey R. The nature of sources of bioelectric and biomagnetic fields. Biophysical Journal, vol. 39, 1982, pp. 309–312.
Plonsey R., and Heppner, D.B. Considerations of quasi-stationarity in electrophysiological systems. Bulletin of Mathematical Biophysics, vol. 29, 1967, pp. 657–664.
Purcell C., Mashiko, T. K., and Ueno, K. Describing head shape with surface harmonic expansions. IEEE Transactions on Biomedical Engineering, vol. 38, 1991, pp. 303–6.
Riera J. Electric lead field for a piecewise homogeneous volume conductor model of the head. IEEE Transactions on Biomedical Engineering, vol. 45, 1998, pp. 746–53.
Salu Y., Cohen, L.G., Rose, D., Sato, S., Kufta, C., and Hallett, M. An improved method for localizing electric brain dipoles. IEEE Transactions on Biomedical Engineering, vol. 37, 1990, pp. 699–705.
Scherg M. Fundamentals of dipole source potential analysis. Auditory Evoked Magnetic Fields and Electric Potentials, Gradori F., Hoke M., Romani GL(Editors), vol. 6, 1990, pp. 40–69.
Schmidt R. F. editor. Fundamentals of Neurophysiology, Springer-Verlag, New York, 1985.
Schomer D.L., and Lopes da Silva, F.H. Niedermeyer's Electroencephalography: Basic Principles, Clinical Applications, and Related Fields, Lippincott Williams and Wilkins, 2011.
Selma S., and Aine, C. J. Magnetoencephalography: From Signals to Dynamic Cortical Networks., Springer-Verlag Berlin Heidelberg, 2014.
Shattuck D. W., and Leahy, R. M. BrainSuite: An automated cortical surface identification tool. Medical image Analysis, vol. 6, 2002, pp. 129–142.
Stratton J. A. Electromagnetic Theory, McGraw-Hill, New York, 1941.
Stufflebeam S. M., Tanaka, N., and Ahlfors, S. P. Clinical applications of magnetoencephalography. Human Brain Mapping, vol. 30, 2009, pp. 1813–23.
Synder W. S., Ford, M. R., Warner, G., and Fisher Jr., H. L. Estimates of absorbed fractions for monoenergetic photon sources uniformly distributed in various organs of a heterogeneous phantom. Journal of Nuclear Medicine Suppl, vol. 10, No. 3, August 1969, Pamphlet No. 5, Revised, 1978.
Tadel F., Baillet, S., Mosher, J.C., Pantazis, D., and Leahy, R.M. Brainstorm: a user-friendly application for MEG/EEG analysis. Computational Intelligence and Neuroscience, vol. 2011.
Tissari S., and Rahola, J. A precorrected-FfT method to accelerate the solution of the forward problem in magnetoencephalography. Physics in Medicine and Biology, vol. 48, 2003, pp. 523–41.
Tobimatsu S., and Kakigi R. Editors. Clinical Applications of Magneto-encephalography, ISBN 9784-431-55728-9, Springer, Japan, 2016.

Vallaghe S., and Clerc, M. A global sensitivity analysis of three and four-layer EEG conductivity models. IEEE Transactions on Biomedical Engineering, vol. 56, 2009, pp. 988–995.

Vanrumste B., Van Hoey, G., Van de Walle, R., D'Havé, M. R. P., Lemahieu, I. A., and Boom, P. J. A. M. Comparison of performance of spherical and realistic head models in dipole localization from noisy EEG. Medical Engineering and Physics, vol. 24, 2002, pp. 403–18.

Van't Ent D., Munck De, J. C., and Kaas, A. L. A fast method to derive realistic BEM models for E/MEG source reconstruction. IEEE Transactions on Biomedical Engineering, vol. 48, 2001, pp. 1434–43.

Weatherburn C. E. Differential Geometry of Three Dimensions, Vol. I, Cambridge University Press, 1927.

Webb A. Introduction to Biomedical Imaging. IEEE Engineering in Medicine and Biology, IEEE press, Hoboken, 2003.

Webb S. editor. The Physics of Medical Imaging, Institute of Physics, Bristol, 1988.

Weinstein D., Zhukov, L., and Johnson, C. Lead-field bases for electroencephalography source imaging. Annals of Biomedical Engineering, vol. 28, 2000, pp. 1059–65.

Wendel K, Väisänen, O, Malmivuo, J, Gencer, N, Vanrumste, B, Durka, P, Magjarevic, R, Supek, S, Pascu, M, Fontenelle, H, and Menendez, R. EEG/MEG source imaging: Methods, challenges, and open issues. Computational Intelligence and Neuroscience, vol. 2009, 2009, Article ID 656092.

Wilson F. N., and Bayley, R. H. The electric field of an eccentric dipole in a homogeneous spherical conducting medium. Circulation, vol. 1, 1950, pp. 84–92.

Williamson S. J., and Kaufman, L. Biomagnetism. Journal of Magnetism and Magnetic Materials, vol. 22, 1981, pp. 129–201.

Wood C. Application of dipole localization methods to source identification and human evoked potentials. Annals of the New York Academy of Sciences, vol. 388,, 1982, pp. 139–155.

Yamatani K., Ohe, T., and Ohnaka, K. An identification method of electric current dipoles in spherically symmetric conductor. Journal of Computational Mathematics, vol. 143, 2002, pp. 189–200.

Yondo R., Andres, E, and Valero, E. A review on design of experiments and surrogate models in aircraft real-time and many-query aerodynamic analyses. Progress in Aerospace Sciences, vol. 96, 2018, pp. 23–61.

Yondo R., Bobrowski, R., Andres, E., and Valero, E. A Review of Surrogate Modeling Techniques for Aerodynamic Analysis and Optimization: Current Limitations and Future Challenges in Industry. (Editors) Minisci E., Vasile M., Periaux J., Gauger N., Giannakoglou K., and Quagliarella D. Advances in Evolutionary and Deterministic Methods for Design, Optimization and Control in Engineering and Sciences. Computational Methods in Applied Sciences, Vol. 48, Springer, Cham, 2019.

Zanow F., and Peters, M. J. Individually shaped volume conductor models of the head in EEG source localization. Medical and Biological Engineering and Computing, vol. 33, 1995, pp. 582–588.

Index

https://doi.org/10.1515/9783110547535-012

De Gruyter Series in Mathematics and Life Sciences

Volume 6
Piotr Biler
Singularities of Solutions to Chemotaxis Systems, 2019
ISBN 978-3-11-059789-9, ISBN (PDF) 978-3-11-059953-4, ISBN (EPUB) 978-3-11-059862-9

Volume 5
S. V. Masiuk, A. G. Kukush, S. V. Shklyar, M. I. Chepurny, I. A. Likhtarov
Radiation Risk Estimation. Based on Measurement Error Models, 2016
ISBN 978-3-11-044180-2, ISBN (PDF) 978-3-11-043366-1, ISBN (EPUB) 978-3-11-043347-0

Volume 4
Sergey Vakulenko
Complexity and Evolution of Dissipative Systems: An Analytical Approach, 2013
ISBN 978-3-11-026648-1, e-ISBN 978-3-11-026828-7

Volume 3
Zoran Nikoloski, Sergio Grimbs
Network-based Molecular Biology: Data-driven Modeling and Analysis, 2013
ISBN 978-3-11-026256-8, e-ISBN 978-3-11-026266-7

Volume 2
Shair Ahmad, Ivanka M. Stamova (Eds.)
Lotka-Volterra and Related Systems: Recent Developments in Population Dynamics, 2013
ISBN 978-3-11-026951-2, e-ISBN 978-3-11-026984-0

Volume 1
Alexandra V. Antoniouk, Roderick V. N. Melnik (Eds.)
Mathematics and Life Sciences, 2012
ISBN 978-3-11-027372-4, e-ISBN 978-3-11-028853-7

www.degruyter.com

Bei Fragen zur Produktsicherheit wenden Sie sich bitte an:
If you have any questions regarding product safety,
please contact:

Walter de Gruyter GmbH
Genthiner Straße 13
10785 Berlin
productsafety@degruyterbrill.com